History of U.S. Agriculture

and Its Relevance to Today

By the same author:

The Day of the Bonanza
The Challenge of the Prairie
Beyond the Furrow
Tomorrow's Harvest
Koochiching
Plowshares to Printouts
Taming the Wilderness

Sonja & Jim Ozbun, may you enjoy this account of our dynamic and rapidly changing agriculture. You, Jim, are one of many who has done so much to bring on the transition.

Hiram M Drache
12-25-95

History of U.S. Agriculture
and Its Relevance to Today

Hiram M. Drache, Ph.D.

Concordia College
Moorhead, Minnesota

Interstate Publishers, Inc.
Danville, Illinois

**History of U.S. Agriculture
and Its Relevance to Today**

Copyright © 1996 by
Interstate Publishers, Inc.

All Rights Reserved / Printed in U.S.A.

Library of Congress Catalog Card No. 95-76604

ISBN 0-8134-3072-0

1 2 3 4 5 6 7 8 9 10   02 01 00 99 98 97 96

Order from

Interstate Publishers, Inc.
P.O. Box 50
Danville, IL 61834-0050
Phone: (800) 843-4774
Fax: (217) 446-9706

Also published in soft cover under the title
Legacy of the Land: Agriculture's Story to the Present

Foreword

M. E. Ensminger

President
Agriservices Foundation

Sir Winston Churchill, the great Prime Minister of England, used history as his compass to shape his own destiny and the destiny of the whole world. Said he:

The further back you can look, the further forward you are likely to see.

Since the roots of the present lie deep in the past, historical events often help to interpret the present and project into the future. So, as we race to the 21st Century and beyond, it follows that each of us has a responsibility for the future, and that we can best understand and chart our course if we have a knowledge of the past. Today's lack of appreciation of agriculture, and lack of understanding of the role that agricultural science has played in transforming America from a nation of food scarcity into a nation of food abundance, stem largely from the lack of knowledge of the history of U.S. agriculture.

In this book, appropriately entitled *History of U.S. Agriculture and Its Relevance to Today,* Hiram M. Drache, one of America's foremost agricultural historians, makes agricultural events of the nation's past come alive as if they happened yesterday. He makes his readers stand tall and be proud of American agriculture.

When the struggling colonists landed at Jamestown on May 14, 1607, they became gatherers and hunters. Planting corn for food, a skill learned from the Indians, saved many early settlers from starvation.

Squanto, an Indian chieftain, showed them how to cultivate corn in the rocky Massachusetts soil; he insisted that each hill be fertilized with three fish heads pointed inward like the spokes of a wheel.

Throughout the colonial period, and during the formative period of the nation, the leaders of America were proud agriculturalists. George Washington, the Father of Our Country and first President, maintained an extensive horse-and mule-breeding establishment at Mount Vernon.

The following excerpt from one of Thomas Jefferson's letters, written while he was President, tells of his love for farming:

> I have often thought that if heaven had given me choice of position and calling, it should have been on a rich spot of earth, well watered, and near a good market for the production of the garden. No occupation is so delightful to me as the culture of the earth, and no culture, comparable to that of the garden.

But, from the colonial period until the 1930's, American agriculture followed the traditional global practice of bringing more land under cultivation to provide increased food and fiber for more people. Until after the Civil War, little was done to develop a science-based, high-yielding, surplus American agriculture. In 1862, President Abraham Lincoln signed three Acts into law which were to change American agriculture forever; namely, (1) the Act creating the Department of Agriculture, which subsequently evolved into the U.S. Department of Agriculture; (2) the Morrill Land Grant College Act, which established Colleges of Agriculture in each state; and (3) the Homestead Act, which made frontier land available to those who committed themselves to living on, and developing, the land.

Upon establishing the Department of Agriculture, President Lincoln said:

> There is hereby established a Department of Agriculture to acquire and diffuse among people of the United States useful information on subjects connected with agriculture.

But implementation of the three agricultural Acts failed to produce magical results. Higher crop and animal yields, followed by increased total production and farm income, continued to lag. So, the U.S. Congress came to the rescue once again. In 1887, the Hatch Act was passed, creating Agricultural Experiment Stations as the research arms of the Agricultural Colleges. During the period from about 1890 to 1910, researchers evolved with improved crop varieties, more productive breeds of livestock, and the control of some troublesome plant and animal diseases. But these discoveries were slow to reach farmers. In order to speed the application of research from the laboratory to the field, in 1914 the U.S. Congress established the Cooperative Extension Service, thereby providing for correlated U.S. agricultural teaching, research, and extension, which is the envy of, and without an effective counterpart in, the world.

Since 1900, the United States has progressed from food scarcity to food abundance.

American agriculture has a past to honor—a future to build. Consider these facts:

- Agriculture is the United States' largest commercial industry.
- The United States provides more than half of the world's shared food aid.
- One American farmer supplies enough food and fiber for 128 people—96 Americans and 32 citizens of other countries.
- Every year from 4,500 to as many as 20,000 new food items are introduced in the United States.
- Americans spend less of their income on food than anyone else in the world—just 9 percent of total personal expenditures go for food.
- Americans have the most abundant and most readily available, the most wholesome and nutritious, and the safest food supply in the world.

Historian Drache conveys the following message loud and clear:

People and food are not only inseparable from history—they're part of it. Without them there would be no history—no humankind.

But this is not finis! Ponder the following challenges ahead:

- More than 700 million people in the world go to bed hungry each night.
- More than 188 million children throughout the world suffer from malnutrition.
- In the next two decades, 2 billion people (the equivalent of two Chinas) will be added to the population.
- Worldwide, the amount of cropland per person has dropped 25 percent since 1950, and it will likely drop another 15 percent by the turn of the century.

The historical achievements of U.S. agriculture and the current biotechnology revolution give assurance that we can look forward with confidence to doing our part, and much more, in meeting the mountainous world food needs in the 21st Century. But, following production, governments must provide for the equitable distribution of foods; otherwise, people in some of the developing countries will starve to death because of population growth outrunning food production. For them, a world average increase in food production does nothing to satisfy their hunger pains when they are far below the average.

Preface

The history of U.S. agriculture is one of the most dramatic stories of the last four centuries. We have evolved from tiny settlements, where virtually everyone had to till the soil, to a country of 260 million people, of whom less than 1.5 percent are employed directly in production agriculture. However, over 15 percent of the labor force is employed in the total food chain, which encompasses all workers from the laborer on the farm to the cashier at the grocery store or the restaurant, making it the largest sector of our economy.

Much of our success stems from the fact that the United States came into existence at the time that our European ancestors were experiencing a release from their traditional ways of thinking, which served to open new methods of commerce, industry, and agriculture. Our farmers were quick to grasp the advantages of the great land bounty that lay at their disposal. Even though the rank-and-file farmers were strong traditionalists, we had leaders who not only developed new methods of farming but also created an agricultural infrastructure that served to make our food chain the most dynamic known to humanity.

After the industrialization of agriculture, we experienced a rapid migration of surplus population from the land to the growing industrial centers, where these workers aided in producing other goods for the betterment of society. At the same time the labor force was being reduced on the farms, the relative cost of food was continuing to decline so that today we spend less than 9 percent of our disposable income for food. This makes our citizens some of the best fed at the lowest relative cost of any nation in the world. Not surprisingly, those who remained on the land and adopted the latest methods of management and technology improved their life style to levels never before attainable by tillers of the soil.

The abundance of free land encouraged many from other nations to come to the United States. This worked to the benefit of the nation and the consumers, for nearly always there was an abundance of low-cost agricultural products. Agriculture also served to finance the growth of our economy. Up to the end of the first decade of the twentieth century, agricultural products annually accounted for more than 50 percent of our total exports.

American farmers functioned in a system that freed them from traditions of the Old World so that they were at liberty to use their imaginative skills to improve their farming methods and to enhance their income. They were benefactors of governmental policies because they were the political majority and because it was beneficial to the nation to have an abundant supply of agricultural goods at relatively low prices. Generally, from the 1600s to the late nineteenth century, agriculture prospered under those policies.

Because farming was so romanticized as a superior way of life for the farm family, it was relatively easy to perpetuate favorable government programs. In later years, those programs made business profitable for the well-managed, commercially oriented farmers, but they meant substandard living for the poorer managers or for those more interested in farming as a way of life. For the latter group, the programs served as modified welfare and prolonged the so-called farm problem.

The government programs and rapidly changing technology have served to widen the gap between the two opposing philosophies, but the myth about farming as a way of life has persisted through most of the twentieth century. The boom of the 1970s and the inevitable crunch of the 1980s have resulted in a more realistic outlook toward those who are the real food producers. Farming, like all industry, is adapting to the most efficient ways of production at a faster pace than ever. Society demands as much. Overcapitalization of human and capital resources has caused a long-term withdrawal from agriculture. This movement is accelerating, and its impact on farm-service villages and the rural areas will continue to be great.

The decline in farm numbers will continue unless individuals are willing to subsidize farming with nonfarm income. Less than 13 percent of our farms today are large enough to support a family solely from farm income. Unfortunately, many who profess to be defenders of small-scale farming are doing so more for the sake of preserving the countryside as it once was than for the sake of maintaining small-scale farmers.

This rapid transition of agriculture, starting in the 1830s and accelerating in the 1900s, particularly after World War II, was caused by a series of mini-revolutions. The initial revolution started by breaking away from the restrictive traditions of Europe while at the same time adapting the latest economic and technological ideas for the enhancement of not only agriculture but all society.

The second revolution appeared after the 1830s when several significant inventions—the reaper, the steel plow, the mower, the grain drill, and the thresher—appeared. The improved market and the labor shortage of the Civil War era hastened the adoption of those machines. By the 1890s, the adoption of horse-powered equipment reached its full stride. Then, the tractor-power revolution commenced, and by the 1950s most horses on the farms were for recreation rather than for work.

In the 1930s hybrid seeds made their breakthrough, followed in the 1950s and after by the increased use of fertilizer, chemicals, and four-wheel-drive power. Biochemistry, plant genetics, computers, intensified use of capital, management, and technology are all part of the rapidly accelerating rate of change. In the decades ahead there will be an even more rapid consolidation and industrialization of the remaining farms. Forces outside of agriculture are aiding this movement.

Restrictions on farming practices, government regulations, environmentalists, animal activists, changing consumer tastes, plus a host of pressures from doomsday spokespersons generally tend to force the traditional farmer to yield to the large-scale, commercially oriented farm entrepreneur who is prepared to cope with the problems and can afford to do so. It is not for us to decide whether that is good or bad, but it is change, and that will accelerate.

This history weaves the story of production agriculture into that of the infrastructure supporting our farmers, together with an account of the food processing and delivery systems. This infrastructure enables farmers to make our food chain the most progressive and efficient system ever created. Once the food and fiber is produced, the food chain takes the goods at the farm gate. It then processes the goods, providing consumers with an abundance of reasonably priced food and fiber. Overall, this massive food chain makes up the largest single industry of our nation.

Hiram M. Drache

Acknowledgments

The origins of this book date back to 1989, when the publishers asked me to consider writing a brief history of U.S. agriculture. As in the past, the Interstate staff has been a delight to work with. Special thanks to Vernie L. Thomas for suggesting the project, to Ronald L. McDaniel for editing the manuscript, and to Kimberly Romine for designing the format.

The facilities of Concordia College, including an ideal office for research and writing, made going to work a complete joy. The able and dedicated Concordia library staff secured every bit of information asked for. Modern technology in the hands of skilled librarians is a boon to researchers, but friendly people are still the key to making a pleasant atmosphere for researching and writing. The library staff was great. Artist George Fredin provided an encompassing cover depicting transitions in the food chain from pioneer times to the present.

The help of individuals knowledgeable in agriculture, such as Joel Belgum, David Danbom, M. E. Ensminger, Gilbert Fite, Tom Isern, and H. R. Lund, made the task easier. These people helped with the outline, made suggestions, or read parts of the manuscript. Thank you.

Once again my wife, Ada, proved to be my most severe and best critic. For the first time we used two computers so that we were able to work independently. This greatly speeded up the process because editing could be done while writing was still in progress. Her technical skills allowed me to concentrate on writing.

Hiram M. Drache

About the Author

Hiram M. Drache's roots are in rural Minnesota. He delivered milk to customers in the village of Meriden in the 1930s, first with a sled or a bicycle with a sidecar and finally a 1927 Studebaker. While working on an uncle's farm, he milked by hand, shocked grain beside his aunt, and did other chores. His first field work was driving a single-row two-horse cultivator. During two threshing seasons he drove a bundle team. With those experiences behind him, he has never understood the strong feelings many people have about labor-intense farming of the past.

Drache's college career was interrupted by service in the Air Force during World War II, after which he received his B.A. from Gustavus Adolphus College, his M.A. from the University of Minnesota, and his Ph.D. from the University of North Dakota. He taught high school, worked in various businesses, and purchased his first farm in 1950. From 1952 through 1991 he was a professor of history at Concordia College, Moorhead, Minnesota, where he specialized in European and economic history. Since then he has been historian-in-residence. Because of several innovations, his farming operation was the subject of many articles in regional and national farm periodicals. In 1962 he developed a computer record system for the farm and his cattle feeding business. After 31 years of farming, the Draches have rented out their farms.

In the past four decades he has spoken to hundreds of audiences in 36 states, 6 provinces of Canada, Australia, and Germany.

This is Dr. Drache's eighth book. In addition, he has been a contributing author to 7 books and has written over 50 articles on contemporary agriculture and/or agricultural history.

Contents

Foreword v

Preface ix

Acknowledgments xiii

Geographical Setting xxiii

Unit I: To 1783 ——————————— 1

Some Events and Technological Innovations
  That Affected Agriculture 2

Introduction 3

Indian Agriculture 4
  Foundations of U.S. Agriculture 4
  Maize and Associated Crops 7
  Wild Seeds, Irrigation, and Hunting 9

Colonial Agriculture 10
  European Customs Are Resisted 10
  Parceling Out the Land 14
  The Growth of the Livestock Sector 16
  Relations with the Indians 19
  Colonial Labor Supply 21
  New England 25
  The Middle Colonies 30

The South   34

Spanish and French Settlements   42

Technology in Colonial Agriculture   45

Power and Early Farm Vehicles   46

Tools and Implements   47

Land and Crops   51

The Need for Markets   55

Challenges to Production   56

The Impact of the Revolutionary War   59

Conclusion   61

References   61

Unit II: 1783 to 1860 — 65

Some Events and Technological Innovations That Affected Agriculture   66

Introduction   67

Early Agricultural Policies   68

The Rise of a Myth   69

Acts Prior to 1800   71

Dispersing the Public Domain   72

Attempts to Create a Department of Agriculture   73

Farming in a New Setting   75

The Philadelphia Society   76

The Agricultural Press   77

Agricultural Fairs   78

Early Agricultural Education   80

Scientific Agriculture   81

Early Protests   87

Westward Expansion 88

    Settling on the Frontier 88

    Women on the Frontier 91

    Forests 93

    The Commercial Livestock Industry 93

    The Cotton Economy 97

    The Mormons in Agriculture 99

    Rivers and Canals 100

    Highways and Railroads 101

Mechanization 103

    The Sickle and the Cradle 105

    The Mower and the Rake 105

    The Reaper 107

    The Header 108

    The Binder 110

    The Fanning Mill and the Thresher 110

    The Plow 112

    Seeders and Drills 113

    The Corn Planter 114

Expanding Demand 115

    Commercial Agriculture 115

    The Food Chain 116

    Food Processing 119

    Changing Work Patterns 120

    The Rise of Central Livestock Markets 122

    The Growth of Foreign Markets 123

Summary 124

References 125

Unit III: 1860 to 1914 — 129

Some Events and Technological Innovations
That Affected Agriculture   130

Introduction   131

Governmental Activities, 1860 to 1914   132

    The Creation and Early Work of the USDA   133

    The Homestead Acts   139

    The Morrill Acts   141

The Rise of the Industrial Society   149

    Exports   150

    Industrial Markets   151

    Regulating the Marketplace   152

    Coping with Increased Domestic Consumption   154

    Growth of the Food Chain   156

Progress in Farming   160

    Muscle to Mechanical Power   160

    Mechanized Farming   165

    Working the Land   170

    Commercial Agriculture   182

    Changing Farm Practices   184

Transportation and Communication   194

    Railroads   194

    Highways   199

    Rural Free Delivery   201

Frontiers   202

    Indians and Eskimos   202

    Cattle Frontiers   204

    The Farmers' Frontier   206

Agricultural Communes   209
Irrigation and Dry Farming   213
The Timber Realm   216
Other Conservation Issues   219

Major Movements, 1860 to 1914   220
The Civil War   220
Migration of Agricultural Workers   222
Agrarian Discontent   223
Farm Organizations   226
The Decline of Agrarianism   231
Cooperatives   232
Changing Rural Life Styles   238

Summary   253
References   254

Unit IV: 1914 to 1954 — 259

Some Events and Technological Innovations
   That Affected Agriculture   260

Introduction   261

Continued Industrialization   262
   The Impact of World War I   262
   The Impact of World War II   265
   The Price Decline in the 1920s and 1930s   268
   Financial Problems   270
   Perpetuating a Myth   273
   The Unstoppable Tide   277
   Changing Roles   282
   Breaking the Isolation   286

Labor in Modern Agriculture   291

The Global Economy   298

Government in Agriculture   301

The End of a Policy of Drift   301

Founding of the American Farm Bureau Federation   302

Attempts to Forecast the Market   304

The Extension Service   306

The Farm Bloc   307

The Agricultural Marketing Acts of 1926 and 1929   309

Cooperative Shipping Associations   311

Cooperative Oil Associations   313

The Rugged Individualist Succumbs   314

Soil and Forest Conservation   320

Social Programs of the USDA   322

The Power Revolution Booms   328

The Tractor Replaces the Horse   330

Machines Outgrew the Farms   333

Cotton Machinery   335

Airplanes and Trucks   337

Productivity Booms   339

Tenant Farming   341

Farm Consolidation   344

Enter Agribusiness   346

The Farm Credit System   346

Livestock Marketing   348

Commercial Fertilizer   350

Hybrid Corn   351

Soybeans   353

Animal Science   354

Chemurgy   357

The Food Chain   358

Summary   362

References   363

Unit V: 1954 to 1994 — 367

Some Events and Technological Innovations That Affected Agriculture   368

Introduction   369

Production Booms   371

Chemicals   372

Biotechnology   375

Energy and Irrigation Use   380

Alternative Products   384

Industrialized Agriculture   390

Agribusiness Infrastructure   390

Professional Management   393

Capital-Intense Farming   396

Integrated and Corporate Farming   398

Farm Tenancy—A Vital Part of Modern Agriculture   402

Surplus Farm Labor   403

Unionism   404

Communal Farming   405

Tomorrow's Challenge   406

Agriculture in a World Economy   407

Chemurgy   412

Cooperatives   413

The Food Chain   415

Farm Programs 427
Changing Structure 433
Environmentalism 438
Fickle Consumers 446

A New Rural Society 449

Accelerating Change 450
Part-Time Farming 453
Changing Life Style 455
Foreign Investments 459
New Farm Groups 461
Declining Rural Communities 462
The Fading Myth 466

Summary 470

References 471

Bibliographical Note 475
Index 477

Geographical Setting

Today the United States is the world's greatest farming nation. It supplies our population of over 260 million people with ease, is the world's leading exporter of agricultural products, and still has an accumulating surplus. A wide variety of crops are possible because of the abundance of good soil and a wide range of climate. This was not the case when farmers were restricted to coastal areas and did not know how to cope with local soils and climate. It took years to learn which were the best crops to grow. Often the crops they knew were not adaptable to the area.

Sometimes first impressions or experiences were misleading. Captain John Smith likened the summers of Virginia to those of Spain and the winters to those of England. However, he observed that the thunder and lightning were more severe than what he had experienced in Europe. Smith also knew that north of Virginia the winters were as severe as in Sweden. The early colonists soon realized that the new land had a hotter sun and rains that were more penetrating than in Europe. Also, the soil was able to produce a wider variety of crops than was possible there. They were able to identify with some native plants and animals, and it did not take them long to learn that Indian agriculture was more productive and less labor intense than traditional European methods.

The coastal plains made up the first major agricultural region. These plains were very near sea level and stretched from Massachusetts to Florida and then to the Gulf of Mexico. Most of this region was covered by forest and contained good, rich soil. It also had some of the nation's largest swamps, such as the Everglades, the bayous, and coastal swamps.

The humid eastern coastal plains give way to mountains, beyond which are the vast subhumid lands stretching from Canada to the Gulf of Mexico. This area is one of the largest well-endowed agricultural regions of the world. It covers from Ohio west to most of Minnesota, the eastern Dakotas, much of Kansas and Oklahoma, and central Texas. Even though drought is a potential handicap, the area is the key to America's dominance in world food production. It has a wide range of summer and winter temperatures and precipitation adequate for a great variety of crops. Most of the moisture comes during the growing season. In its native state the largest part of this region was covered with prairie grass that ranged from 3 to 10 feet high.

To the west of the rich subhumid region lies the Great Plains, a belt of land varying from 300 to 400 miles wide and extending from Canada on the north to the Rio Grande on the south. The elevation varies from 3,000 feet on the east to 4,000 feet on the west. Rainfall ranges from 12 inches in the north to 20 inches in the south, but there is a wide range from the driest to the wettest year. The region is subject to rapidly changing temperatures, with extremes from −63°F in the winter to 117°F in the summer. The Great Plains are a very risky agricultural area, but they have the potential for excellent crops when conditions are good. The region is ideal for range cattle and sheep. In recent years extensive irrigation projects have altered the economy of some areas.

The western edge of the Great Plains blends into a large arid region that averages about 600 miles in width and stretches 1,100 miles from Canada to the southwestern border of Texas. Lying between the Great Plains and the crests of the Rocky Mountains, covering part of 11 states, this arid area is extremely challenging to agriculture. Generally, crop production is not feasible without irrigation. Fortunately, heavy snow and rainfall in the mountains provide a large supply of water for irrigation.

West of the crest of the Rocky Mountains to the Pacific Coast lies a vast summer-dry area extending from Puget Sound to Baja California. This region includes the western part of Washington and Oregon and most of California. Except for an occasional local thunderstorm, the region experiences less than 10 percent of its precipitation during the summer growing season. Its mild winters seldom have temperatures low enough to harm winter-growing crops.

The early Spanish settlers to California brought fruit trees and grain crops that flourished there. The Catholic missions, which developed a high state of agriculture, raised livestock and grain, in addition to figs, olives, grapes, and a wide variety of garden crops. Unfortunately, this excellent agriculture declined after the 1820s, when Mexico became independent from Spain. It partly revived after gold was discovered in 1848. Cattle raising, which had survived, gave way to wheat production in the 1860s and 1870s, followed by wool in the 1870s and 1880s, fruit in the 1880s and 1890s, and dairying from the 1890s onward.

The pattern was different in Washington and Oregon, where trappers first entered the area, followed by traders with the Indians, missionaries, frontiersmen, and later the settlers. The settlers brought the practices of midwestern agriculture with them. The short summer-dry season and heavy winter rains with a relatively mild climate provided a solid base for agriculture. In contrast to California, which must be extensively irrigated, there is little need to do so in Washington and Oregon.

Unit I

To 1783

Some Events and Technological Innovations That Affected Agriculture

- **1525** Spanish introduced cattle to North America
- **1609** First corn grown
- **1611** First wheat planted in Virginia
- **1612** First tobacco planted by English settlers
- **1618** Headright system implemented
- **1619** Blacks arrived as indentured persons
- **1646** Scythe invented
- **1660** Pork cured and packed for export
- **1713** British funded development of naval stores industry
- **1775–1783** American Revolution

Introduction

The progress of the human race was extremely slow until it learned how to produce an adequate supply of foodstuffs. Living off nature makes people physically strong, but it does little to stimulate their mental abilities. Eking out the essentials of life left little time and energy for other pursuits. Societies, therefore, remained primitive until they learned how to produce the necessities of life, especially food. Nearly everyone had to be involved in the production of food. Today the smaller the percentage of people directly involved in producing food, the more progressive that society is.

Throughout most of history farmers were almost always involved in other occupations to supplement their meager farm income. Unfortunately, these secondary occupations often interfered with good farming practices. The most common secondary occupations were fishing, hunting, gathering sweeteners and salt, making potash, and lumbering.

These secondary occupations provided the farm family with goods necessary to barter for supplies not produced on the farm. The Europeans who settled on the North American continent to farm were seldom totally self-sufficient. Nearly all of them relied on goods or services of others in nearby communities. Very few were so isolated that they basically functioned without help from others. Those families survived under extremely primitive conditions and had a very low standard of living, so whenever possible the immigrants settled along rivers and as near to existing communities as they could.

One objective of this unit is to show how the Indians, i.e., Native Americans, survived with their primitive agriculture and to observe how newcomers adopted Indian farming techniques and crops. The Europeans also brought along their crops and methods of farming, but even in the late 1700s agriculture was still very primitive. It resembled the agriculture of biblical times more than the profession as we know it today.

Indian Agriculture

Over the centuries the Indians learned to use nature in such a way that enabled them to survive under the best conditions possible given the technology available. The figures vary greatly as to how many people existed in what is now the United States prior to the advent of Europeans, but one thing is certain: they never had an adequate food supply. We also know that most of their time and effort was devoted to acquiring food.

Foundations of U.S. Agriculture

The Indians made many contributions to U.S. agriculture. They propagated and controlled wild plants, which increased the available food supply without the need for additional cultivation. Practically every known plant with sufficient seed to justify gathering was used. Many poisonous plants were prepared in such a manner that they were safe to use. The Indians practiced plant breeding by making selective seed selection. They learned the need for spacing seeds to admit sunlight to the plants, permit cultivation, and secure soil nourishment to a greater extent than in wild conditions.

Once the Indians became more sedentary, they practiced using various forms of fertilizer. Those who lived close to the water had an advantage, for they could use fish and shells as fertilizer. Those further inland used roots, grasses, ashes, and other organic matter, including manure, to improve production. They broke and pulverized the soil surface to make more plant food available without losing excess moisture. They knew that it was essential to control weeds. Multiple cropping was practiced to gain the most from the soil and the labor needed to produce crops.

The Indians learned very early that silt soil, although not as rich as the lowlands, was the easiest to work. This was critical in the days when all power was produced by humans, working with three-sided stone hoes to break the ground. Sometimes the shoulder blade of a deer or of a moose was used as a mattock. Hoes for weeding were often nothing more than large clam shells attached to a wooden handle.

To clear land for crop production, the Indians girdled the trees. After the trees died and dried out, they were burned. This was a wasteful process as far as the trees were concerned, but it was the most labor-efficient way to clear the land. It made little difference that the trees were lost, because the Indian population was sparse and except for firewood had little use for timber.

Corn cribs were made to store and dry corn. Tobacco was cured by artificial heat, and syrup and sugar were made by using heat to evaporate the sap. The Indians preserved fruits and berries with syrup, sugar, and honey. Meat was preserved by drying and smoking. Sun and air drying were used to preserve vegetables, fruits, berries, and meats. Vegetables, fruits, and meats were protected from rotting bacteria and fungi by burying them in the ground. Oil was extracted from nuts by boiling, and paints, dyes, and stains were extracted from plants by various means. Fibers from vegetables and cotton were adapted to spinning and weaving. Through the process of experimentation, the Indians learned how to cook, process, and mix various foods to gain the greatest benefit from them.

Better crops were developed through constant experimentation. Women led in this work because they were in the best position to observe how the crops performed under given conditions. This was especially true of corn (commonly called "maize"), which was the Indians' major crop. Experiments led to the development of new strains of dent corn, flint corn, sweet corn, popcorn, tobacco, beans, pumpkins, and summer and winter squash. The Indians had advanced enough in crop production so that in some regions they lived almost totally from cultivated plants. In the eastern and coastal areas, maize was more important as a food crop than wild game.

Because of a lack of knowledge regarding soil fertility, villages relocated every 10 to 25 years as the soil became exhausted. The length of time between moves depended upon the natural fertility of the soil. In time wild game and firewood became more difficult to secure, providing another reason to move.

The villages contained between 10 and 50 families, and the amount of cultivated land per village ranged from 20 to 200 acres. However, a few isolated villages had 1,000 or more acres in cultivation. In addition to the corn field, which was nearly always common ground of the village, there were also individual garden plots. Each family took care of its own vegetable plot, which in most cases was probably less than an acre.

In many tribes women ruled the household, and it was not uncommon that if the husband got out of hand he was ejected from the lodgings. There were incidents where early explorers found entire tribes ruled by women. According to the Constitution of the Iroquois League, women owned the land. Their society was based on maternal lines. This probably came about because as early as A.D. 1000, women's agriculture was considered the mainstay of the Iroquois economy. The women who owned land elected a matron to inspect the fields and direct the work schedule for the fields.

Women did most of the farm work because the men spent their time hunting, fishing, trading, and fighting wars. Farm work was not looked upon as merely "women's work," for it was considered every bit as important as housework and raising a family or any of the other tasks that had to be done. It was absolutely essential that the women did this work, for without their efforts the Indians could not have survived. Men had to stay agile for hunting and fighting, which often required feats of strength and endurance. If men had farmed they might have lost some of their fleetness and dexterity. However, when men had free time, or if the task was particularly difficult, they took part in farming.

Indian women did not feel that their lives were unhappy. Some accounts indicate that they looked forward to spring and working in the fields and gardens. They took pride in having clean gardens and good crops.

Other than domesticated turkeys of the western pueblo dwellers, the only animal domesticated on a large scale was the dog. This was a major obstacle to progress in food production because it limited the power available for farming. Because the Indians had no domesticated cattle, they needed no forage, so they frequently burned the meadows. After doing this a new growth of grass appeared, which attracted deer, elk, or buffalo. Was this a form of pasture management, or was it a method that they practiced because it drew the animals?

Except for scarecrows to frighten birds, the Indians had no means of protecting their crops from natural pests or diseases of the crops or soil. Plagues of any nature were looked upon as punishment for some wrongdoing.

It is estimated that as recently as 1992 nearly 60 percent of U.S. crop production came from plants bred and enhanced by Indians. The major ones are corn, squash, beans, pumpkins, white and sweet potatoes, tomatoes, blue grapes, peanuts, sunflowers, and cotton.

Maize and Associated Crops

The Indians made great strides in breeding maize, their "king of the crops," which proved to be a boon to later immigrants. Indians had farmed as early as 7000 B.C. and by A.D. 1000 were producing maize. Beans and squash were produced in the Ohio area as early as 650 B.C., and by A.D. 800 agriculture was advanced far enough so that it was no longer strictly dependent upon garden plots but was expanded to an extensive field system.

Indians of the East Coast and the Midwest likely had contact with Mexican and southern maize growers even though there was a great distance between their homelands. Because maize was one of the most efficient crops in terms of labor and land use, its production was extended as far as climatic limits made it economically feasible.

In the northern sections of the maize area, squash, beans, and maize were all planted in the same field. The squash vines shaded the ground and minimized weed growth. The beans climbed the cornstalks to gain maximum sunlight. In southern sections, maize fields were also planted to millet, melons, sweet potatoes, and gourds. As in the North, these crops were complementary to maize in terms of labor and ground needed to produce the crops.

Maize was planted in April, May, and June, which spread harvesting of the mature crop through August, September, and October. Over the centuries the Indians adapted maize to the areas where it was grown. This meant that they developed plants that matured in from 90 days in Canada to 150 days in the South. When virgin land was planted, maize yielded as high as 30 bushels per acre, with 15 to 20 bushels being a normal crop. When production fell under 10 bushels per acre, fields were abandoned for new locations.

Maize was dried in aboveground cribs, after which it was placed in underground pits for future use. Dried grass was placed between the layers of maize to enable air circulation for continued preservation.

Everywhere in the maize area, domestic crops were supplemented in the diet with whatever nature had to offer. In the North, wild rice was a staple. Maple sugar, which required processing, was another major natural food. Naturally grown crops were far more abundant in the southern regions of maize production than in the northern ones. Throughout the maize region the domestic and wild crops were supplemented by deer, turkey, small game, and fish.

Tobacco was produced wherever it would grow. It was considered essential for good health as well as for religious and political rites. In many of the tribes tobacco production was handled by the men rather than the women, but the reason for that is not clear. Unlike most other crops, tobacco was not grown within the fields planted to maize.

In the rugged terrain of the extreme Southwest, the Indians practiced intensive agriculture based on irrigation. North of the area of irrigation and west of the crest of the Rocky Mountains was an extensive wild-seed area. Under natural conditions there were ample berries, nuts, and seeds to supply the needs of a sparse population of gatherers. To the north, in what is called the Pacific Northwest, where native vegetation was not so abundant, the Indians practiced salmon fishing for their livelihood.

The pueblo dwellers of the intense crop area of the Southwest produced maize, beans, melons, squash, sunflower seed, tobacco, and cotton. Those crops were supplemented with onions, chili peppers, and other wild plants of the area. Maize bread, which was processed into thin sheets, was the dominant food. About the only game found in the area were rabbits and turkeys. Turkeys were domesticated.

Wild Seeds, Irrigation, and Hunting

In western areas beyond the region of good maize production, Indian agriculture changed to match the environment. Maize was planted along the Missouri River, but sunflowers were increasingly important. They worked well in the shorter growing season because they could be planted before the soil was warm enough for other crops and did not have to be harvested until after freeze-up in the fall. Seeds harvested after frost produced an oily meal, which was treasured. It was used separately or mixed with beans, squash, or maize. Balls of this mixture were carried by warriors or hunters for quick nourishment.

In the South, common cane was widely used. Bread or gruel was made out of the grain of cane, and the young shoots were eaten much like asparagus.

In the Southwest, the fruit of cacti was a favorite. It was called "prickly pear" and was unique in that the fruit could be preserved in its own juice. It was one of 14 kinds of roots and 42 kinds of fruits from trees and plants consumed by the Indians. The meats of many of these plants were dried and then pounded into powder, which enabled them to be preserved indefinitely. Later the meats were mixed with gruel from cane or maize for eating.

Irrigation commenced about 100 B.C. in the Salt River Valley of Arizona. As many as 250,000 acres were irrigated from reservoirs and long ditches, some of which were carved out of solid rock. In areas of sandy soils, the ditches were lined with clay to hold water. The Pueblos also irrigated in the Rio Grande Valley in what is now New Mexico.

Soon after the Spanish explorers discovered these systems in the sixteenth century, that knowledge was transferred to California.

There was less agriculture in the form of tillage as we know it along the Pacific coast, because of the access to seafood, wild fruit, nuts, grains, and wild animals and fowl. Because of the relatively ample supply of staples, the coastal settlers generally maintained permanent villages. Berries and salmon were abundant, so there was little need to move.

Hunting was a risky business because weapons were primitive and a great deal of skill was needed to bag wild game. It is estimated that in any settlement of 100 persons, not more than 5 or 10 were able-bodied hunters. This put a considerable burden on them, for they had a daily quota to fill in addition to keeping their hunting equipment in repair.

To keep a settlement of 100 people supplied, at least 100 deer skins were needed annually for tepees and clothing. Each hunter had to bag a deer every other day to meet the food needs of the group, which consumed 4 pounds of meat per capita daily. A dressed bison yielded about 1,000 pounds, a moose 800 pounds, an elk 350 pounds, and a deer 100 pounds. Venison was the favorite, but deer were elusive and hard to hunt. Bison were the most dangerous to hunt, but often they required the least effort.

The accomplishments of the Indians regarding permanent agriculture were significant. Their efforts greatly lessened the hardships of the immigrants in later years.

Colonial Agriculture

European Customs Are Resisted

Soon after European immigrants arrived on American shores, their customs and habits came into conflict with the new environment. But the newcomers clung to their traditional life styles, legal practices, and farming methods until they were forced to alter them to conform better to the new conditions. Events moved slowly in those days, and from 1607 to nearly the 1800s, agricultural practices remained basically unchanged.

Land, the magnet that drew so many to our shores, could be acquired so easily that individuals took great risks to come to the New World. Almost anyone who desired land was able to secure it at a reasonable price. Those who did not want to pay were free to enter the wilderness and establish a farm to their liking. A heavy price was paid for their independence by living away from others and under very primitive conditions. Peasants in Europe had suffered under harsh conditions for centuries, so the price of freedom was not a particular hardship for them.

Most countries of Europe had experienced agrarian revolts. What the peasants learned from those events influenced their expectations of the new land. In seventeenth-century England, small farmers were under social and financial pressure to sell their farms, and the remaining tenants were placed under more rigid regulations. Small farmers were driven from the land as farms were consolidated to take advantage of more profitable farming procedures.

Agricultural pressures in England and the hunger for land made for ideal conditions that had a direct bearing on what happened in colonial America. Fortunately, England realized that to achieve successful colonies, agriculture had to be permanently established and encouraged. France and Spain had failed to recognize that fact. Unfortunately, England attempted to establish the English land system in the colonies. This displeased the immigrants, who wanted nothing less than actual ownership. When Governor Thomas Dale attempted to assign 3 acres of cropland and 12 acres of woodlot to each newcomer, it caused sufficient unrest to lead to his recall.

The next governor removed Dale's restrictions. He allowed settlers freedom to produce more tobacco and granted them up to 100 acres each. After he abandoned the communal policy, the colony experienced an influx of immigrants. This was a clear sign to the leaders of what was necessary to get people involved in agriculture.

The introduction of slaves and the importation of young women in 1619 were, in part, moves to aid settlers. The slaves were destined to make up for the constant shortage of farm workers that resulted from most free Europeans wanting to farm for themselves rather than work for another. Each settler who married one of the women had to pay a sum of 120 pounds of tobacco to cover the cost of her transportation.

After the initial protest against communal farming in Jamestown, actions were taken that helped to shape the laws governing agriculture and land ownership. The law of primogeniture was abandoned in New England and was replaced by acts that provided that all heirs should receive equal portions of an estate. The one exception was that the eldest son received a double portion. This law was passed because land was basically the only source of wealth.

In Maryland tenants rebelled against the manorial system because they wanted to be landowners, not tenants. A 1634 revolt in Massachusetts Bay Colony was directed at the removal of privileges. Bacon's Rebellion of 1676 occurred because farmers felt that they were overtaxed, they resented not having a voice in government, and they were discriminated against by large planters who received priority over small holders when shipping tobacco. When trouble arose with the Indians and the government did not respond to calls for help, the settlers rebelled, which resulted in reforms in 1679.

Quitrents, which were levied from New York south, were a constant source of irritation to the settlers because they were collected by nonresident landlords or they were not used for valid governmental purposes. Farmers resented them because they were taxation without representation and they showed the inequities in the land rental system. Even proprietors who lived in the colony, such as William Penn, found quitrents difficult to collect. Collecting quitrents proved so challenging that many leaders gave up trying to do so. But since quitrents came with the land and they might be collected forever, they had a negative impact on the sale of land. Farmers found methods to evade paying them.

Colonists were encouraged by the changing conditions in Europe relative to landholding and rental payments, so they stiffened their resistance against whatever they did not feel was reasonable. Quitrents headed the list. Eventually the courts generally sided with the settlers in suits involving quitrents. This was the beginning of quieting the issue, although it took the American Revolution to finally do away with quitrents.

The Dutch had one of the most rigid landholding theories, under which they attempted to establish the patroon system. Royal favorites received large tracts of land and were encouraged to perpetuate the feudal system as it was known in Europe. Local settlers who rebelled against the patroon system and its attempts to collect quitrents suffered severe penalties for their efforts. The large landholdings persisted even after the British acquired New Amsterdam. It was not until the 1880s that the last problems regarding patroon landholdings eventually faded away.

Probably the most democratic of the early landholding systems was the village system in New England. It was devised to create a compact settlement within a township six miles square for protection of settlers as they faced the potential hazards of the unknown frontier. Ideally, each settlement was to include at least 20 fighting men. Each individual grant provided for a small lot within the village, use of the commons land, and an acreage for farming in the outlying area.

Theoretically these were to be homogenous racial or religious settlements. Use of the commons land produced an immediate and constant dispute. The lack of crop protection for the outlying fields, the inconvenience of going back and forth to the distant fields, and the restrictions on crop rotations became some of the basic causes of dissatisfaction with the village system. By the end of the eighteenth century, most of the commons land had disappeared, even though the New England village itself has persisted to this day.

Although the colonies had liberal land distribution policies in relation to those of Europe, there were always tenants. This was due to the fact that a great amount of land was taken by speculators and the prices they asked did not seem reasonable to some buyers. Other available land was isolated from the settlements and subject to attack.

Some farmers preferred renting improved land rather than facing the task of improving land on their own.

The movement of frontier farmers was part of a trend toward the decline of feudalism and the rise of capitalism. It was a strong factor in the cause of the American Revolution, because farmers felt that they had suffered more than their share from royal decrees and parliamentary laws. This was particularly so in the realm of taxation, because land was the most accessible source of funds.

Parceling Out the Land

Once the leaders realized that land was the magnet to get settlers to the New World, steps were taken to make it readily available. In 1616 the London Company declared that settlers who paid their own way should be given 100 acres of land each. Shortly thereafter each stockholder in the Company was given 100 acres for each share of stock owned. Later a second 100 acres per share was given.

The next major step taken was the allocation of a headright of 50 acres for each person who came over at his or her own expense and another 50 acres for each person he or she brought over. Indentured servants spent four to seven years to pay for their passage. Frequently at the end of their servitude they secured land, the amount of which varied by colony.

The purpose of the headright system was to stimulate immigration. Communities in England saw it as a means of getting rid of paupers and criminals. This provided for the rapid growth of the colonial population and the number of farms. Because additional land was granted to those who paid the way of others, the headright system became a means of acquiring sizeable blocks of land by some individuals. Considerable corruption existed under this system, for sailors' names were used, headrights were filed for in more than one county, names were falsified, or headrights were wrongfully purchased from colonial officials. Some planters paid for the passage of more workers than they needed simply to acquire more land. In some colonies more land was distributed via the headright than by public sale, for its chief aim was to attract settlers, not to raise revenue.

The motives in New England were different from what they were in the Middle Colonies and the South. Soil conditions and climate there were not as attractive for acquiring large blocks of land, but people were still interested in owning land. In Plymouth every person received a share of stock and 20 acres of land. At first land was worked in common, but too many shirked their duties. After 1623 everyone had an acreage and production improved.

The New England township described in the previous section soon became modified as more settlers moved into the township and the land had to be redivided. Some of the farmers became carpenters, cobblers, millers, or merchants and gladly sold their land to others who wanted to increase their holdings. Those who retained their original small farms frequently turned to trapping, fishing, or logging to enhance their income. There were virtually no large farms, because there was no crop that encouraged expansion. This meant that farms were mostly owner operated and tenantry was almost unknown.

Pennsylvania soon became the breadbasket of the colonies because of its fine, rich soils, its good climate, and its central location. William Penn wanted it to remain an area of small owner-operated farms, so he sold 100 acres for 2 pounds sterling and a very nominal annual quitrent. Individuals too poor to buy land could rent for one shilling an acre per year.

Thousands indentured themselves to come to Pennsylvania. Some even indentured their children. Once these people were freed from their obligations, they went to the unsettled areas to start their own farms. Penn appealed to Germans, who came in large numbers and became good farmers on the rich loam soil. Most of them raised livestock in addition to growing wheat, hemp, and flax. To this day Pennsylvania has remained a state of small farms with intense livestock, poultry, and vegetable enterprises.

The Growth of the Livestock Sector

It was obvious to those who planned the colonies that to have a successful agriculture, livestock had to be produced. Hogs, cattle, sheep, goats, and horses were on the first vessel to arrive at Jamestown in 1607. Unfortunately, those animals were all consumed during the first winter. In 1611 a second shipment arrived, which became the basis of a permanent livestock economy for Virginia and the South.

Because of constant loss to Indians, hogs were placed on Hog Island in the James River, and the other animals were kept in a stockade. Soon there were 200 cattle, 200 goats, and more hogs and horses than could be cared for. The hogs and horses were allowed to roam and fend for themselves. Despite hunting by the Indians and predators, their numbers continued to increase, and before long, plantation owners reported damage to crops from the livestock that ran wild. The famous zigzag rail fence, which was more picturesque than effective, soon became a popular sight.

Additional shipments of cattle added to the growing numbers. By 1627 there were between 2,000 and 5,000 cattle and innumerable hogs. Sufficient livestock was available in Virginia by 1633 so that ships could be provisioned before returning to England. It then became necessary to search for a market for live cattle.

Livestock problems were handled differently in New England, where livestock was far more critical for the survival of farming. Many of the earliest town regulations grew out of agricultural requirements, especially grazing rights. Stallions, bulls, and rams were particularly controlled to avoid indiscriminate breeding. Those not kept specifically for breeding were required to be castrated upon reaching a specified size.

A "no fence law" was the vogue, which meant that livestock had the freedom of the "open range." Some colonies specifically required that cropland be fenced to protect it from damage by livestock. Meadowland was also protected. After the cropland and meadow were harvested, a restricted number of animals were allowed to graze. Frequently townships had fence viewers, whose duty it was to see that fences were in repair to protect the crops.

The cost of fencing, in comparison to other farm expenses, was excessive. Fortunately, it was somewhat eased in New England settlements, where everyone lived in compact communities. The same applied to the areas of Dutch settlements. However, community fencing was not possible in the Middle Colonies and the South, where farms were more isolated.

Another benefit of living in compact communities was that herders rotated the task of caring for the animals. The young stock, sheep, and hogs were pastured farthest out in the remote, uncleared areas. A shepherd stayed with the sheep the entire season that they were on pasture. For added protection the sheep were enclosed nightly to avoid loss to predators, especially wolves and bears. Sheep were difficult to raise to maturity, but they were essential to the economy because their wool was needed for personal clothing and by the budding textile industry.

Originally, throughout the colonies, bounties were paid to the Indians for the capture and elimination of predators. Wolves, bears, cougars, panthers, foxes, and wildcats topped the list. Most of the time the Indians were given a cow, a hog, or a sheep as bounty in exchange for a specified number of predators.

Hogs and goats did not need as close attention, for they were more capable of staving off or escaping from predators. Because hogs were the most troublesome, some colonies required that they be pastured at least five miles from planted crops. Free-roving hogs were caught and ringed and sometimes had wooden yokes placed around their necks to prevent rooting. Hogs were also the best scavengers and fed on clams, shellfish, and acorns and other nuts. Supplemented with corn, the above ration enabled hog numbers to increase rapidly enough so that by 1660 it was possible to export salted pork to the West Indies.

Oxen, horses, and milk cows, which were needed on a daily basis, were pastured nearer to the village. The cowherd, who was considered a village official, took all the cows of the community to the pasture every morning after milking and returned them for evening milking. Village boys and girls sometimes were required to help the herder. The number of cattle each farmer was allowed to graze was determined by town authorities, based on the individual acreage tilled.

In spite of the attention that the animals were given, they did not do well. The natural food supply of the forest and the meadows was not always sufficiently nutritious to maintain good health. The animals were not housed during the winter and often were not provided with the additional feed necessary for body heat during cold weather.

The farmers initially did not realize what a severe toll the harsh New England winters would have on livestock. Also, they often did not have the means with which to provide the shelter needed. It was several decades before tame grasses and root crops were provided for winter feed. Such crops demanded more of the farmers' limited time. This competed with the time needed for their secondary occupations, which were essential for the purchase of goods not produced on their farms.

As the population of the colonies increased, livestock numbers rose proportionately, and the problem of keeping a proper balance with the available commons land intensified. It was a regular practice to overgraze the commons and the pastures nearest to the village. This often meant that some of the inhabitants had to establish another village. After the American Revolution, cattle were no longer herded together, and branding was discontinued. Individual farmers then became responsible for their own livestock, and the common pasture became the village green.

The other alternative, especially in the South, where natural conditions were more favorable, was to turn the livestock loose. This meant that the animals would mix with other herds. As late as 1710, Virginia forbade settlement further west than rangers could protect the herds. However, after 1720, when the rich prairie grasslands of the Shenandoah Valley were discovered, there was no stopping the influx of settlers. Under such conditions livestock was not closely watched, and the obvious outcome was stealing by individuals waiting for the opportunity. When branded cattle were missing, it was often rumored that a neighbor had been in the hills butchering.

One of the reasons livestock multiplied as rapidly as it did in the early years was that the animals lived in a pest-free environment. As long as there was an abundant supply of native grasses, they thrived; but that soon came to an end, and livestock frequently died from starvation. In a semi-starved state the animals also were easier targets for disease.

It was not until cattle were exposed to the low, swampy, fly-infested country in Georgia that other problems occurred. The streams abounded with leeches that fastened themselves onto the cattle, causing water-rot or scald, which frequently resulted in death. The first reported cases of rabies appeared in Virginia in 1753, and around Boston in 1769. The disease was known in Europe and was quickly recognized by the colonists.

Herding organizations employed cowboys, who erected temporary fenced enclosures called "cowpens" to hold, brand, or otherwise round up the cattle for a drive. Sometimes crude shelters were built for the cowboys, who lived on the frontier with the cattle. Once cotton became more significant in the back country, livestock freedom was greatly reduced.

Relations with the Indians

By the time the first Europeans arrived in the New World, the Indians had developed a relatively high and stable culture based on hunting and crop raising. Columbus acknowledged as much when he landed in the West Indies in 1492. In 1586 a colonist at Roanoke, in what is now Virginia, recorded that the settlers had no corn for seed, yet they were aware that the Indians had ample food supplies.

In 1609 Governor Dale, of Jamestown, contracted with the Chickahomineg Tribes to secure one bushel of corn for each English bowman. This not only saved the colony from starvation, but it also acknowledged that the settlers knew of the success that the Indians had in growing corn.

At a later date, Captain John Smith kept Jamestown settlers alive by purchasing corn from the Indians. He noted that there were fields of up to 100 acres and reported that at least 3,000 acres were planted in what is now Virginia. Smith knew that generally each stalk had two ears, sometimes three, signifying a relatively high degree of production.

After the settlers of Massachusetts Bay Colony nearly starved to death, Wampanoag tribal chieftains Squanto and Massasoit aided the Pilgrims by showing them how to plant corn using fish heads for fertilizer. That corn produced an ample crop that was harvested and served as the basis for the November 1621 Thanksgiving. The Indians also showed the settlers which native plants were edible and could be cultivated for crops.

The axes, shovels, hoes, and saws used by the settlers were so prized that the Indians readily traded agricultural land for them. They also treasured other tools, weapons, clothing, jewelry, and goods of the settlers and gave land to secure them.

Prior to the advent of Europeans, many forests had been cleared for hunting and for farming. That made conditions easier for the newcomers, for it would have taken at least a generation to prepare that land for cropping. Ironically, the reduction of the Algonquin Tribe in New England from smallpox made a considerable acreage available to the immigrants without any opposition.

Unfortunately, life between the newcomers and the Indians was not without conflict, and a state of tension seemed to be almost constant. In 1622, 347 Virginia settlers were killed as a result of an attack. Unrest continued until 1644, when the English became powerful enough to gain the upper hand. This prolonged conflict was caused by settlers who stole corn and took land from the Indians.

Similar conflicts took place throughout the early colonial period, and in nearly every case the wrongful seizure of land or of crops was the cause. Destroying crops, cattle, storage buildings, and homes seemed to be the favorite tactic of both sides, for that caused the greatest suffering. Families living on isolated farmsteads were the easiest targets because they had only limited protection and were susceptible to ambush. Some colonies directed that at least one-fourth of their male inhabitants be under arms daily as a precaution against trouble.

Much of the problem was caused by the fact that neither the French nor the Spanish recognized that the Indians had title to their land. Therefore, they rarely were compensated when others occupied it. Initially, English settlers adopted the same policy, but eventually they recognized the rights of the Indians. Indians did not place a value on real estate as Europeans did. Their interest in the land was in having the freedom to roam and hunt, not in owning it.

In 1634 Lord Baltimore probably was the first European to purchase land and houses from the Indians. Their houses were not lavish, but they were better than what most of the early colonists built for themselves. That first spring the colonists were helped by the Indians to plant corn, tobacco, and garden crops.

The previous year the General Court of Massachusetts declared that Indians had rights to the land that they lived on and had improved. Colonists were required to obtain permission before they could purchase land from them. It was clear that the New England authorities recognized the need for protection of the Indians against wrongdoing of the newcomers.

Colonial Labor Supply

To a great extent the abundant supply of cheap land resulted in a scarcity and high cost of labor. Early legislatures attempted to solve the problem by establishing maximum wage rates. This was not a solution, for such laws were extremely hard to enforce. As long as land was available, a large portion of the free workers made every effort to acquire their own farms rather than work for others. Most attempts to employ Indians had proved unsuccessful.

Four ways were open to solve the labor shortage: (1) have large families; (2) import marriageable women; (3) import indentured workers; (4) import Africans as slaves.

Large Families

Large families were the rule in New England. However, they were only a partial solution to the labor shortage, for in most cases, as soon as the children were of age, they left home. So many opportunities were open to them—either working for others or securing land of their own.

Families of 10 to 12 children were not uncommon in the South. The need for companionship because of the isolation on the frontier and on the plantations was strong, but children were also looked upon as defenders and workers. Often a woman died in childbirth after already having a large family, and the widower remarried and had a second family. As in the North, children left home as soon as they could, but until that time they were put to work as quickly as they could lift a hoe.

Importation of Marriageable Women

Soon after the first colonies were settled, stockholders realized the need for women as workers but more importantly as a means of stabilizing conditions that would result only after families were established. In 1619 it was recommended that 100 women be imported to Virginia. They could be married only to freemen who had the means to support them. The following spring 90 women arrived, and in 1621 another group of 38 arrived. By 1622 all of the women had married, but only 35 were alive in 1625 after an Indian massacre.

Several colonies passed legislation granting land to women just as they had to men. Other women came as indentured servants and generally could not marry during their time of servitude, or they were restricted from marrying until later than normal age. Late marriages, high infant mortality, and a high male-to-female ratio meant that the labor shortage persisted. Women were reluctant to marry men who did not own land, which further limited growth potential of the labor supply.

Indentured Workers

Securing indentured workers was the next obvious alternative. This labor supply came from two sources—prisoners who were delivered to our shores and sold for transportation costs, and individuals who voluntarily indentured themselves. Indentured persons could be either female or male. The Middle Colonies, in particular, received a large immigration of English, Scotch-Irish, and Germans, who came to labor as indentured workers for four to seven years.

The goal of indentured persons was to work out the period of servitude to pay for their transportation, learn the customs and the language, and learn how to farm. After they were free, they worked for others. The relatively high labor rates were attractive, and within a few years they could accumulate enough cash to start farming.

It is estimated that about 50 to 75 percent of the immigrants who came from England between 1630 and 1680 were indentured. Males outnumbered females about six to one in the early period and three to one in later decades. Most of these persons were youth with no skills, and probably not more than 25 percent of them had any farm background.

In the long run they did not greatly reduce the demand for labor because of the high death rate from malaria, smallpox, or influenza. Indentured immigrants did not replace themselves in the population. Because of restrictions, most did not marry until they were into their thirties. With the high mortality rate of those days, it was common that one or the other of the spouses died within 9 to 13 years.

England prohibited emigration of skilled workers to the colonies to guard against the transfer of any knowledge of industrial value. The Germans did not care to go to the South because of the domination there by aristocrats and the negative attitude against manual labor. As conditions improved in England, individuals were less willing to volunteer to become indentured. This caused the period for indentures to be reduced to four years. Americans became less willing to purchase indentures because of the cost for the time involved.

Initially, indentured individuals made up a large portion of the workers of Virginia, far outnumbering Africans, Indians, and even the plantation owners. In Maryland, there were six indentured persons for each free person. In New England, the percentage was never high, for the economic opportunity to use the indentured in agriculture was not that important.

The indentured system was partially satisfactory in some colonies until about 1680. By 1700 trade in indentured people had dropped drastically, even though the system persisted until prior to the Revolution. Indenturing was a positive factor for many English criminals, as it opened the way for a new life. It also gave opportunity to other Europeans who wanted to improve their conditions.

Slavery When European indentured persons of any nationality were no longer available, planters turned to buying Africans as slaves. Some planters were reluctant to do so, while others eagerly awaited the slave ships and purchased Africans to speculate on increasing prices.

The first Africans arrived in 1619 and were received as indentured servants. Some of them may have acquired their freedom after their indenture expired. After 1640 most of the Africans who arrived were made bondspeople for life. By 1660 there were 1,700 transplanted Africans in Maryland and Virginia. In 1661 the uncertainty as to their status was ended when the Virginia legislature recognized slavery. In 1698 Parliament opened slave trade to private merchants, which increased trading activity, and by 1700 about 3,000 Africans were enslaved, a number equal to the total that had been enslaved in the previous 20 years. By 1768 about 6,300 slaves had been imported.

Initially, planters had preferred white servants to black slaves because they were English speaking and were thought to be more productive. African women were just as involved in field production as the men, for it was obvious from the beginning that the cotton, rice, and tobacco economies could not survive without their input. It was sometimes argued that slaves were less expensive than indentured persons because indentured women did not work in the fields. In any case, if the planters wanted to continue raising plantation crops, their hands were forced.

If only tobacco was grown, after three crops the land had to be idled for 20 years. When other crops were produced in rotation on a plantation, at least 50 acres were needed for each field worker. One worker could handle about three acres of tobacco a year. Cotton, another leading plantation crop, was so labor intense that its production was on the decline in the late 1700s because cotton goods became too costly.

Because of the capital investment in necessary slave labor, plantation ownership by former indentured persons was basically out of the question. Therefore, plantations were limited in number and generally were continued through inheritance. The problem of financing was further complicated by the fact that tobacco was extremely hard on the soil, so if a high level of production was to be maintained, new land had to be opened constantly. The best alternative to moving west was to have enough acres to enable a constant rotation plus some idled acres at all times.

Pressures combined to entrench the system of slavery as English authorities realized that the greater the population growth, the greater the volume of tobacco and the greater the tariff revenue. Invention of the cotton gin and the Industrial Revolution in England added to those pressures.

New England

Initially, settlers in New England experienced a life not much different from that of the Indians. They attempted to live by hunting and collecting, and even though natural food was abundant, they verged on starvation much of the time. Hunting and gathering food from nature was difficult.

The first major task was building shelter, followed by clearing land. Only limited clearings existed for good grazing, hay meadows, and, more importantly, cropland. Some of them were the result of previous work by Indians. At first the settlers felt that they had to have clear fields just as they had in England. They were reluctant to adopt hoe culture, which simply meant planting crops among the fallen trees.

The leaders of Plymouth Colony were not knowledgeable in agriculture, but, fortunately for them, Squanto, who could speak English, was able to advise them on corn culture. The settlers secured seed from an underground storage left by Indians who had previously farmed the land. This provided seed for 20 acres, which they planted near Cape Cod. On another six acres they planted English barley and peas. The corn was fertilized with fish and was the key to the survival of Plymouth Colony. Neither the barley nor the peas survived.

In the first two years, crops were raised by the communal method. However, starting in 1623 each family was assigned one acre per person, and production improved immediately. From that time on, the settlers seldom lacked for food and soon were trading corn for beaver skins from the Indians. After seven years, each family was granted 20 acres. Only the meadowlands remained in common.

Oats and barley were widely grown throughout New England and the Middle Colonies. Barley was used chiefly for beer, and oats were used primarily for horse feed but sometimes for human food. Peas and oats or rye and oats were frequently sown together and were used as cattle feed. Peas sown alone were used solely as human food. Sometimes wheat and rye were sown together, and the grain was used for making bread.

Apple, pear, plum, quince, and cherry trees were introduced from Europe and successfully grown in New England. The apple was by far the most popular. Cider replaced beer as the dominant beverage.

Some colonies required the production of flax or hemp for making cloth. Because it was difficult to make flax into linen, the colonists resisted its use until after the original imported clothes were worn out or the fur trade failed. Once the spinning wheel for linen was developed, flax became more important. Later, sheep wool became the major source of material for clothing.

The Indians had no need for forage crops, so they had not developed them. Because the native grasses were not satisfactory as hay, New Englanders had to import timothy, bluegrass, red clover, and white clover seed from Europe. It was not until the mid-1600s that hay production reached a volume sufficient to prevent animal starvation during dry years.

Establishing an adequate agricultural base was a long and difficult task. At first, new colonists were advised to bring at least one year's food supply with them. Later, until 1650, even those intending to farm were encouraged to have two or three years' supply with them.

Massachusetts Bay Colony, which had a fairly easy and rapid start in producing food, had major supply problems until 1640. In 1648 it prohibited all grain exports because of a local shortage. Other colonies had problems producing ample food until the 1650s.

When the populations of the original communities became sufficiently large, new communities were established by splitting the older settlements. Land was allotted to each family based on the investment made and the ability to use the land. Family size was generally the major factor in determining ability. Average-size holdings were about 25 acres, but they varied from 10 to 1,000 acres. Individual allotments were not necessarily made in contiguous blocks.

By 1650 the general court of one colony expressed concern that there was considerable consolidation of land. Towns attempted to prevent this trend out of fear that they would lose population. The sale of blocks of land required the consent of the town, and if the planned sale was to a nonresident, the town reserved the right to preempt it. This was probably the first effort within the contemporary United States to restrict the size of farms. Persons with larger-than-average-size farms were often referred to as being "land poor," because they had more land than they could properly work. This was true in part because they had little capital and were short on labor, so they practiced extensive agriculture utilizing land, the cheapest commodity in the production chain.

One of the chief problems in food production, which continues to this day, was overcoming problems inherent in nature. Weeds were a relentless plague, and they often choked out the planted crops. Drought and frosts were constant threats over which the farmers had no control. The most pathetic of all losses, however, was the destruction of crops by enemies or by animals.

The destruction of farm animals by predators, climate, or sheer neglect was enormous and never ending. Sheep were by far the most vulnerable. They were easy prey for wolves or any wild creature that cared to attack them. Even less than other livestock, they could not endure the winters without shelter. It was not until the mid-1700s that shelters were commonly provided for animals. With the advent of woolen mills in the 1660s, the market for wool was more important than that for lamb or mutton. Colonies passed legislation virtually mandating the production of sheep. Once the textile demand became stable enough so that farmers could rely on a market and better prices, most problems were overcome, and sheep numbers grew rapidly.

Early American farmers were fortunate that they could draw on European breeding stock to build their herds and flocks—sheep from Spain; horses from throughout Europe; hogs, except for Poland China, Duroc-Jersey, and Chester White, which are of American origin, from several countries; plus beef and milking breeds of cattle chiefly from England and northwestern Europe.

By 1632 cattle numbers had become so great in Massachusetts that new communities had to be created because the cattle had outgrown the available pasture, meadow, and hayland. By then, Massachusetts Bay Colony was considered quite prosperous, because the 4,000 residents owned 1,500 head of cattle, 4,000 goats, and innumerable swine. By 1700 the production of cheese and butter was sufficient to enable exportation to the West Indies. Commercial dairying developed around the larger towns as soon as the market could justify such ventures.

The sandy soils of New England were never very productive, and after a few years of continuous cropping, yields declined rapidly. This was especially true for small grains. Fortunately, corn, beans, squash, garden crops, and orchard crops did fairly well and provided for a greater variety of diet than the settlers had experienced in England.

Nearly all the food and clothing used by New Englanders was produced on the farm. Initially, the only food purchased was salt. Soon brown sugar, green coffee, pepper, spices, rice, raisins, tea, and molasses were added to the list. It was not until after the Revolution that tea,

Hand-carved washbowl and scrub board with bar of homemade lye soap. (Peet Museum)

coffee, and chocolate became widely accepted. Purchased tableware and kitchen utensils were treasured and were handed down to succeeding generations. Nearly all were made of wood.

The isolation of the early settlers becomes very evident when one realizes that the first post office in New England was not established until 1639. It was not until 1693 that an effective postal system was commenced, and this was restricted to the larger communities.

In addition to isolation, New England pioneers suffered from less than adequate housing in what could often be a hostile climate. A log cabin was neither a comfortable nor a healthy dwelling. It was a dark, dirty, and dismal abode, which generally had to be lived in 10 to 15 years before the family could afford an improved home. In much of New England, pioneers lived in log cabins 25 years after the first settlement. In one Vermont community settled in 1775, people still lived in log houses in 1865.

All members of a New England farm family were expected to do their share in making the farm produce. Boys began working as soon as they were physically able. Starting at about age five, girls were expected to help their mother do some of the work that they had observed since birth. Churning, spinning, sewing, and reading to the younger children were the initial jobs. Other jobs were added so that by the time they were 10 or 12 they helped milk, cook, bake, and wash clothes. They were full-scale apprentices to their mother.

After daughters were fully trained, it was customary for them to work for others. When they hired out they were expected to spin, sew, weave, milk, and help with extra chores. When a girl married and with her husband became part of a hired couple on a farm, she was expected to work for no additional pay, only for food and quarters. Sometimes she was assigned specific duties, such as harvesting, dairying, caring for animals, or doing household chores. Hired couples lived a very nomadic life.

If a woman's husband died, she generally received one-third of his estate, including real estate, for life or until she remarried. She also received his personal estate. Unfortunately, in most cases the real estate probably consisted of a farm, which lost much of its value without added labor to operate it.

In spite of the struggle to produce food and the colonists' life of constant drudgery, the population grew steadily. The Pilgrims landed in 1620 with 101 individuals. They numbered only 300 by 1630, but by 1660 there were 3,000. Also, other colonies had been created from the original one.

The Middle Colonies

The Middle Colonies consisted of New York, Pennsylvania, Delaware, and New Jersey. They quickly became not only the fastest growing area from a population standpoint but also became the most prosperous. Location played an important role, but more significantly this prosperity was due to the rich soil farmed by immigrants who were more knowledgeable about agriculture than those who went to the other colonies.

The Dutch, who settled in present-day New York, probably were the best of the farmers who came to the colonies by the early 1600s. In their first year they produced excellent crops of wheat, rye, barley, oats, buckwheat, canary seed, beans, and flax. All crops, plus a good mixture of garden produce, were harvested by the middle of August 1626.

In 1625 they received 103 head of stallions, mares, bulls, cows, hogs, and sheep. The Dutch black-and-white cattle were of the greatest importance. By 1641 interest in livestock was sufficient to hold an annual fair for hogs and another for cattle. The work of the Dutch in livestock provided the basis for some of the colonial foundation stock and horses.

The Dutch were the first farmers to grow European grains on a profitable commercial basis. By 1645 they exported large quantities to the West Indies. They also excelled in the development of apple, pear, cherry, and peach orchards. Because of their success in farming, the Dutch were able to build substantial homes in contrast to the settlers in most other colonies. Those houses included cellars of 6 to 7 feet in depth.

Slaughterhouses were a conspicuous part of the economy in the colonies. An early thriving trade in livestock and livestock products was developed with the West Indies. Dairy products were the only livestock commodities that were not abundant in any of the colonies except Pennsylvania.

The greatest drawback to more development in New York was the patroon system. The patroons attempted to run the colony as if it were their property. This was contrary to the democratic inclinations of most of the settlers. This weakened the colony and made it easy for the British to acquire it in 1664.

Although it was not founded until 1681, Pennsylvania was recognized early as the breadbasket of the colonies. In 1643 Swedes from Delaware had moved onto the rich land in the southeastern portion of Pennsylvania. They prospered and boasted that there were no poor in their country. Prior to 1665 each freeman who came to Pennsylvania or New Jersey and provided himself with a musket, ammunition, and six months' provisions received 150 acres of land. Any master who provided an immigrant male servant received 150 acres of land. At the end of his service the servant received 75 acres.

When William Penn received his patent in 1681, Pennsylvania was already on solid footings, and within a year more than 2,000 settlers arrived. Penn was liberal in granting land but was careful to make only small grants. The colony's eventual good fortune was due to several factors. It had a large area of fertile soil, an industrious population, a supply of excellent livestock and work animals, good local and foreign commerce in foodstuffs, and the ability to avoid destructive wars with the Indians.

Pennsylvania profited from the early Dutch and Swedish settlers in New Jersey, New York, and Delaware who sold horses, oxen, dairy cows, and hogs to newcomers of means in the colony. Pennsylvania was unusual in another respect, for many of its early settlers had more capital than those going to other colonies. Many who came had adequate funds to pay for their fare but decided instead to indenture themselves and save the money to buy land after their indenture was ended. They gained from their experience during indenture and were better prepared once they became freeholders.

As more settlers with capital arrived, the original settlers often sold them their partially improved farms and moved west to start other farms. This provided Pennsylvania with a compact rural population of land-owning farmers and was in line with Penn's idea of having a colony in which most of the settlers were small landowners. His early goal was to settle groups of 10 families in townships of 5,000 acres. In a township the settlers were to live in a central village of 500 acres, with each family having an additional 450 acres in the immediate outlying area.

Pennsylvania was also blessed with a larger acreage of higher-quality soil than any other colony. That fact, combined with experienced progressive farmers, a good supply of free labor, a preference for crops capable of intensive agriculture, and yields in both field and garden crops that surprised everyone, established Pennsylvania as the queen of colonial agriculture.

Penn was a progressive agriculturalist, and his secretary, James Logan, was probably the first scientific agriculturalist in the nation. Logan made early accurate assessments of what crops would do well in the colony. Religious freedom, good land that was reasonably priced, and progressive government attracted high-quality settlers. By 1685 over 10,000 settlers had arrived, and 15 years later the predominately German population produced sufficient corn, wheat, and rye to provide for Philadelphia and also to export to the West Indies.

Complementing Pennsylvania's excellent agricultural base was its location near the coast. Whales were caught to provide for a profitable business in oil. The rivers were filled with an abundant supply of fish, which provided an addition to the diet as well as an export commodity.

Livestock was confined almost from the start, especially by the German settlers. They prided themselves on their fine barns, which often were built before satisfactory houses were erected. Penn imported good seeds, and within a few years livestock was supplied with quality hay during the winter season to keep the animals in condition. Hay was supplemented in the ration with turnips and other root crops.

Providing housing for livestock at an earlier date than in most other colonies, plus better rations, caused a faster increase in the number of top-quality animals. Soon beef, pork, cattle, hogs, butter, and cheese were exported to the West Indies. By 1700 all of agriculture, but specifically dairying, had advanced beyond that of any of the other colonies. With food supplies assured, Pennsylvania experienced a rapid increase in population.

It is estimated that 98 percent of the colony was originally covered with forests that were above average among the eastern states in commercial value. They provided the settlers with building materials, meat, clothing, maple sugar, and, as the colony became urbanized, cash from the sale of timber. On the other hand, the forests had to be cleared for farming. Forests provided shelter for livestock predators and for crows, blackbirds, and squirrels, which damaged field crops.

Until 1735, when restrictions were first instituted, the practice of firing the forest was used without limits. Because of the potential danger to other settlers, in 1794 laws were strengthened by increased penalties. It was not until 1840, however, that protective measures were passed to assure the need for conservation of the forest.

Weeds imported with grain and hay seeds were different from those already in the country. Black stem rust, or blast, which was first noticed in 1664, was traced to the barberry bush. Another unwelcome newcomer, the Hessian fly, appeared in 1768. It had entered the country on wheat grown in Asia but imported from Europe. The Hessian fly was very destructive to wheat and moved west rapidly. Because of it, by the 1790s some eastern farmers had abandoned wheat and turned to livestock and poultry production.

In New England women did dairy, poultry, and garden work but were in the fields only during harvest. By contrast, in New York and Pennsylvania, especially in the German settlements, women worked in the fields on a regular basis. As the urban areas grew, there was a demand for more products for local consumption. The farm wife, who once produced textiles, found a more profitable source of income in dairying. Butter that was not consumed in the local market found its way into export channels and by 1776 entered China. Generally, women were in charge of the butter-making process. Once commercial herds were established, the less work children could do, and adult help had to be employed. This meant that girls were freed to attend school. At least one-third of the families in the dairy area had live-in hired help, most of which was female.

The South

The ill-fated Roanoke Island settlement was founded in 1585 as the site of the first English colony. The French twice attempted to settle the coast of Carolina, first in 1562 and again in 1629. Settlement was delayed due to frequent violent coastal storms. The area had to wait for settlement from overland migration.

In 1607 colonists landed at Jamestown, and within two weeks they had cleared land for wheat and a garden. In addition to wheat they planted melons, pineapples, and oranges, none of which were indigenous to the area around Jamestown. Later, Mediterranean fruits were attempted, but only the fig was successful. That must have been discouraging to the settlers, most of whom had little knowledge of farming and had no intention of learning about it. They preferred to barter with the Indians for food, but by 1609 the Indians were becoming reluctant to trade.

Fortunately, by then 40 acres of corn were planted as a communal project. Unfortunately, most of it was eaten as soon as it was ripe, so little was saved for winter food. The colony's fate appeared to be sealed, but in 1611 a turning point was reached when Governor Thomas Dale arrived with 100 cattle and 200 hogs. Because corn had a natural advantage over small grains as human and animal food, it was the obvious crop to raise. Dale put everyone to work planting corn and garden crops. By then it was clear that without an agricultural base the colony could not survive.

After the near failure of the colony, Dale rented three-acre plots to the most productive settlers. They were free to work for themselves but had to make a payment to the colony of two and one-half barrels of grain a year. Production improved, but there were still problems. Later, each new family was given a house and 12 acres, rent free for the first year. Eventually they were free to farm as they wished.

A milestone occurred for Jamestown in 1612 when John Rolfe introduced tobacco. Tobacco was in demand in Europe, and its high price as a luxury gave potential for great profit. This endangered the growing of food crops, so Dale decreed that no one should plant tobacco until they first had two acres of grain seeded. Later, it was decreed that each settler had to store enough food for the winter rather than rely on the colony.

The reluctance of the settlers to farm caused continued suffering because of the constant food shortage. The colony was saved by the annual flow of new immigrants, who brought food and supplies. By

1622, 6,000 settlers had arrived, but only 2,000 survived. A few returned to England, but most died from starvation, disease, or Indian wars.

The future of the colony remained uncertain until the 1640s, when subsistence level was finally reached. From then until 1780, production grew more rapidly than population, so there was a surplus for export. In the early decades the settlers tried crops that were grown in similar latitudes with little success. Tobacco was important because it provided the cash necessary for developing the rest of the economy, including clearing land for further food production.

By 1620 the export of tobacco was sufficient to convince the stockholders that it would be a profitable crop. Despite a massacre in 1622 that greatly depleted the population, tobacco exports reached a half million pounds by 1628. By 1669 the figure rose to 9 million pounds. Then tobacco became such a glut on the market that it was practically worthless. During the 1680s, vigilantes invaded the tobacco fields, destroying tobacco plants in an attempt to reduce production. Such activity took place several times in an effort to keep production down.

Legislative action followed that resulted in the crop being inspected with the intent of destroying the poor-grade tobacco. Later action mandated that one-half of the good crop be destroyed. By the 1680s the price had dropped to a half penny a pound, less than 1/50 of the price when Virginia first started growing tobacco. However, the cost of production did not drop, because the methods used remained the same. Tobacco was hard on the soil, but yields per acre held constant as virgin land was constantly being opened.

In 1632 Maryland was settled for the purpose of granting religious freedom to Catholics, the first of whom arrived in 1634. In some respects it was a mix of the Middle Colonies and the South, both socially and agriculturally. Lord Baltimore initially attempted to create a feudal domain of 60 manors of about 3,000 acres each, with quitrents on a declining basis. However, only a few large estates were founded, and they were worked by indentured servants.

Fortunately, Baltimore altered his plans and offered headrights of 100 acres for each colonist plus 100 acres for each servant and family member over 16. For those under 16 the grant was 50 acres. A small quitrent was assessed annually. Many individuals indentured themselves to leave Europe because of the ease with which they could acquire land once their servitude ended. Maryland was settled rapidly because of its generous land policy, religious freedom, a considerable amount of self-government, plus the knowledge that tobacco could be raised for cash income. The newcomers knew of the experiences of the earlier colonists and went directly into farming for food production prior to raising tobacco.

A series of very cold winters in the first years after settlement caused the loss of much of the livestock. Ironically, horses, which had been allowed to fend for themselves, later caused serious damage to the crops.

In 1663 the Carolinas were granted to eight English lords in an attempt to establish the feudal system. That action called attention to the area, and English planters from the West Indies soon arrived. They introduced sugar, rice, and indigo and brought black slaves to work the fields. This became the foundation of the economy.

Commencing in the 1660s, scattered settlements of squatters produced limited quantities of tobacco. They marketed their tobacco through New England merchants whose vessels traveled along the coast. That enabled the producers to evade English customs. Efforts by authorities to collect those intercolonial duties led to the Culpepper Rebellion of 1677. In 1679 Virginia forbade the importation of North Carolina tobacco because it was of an inferior grade. This policy, which continued until 1731, forced the Carolinians to market their tobacco to New England or Newfoundland.

North Carolina produced tar, pitch, turpentine, masts, and other forest products. Unfortunately, the export of those products was handicapped because they had to be marketed through South Carolina ports, where high duties were placed on them. Such geographical and commercial obstacles delayed the settlement of North Carolina. However, the lack of good ports on its coastline did not prevent pirates, privateers, and debtors from seeking refuge there. Many of the early settlers to North Carolina were poor economically and frequently not in good health. This was a serious handicap in a desolate country subject to Indian attacks. Many settlers came because of conflict with laws or religious beliefs in other areas and sought freedom from all restraint, which was possible in the remote Carolinas.

The lack of good transportation routes was not a deterrent to those who wanted to raise livestock. Wild cattle and horses, offspring of animals that escaped from the Spanish, provided the basis for a lucrative livestock industry for both Carolinas. An abundance of good grass gave excellent grazing for cattle and horses, and the hardwood forests had a good supply of roots and nuts for hogs. Mild winters enabled year-round grazing and did not take the toll of livestock like those of the New England and Middle Colonies. Large herds of cattle, hogs, and horses ranged the interior without restraint.

After 1700 many substantial agriculturalists came to North Carolina to capitalize on its natural assets and quickly developed a large livestock economy. But they did little to improve the area. Towns and churches were virtually nonexistent. At the other extreme a few marginal farmers produced corn, wheat, rice, buckwheat, beans, and peas for local consumption. Beans and peas were the only commodities produced in surplus for export.

Livestock numbers increased in North Carolina, and the cattlemen pushed against the Indian settlements as they moved into the grasslands of South Carolina. Two crops of corn could be grown annually along with all other grains, vines, fruits, vegetables, tobacco, indigo, and cotton. Unfortunately, the Indians were relentless, causing colonists to move elsewhere and delaying the development of the area.

Although settlers into the Carolinas realized that a wide variety of crops could be grown in the area, not much was done to get an economy started. Prior to the 1700s there were not many people of wealth or many plantations. A few industrious small farmers of European background planted a wide variety of crops, but there was a far greater number of less progressive and thriftless families who lived in crude structures and existed by herding cattle and by hunting and fishing. They made a bare living in the mild climate.

The proprietors of the Carolinas favored large estates but not necessarily a system relying on slavery. They preferred white indentured servants who would become tenants when their period of servitude expired. But most Europeans did not care to immigrate to the South, and most white settlers who came did not want to farm.

Several attempts were made in the 1660s to produce rice in Virginia but without any lasting success. However, rice culture developed in South Carolina almost by accident. A ship captain gave a Charleston official a bag of rice, which was planted and yielded excellent results. By 1677 a rice economy was developing. Varieties from several countries were experimented with, and seed from Madagascar proved to be superior and the most adaptable to the area.

At first, rice was grown chiefly for local consumption. It was the staple crop, supplemented by herding, lumbering, and Indian trade. By 1710 a small aristocratic class had developed, and South Carolina was ready to emerge from pioneering. Rice exports grew from a few hundred tons annually to 12,000 tons by 1739 and 39,000 tons by 1775. A volume of that scale required at least 60,000 acres and 20,000 slaves, which meant that rice became a major part of the economy.

Once the British realized the success of rice culture, they passed legislation to promote its production. Only tobacco was more highly regulated and encouraged. Demand for rice grew steadily, except during periods of European wars. Efforts were made to expand rice production in Georgia and Florida, but naval stores and indigo were more profitable there.

Initially, rice was grown in South Carolina along the coastal islands, but by 1724 irrigation was applied by using freshwater swamps. Cyprus swamps, which were fed by springs or from water stored behind specially constructed dams, were preferred. By the late 1750s, when all the land in the eastern inland swamp area was occupied, rice production moved to the tide swamp lands. Rice was a very labor-intense crop, but because of the ease with which workers were secured there was little effort to find ways to reduce the need for hand labor. Horse power was not applied to rice production until shortly before the Revolution. Harvesting was done with a sickle until after the Civil War.

As early as the 1680s several attempts were made to introduce the production of indigo to South Carolina. Much of the work was conducted by one person, Eliza Lucas (later Pinckney), who had lived in the West Indies and knew that the plant thrived in South Carolina. In 1744 the colonial government offered a bounty to develop the industry, and the following year the British also added a bounty. Sugar cane was replacing indigo in the West Indies, and the British did not want to have to rely on the French islands as their only source of supply. The bounties made indigo production very profitable.

To achieve improved quality and to accelerate production in South Carolina and Georgia, the British increased the bounty in 1748. Production rose in South Carolina but not in Georgia. Because of the British need for indigo, production was extended to Florida, North Carolina, and Mississippi. The southern colonies, however, were at a disadvantage with the West Indies in indigo production because they were limited to two cuttings annually, while in the Indies three, four, and, in extreme cases, eight crops could be harvested. The quality of West Indies indigo was also superior, but because of demand and competition from more profitable crops in the West Indies, the southern colonies continued to produce it.

Georgia was established in 1732 to create an agricultural peasantry from among the poor and the debtors from England. The immigrants were expected to develop industry and agriculture. Most of them could not pay their way over and had no funds to become established. They were not indentured, but they were expected to clear the land and construct facilities. In addition, they were asked to serve in the military to provide a buffer between the Indians on the west, the Spanish colonies on the south, and the Carolinas.

Grants of up to 500 acres were made to individuals who would transport 10 persons to Georgia. Most of those ventures were unsuccessful.

Considerable help came to those in the Georgia settlement from people of South Carolina in the form of gifts of seed and livestock. By 1739 it became clear to the Georgia trustees that their policies only encouraged people to take advantage of the charity provided. When the public store was closed, many colonists left the project.

Rice and indigo were the staple commercial crops that supplemented corn, beans, peas, and other vegetables. Livestock was important in the uplands away from the swamps and the coastal areas. Cattle from those herds provided the foundation stock for the cattle industry of the Great Plains and the Southwest.

The other bright spot in the early days of the Georgia settlement was the arrival of the Saltzburgers, a German religious sect, who became active in horticulture and were the leading agriculturalists. They established grape growing and silk production. The Saltzburgers were the key in the silk culture and developed silk into a major export crop. But Georgia's economy grew slowly. It was not until after 1750, when South Carolina planters succeeded in forcing the British government to allow slaves, that population growth became significant.

Despite all the efforts to establish a society of free people, a constant struggle took place in the colonies because of the labor shortage. Northern farmers were less concerned about producing more than their needs because the market was so limited that little profit could be derived from sales. It is difficult to determine what portion of both northern and southern farmers took agriculture seriously beyond the opportunity of improving the land for future sale at increased prices.

In the South, selling commercial crops for profit presented greater opportunities than in the other colonies. The combination of inexpensive land and commercial crops that had great profit potential motivated individuals to seek a labor system that would make profits a reality. Many white servants in the South had voluntarily indentured themselves. Sometimes they renewed their terms when the first period expired. However, during much of the 1600s most of the individuals who were kidnapped from various countries and sold here left their bondage as soon as they could, resulting in a constant shortage of labor. Because of that, slavery became the solution, and the system grew.

Ironically, as early as 1640 the costs of credit and marketing were so large that they consumed most of the profits and at an early date had put planters in a position of permanent indebtedness. Extravagant life styles helped to create an aristocratic element and at the same time increased their financial burdens. It was alleged that by 1738 two-thirds of the Virginia planters were so heavily in debt that they could not change factors.

However, by 1850 there were 101,335 plantations, which amounted to nearly one-fifth of all the farms in the South. Despite the image that plantations were large enterprises with hundreds of slaves, about half of them had fewer than 20. These small planters were still part of the system, and all else being equal, they were at a disadvantage because they could not compete in the market with the larger planters.

Spanish and French Settlements

Although the Spanish and the French were reluctant to establish purely agricultural colonies, they were aware of the need to transplant their technology, animal breeds, and plant varieties to the New World. The Spanish were active in American agriculture for three centuries and made intelligent and efficient contributions to it. In 1494 they sent 6 roosters and 100 hens to the West Indies. In the following year they sent jacks, jennets, mares, cattle, pigs, sheep, rice seed, millet seed, farm laborers, gardeners, a millwright, and a blacksmith. They continued to provide materials and agricultural knowledge, including their ideas of a generous land system. By 1500 the government decreed that Indians could be compelled to work but should be paid a fair wage.

By the mid-1500s ships leaving Spain for the New World were required to bring seeds, plants, and domestic animals to the West Indies. The Spanish settlers raised cotton, corn, and cacao, which were indigenous to the islands. They introduced wheat and sugar cane, which became significant crops. By 1535 the Spanish settlements in Mexico exported wheat to the West Indies to supply those producing commercial crops for the European market. By 1546 sugar cane production in the West Indies was sufficient to keep refineries operating. Ranches and plantations were worked with the aid of horses and mules that had come earlier. Irrigation systems fed by windmills were in operation.

The Spanish settled Florida in 1565 for strategic reasons and had little intention of using it for agriculture. They planned to supply Florida with food from the West Indies. Before long, individual soldiers raised crops for survival, and it was realized that the area was capable of producing a wide variety of crops. The biggest handicap to Florida agriculture was that the Spanish government was determined to institute a feudal system. It was not until after 1763, when Florida was acquired by the English, that much was accomplished there in agriculture.

Spanish missions were in operation in Texas, Arizona, New Mexico, and California by the late 1600s. These missions were based on a livestock economy, and because there was a lack of Spanish workers, the missionaries taught the Indians how to develop agriculture. Sheep raising was the principal industry of the Spanish in New Mexico, where production was adequate to provide 500,000 head for export. Cattle and horses were the chief animals in the other areas.

California was a stockman's paradise. The missions prospered as horses, mules, cattle, sheep, goats, and hogs multiplied. Large ranches with thousands of head of livestock were in operation. One of the largest mission ranches had 76,000 cattle, 79,000 sheep, 310 yoke of oxen, and 6,000 horses. The chief export products were hides and tallow, which were controlled by the Boston market. Each vessel that brought horses or livestock took hides and tallow on its return trip. Most of the meat was wasted because local demand was limited and no satisfactory means of processing and transporting it to foreign markets was available.

The French made several explorations in the lower Mississippi Valley between 1660 and 1682 and were well aware of the agricultural potential of that area. They also explored the coastal regions and inland as far west as New Mexico. In 1717, in an effort to develop the lower Mississippi region, they sent out 7,020 persons, of whom 2,462 were indentured servants, 1,500 were slaves, and 1,278 were criminals or other exiles. These persons were immediately put to work building levees and drainage ditches to protect farm land and communities.

Those efforts paid off, for by 1720 rice was grown at several sites along the Mississippi River. Each spring the levees were opened to flood the fields, and then the water was removed through drainage ditches. In 1721 a price floor was established in an attempt to increase production. That proved successful, and by 1728 enough rice was grown to export to the West Indies. However, rice was never a major crop in the region.

Corn was cultivated in Louisiana before 1710 and quickly became the major food crop for the slaves. It was also the chief crop in domestic and export trade. The French preferred wheat as their food crop, which flourished in the lower valley until it was hit by rust. They introduced a Turkish variety but also used wheat grown by the Indians. To avoid the problem of rust, the French traveled to the drier regions of the upper Mississippi and by 1721 grew wheat as far north as Illinois. By 1725 wheat was sent downstream in sacks made of deerskins.

Because of the favorable southern climate, a wide variety of vegetables, fruits, and specialty crops were grown. Some, such as oranges, suffered from occasional frost. Because of greater heat and more moisture than they were accustomed to from the land of origin, some fruits grew too fast, causing them to lose their flavor and become woody.

In 1711 tobacco was first grown in Louisiana, and by 1719 a settlement was established purposely for its growth and manufacture. However, an inadequate labor supply prior to 1740 restricted tobacco production. Then the French government established a price floor, and production increased rapidly until the late 1700s, when tobacco was replaced by cotton and sugar.

Indigo was the third major crop in Louisiana to be produced for the commercial market, after corn and tobacco. It was first grown in 1722, and by 1728 much of the production was exported. The Jesuits processed indigo into a dye that became a major export item. The British bounty made it a very profitable commodity for smuggling into the trade channels.

The original intention of the French was to establish large plantations. The size of the concessions was dependent upon the number of people brought into the colony. Some of the grants contained at least 70,000 acres. The concessions were successful in getting settlers to Louisiana, but they soon failed. The wilderness conditions and their large size made them difficult to manage. As was true of many of the English colonies, too few people were knowledgeable about agriculture or sufficiently interested in it for the plan to succeed.

In the case of both the Spanish and the French, the development of commercial agriculture was hampered by their restrictions on slave trade, which kept labor in short supply. Later, they encouraged small holdings, but little progress was made on them, for most of the farmers were not overly progressive. They preferred diversified farming, which enabled them to maintain a subsistence without too much effort. Most of them made very little money because they were not active in either the domestic or the export market.

Technology in Colonial Agriculture

"Technology," as used here, means the tools, the seeds, the livestock, and the knowledge involved in the practice of farming. Agriculture, as was true of most other economic activity, had not progressed much from its origins. Agricultural progress is most critical to human survival and growth, for little can happen until there is adequate food and fiber. The simplicity of agriculture just three centuries ago is difficult for individuals to comprehend who were born in the closing decades of the twentieth century, when technological explosion after explosion has taken place.

Power and Early Farm Vehicles

Human beings provided the chief power on colonial farms for all work, including seeding, cultivating, and harvesting. Except for an elite few in the South, no family member was excluded from even the heaviest work.

Oxen and horses were used to plow, to harrow, and to pull skids or carts for hauling grain or hay. Oxen were the preferred draft animals because they had more power, which made them steadier under heavy pulling, such as that required when plowing through rooted soil. Oxen were also easier to care for since they were less vulnerable to disease, less expensive to provide for, and less costly to purchase. A final important factor in favor of oxen was that once their working days were over they could be eaten.

Because oxen traveled only slightly over one mile per hour in contrast to horses, which had a gait of up to four miles per hour, horses were preferred for riding and for road work. Even though they were more expensive to purchase than oxen or other farm animals, horses probably suffered more from neglect because they had no food value.

In spite of the disadvantages of horses, more farms had them than oxen. But it must not be concluded that all farms had either oxen or horses. Plowing, which was the heaviest work done on farms, was not universally practiced until the 1650s. Little land was cleared for plows, and few plows were available. Much of the heavier work was done cooperatively or by a custom operator, who probably had a livery barn or maintained animals specifically for hiring out. This was especially true in New England, where farms were too small to justify owning horses. There were exceptions, for some New England farmers raised horses to export to the West Indies. In the southern colonies, horses ran wild and were so plentiful that nearly everyone rode. The upkeep was less there because year-round grazing was possible.

For moving heavier loads, early farmers used a vehicle, often called a "log boat" or a "log sled," that was simply the crotch of a small tree. A more advanced version, used only by the wealthier farmers, was nothing more than several planks fastened together and dragged over the ground. This was called a "stone boat" and was used to haul manure from the barns to the fields and hay from the fields to the barns. Farmers who could not afford such a rig let the manure accumulate. Another rig consisted of two poles, with one end fastened to each side of the horse and the other end left to drag. It was similar to the travois commonly used by the Great Plains Indians.

Wheeled vehicles first appeared on farms in New England about 1670. These crude two-wheel carts, with solid wooden wheels made from cross sections of hardwood trees, were not in common usage until after 1700. As was true of most tools or implements used, the carts were nearly always made by the farmers who used them. Only the wealthier farmers could afford tumbrels (tip carts), which were more advanced than regular carts because they could self-dump their loads.

The four-wheel wagon, which was the natural next step, did not appear until 1740 and was not in common usage until after 1800. The most used early four-wheel wagon, called the "Conestoga," was introduced in 1750. However, this wagon was too costly for most farmers until later.

After roads came into existence and the Conestoga wagon was developed, the Conestoga horse was bred especially for use with the wagon. This was the only widely known native draft horse produced in America. It was a good horse for teamsters, stage drivers, and mail carriers who traveled between the Atlantic seaboard and the Ohio River. Even though it could travel faster and longer than any other breed of horse, it was not widely adopted for farm use and died out about 1850. In the South the most prized horse was the Chickasaw horse, which was adopted from the Chickasaw Indians. This small, active, and beautiful horse came from the Spanish herds of the late 1500s.

Tools and Implements

The necessary equipment for a family of five on a 100-acre farm, as defined by an authority of the 1600s, consisted of the following: a plowshare and colter, a plow chain, two scythes, four sickles, a horse collar, some cordage for harness, two stock locks, two weeding hoes, two grubbing hoes, a crosscut saw, two iron wedges, an iron pot, a frying pan, two felling axes, a broad axe, a spade, a hatchet, and a splitting tool to cleave clapboard, shingle, and cooper's timber.

Most of the tools were handmade, heavy, clumsy, and poorly suited for the purposes intended. They were a clear indication of the labor intensity of farming at that time. Another listing relating to New England in 1638 pointed out that a family of six should have five broad hoes, five narrow hoes, five felling axes, and assorted hand tools, such as spades, chisels, hatchets, pickaxes, and pitchforks, plus nails, pins, and spikes. This was reputed to be more than the average inventory of most settlers.

In both cases, forest tools were listed because they were needed to clear land and because most New England farmers also practiced forestry to supplement their income. It was generally recognized that a strong man could carry all the implements used in seventeenth-century agriculture on his shoulders. It was many years before the plow, the harrow, and the cart came into use to make this statement incorrect.

For the first 12 years after landing, the Pilgrims had no plows and broke the soil with hoes and mattocks. The wooden plows of that day were quite useless in both the rooted forest land and the rocky soil of New England, so there was no strong desire to secure them. Only 16 out of 58 estates probated in one county in Massachusetts between 1636 and 1664 had plows. In 1650 there were fewer than 200 plows in all of North America, and most of them were in Virginia.

The early wooden plows did not turn the soil over. They did not scour, which made them pull hard, and they required someone to scrape the moldboard constantly. Deep plowing was virtually impossible. Four to six oxen were required to pull a plow, and even though two men could manage, it was better if a crew of three men and a boy was used—one man to hold the handles, one to press on the beam, another with a shovel to scrape the moldboard, and the boy to drive the oxen. Such a crew could do one or two acres a day, but it was more like scratching the soil, not deep plowing. The need for so many oxen and such a large crew made custom plowers essential. Some towns paid a bounty to farmers who purchased plows and kept them in repair for custom work.

The only other early farm implement that required animal power was the harrow. Often this was nothing more than a tree branch or an A-frame wooden implement with either wooden or iron pegs that was used to pulverize the soil. It was used again to cover the grain after it had been hand seeded. The purpose was to keep the birds from eating the seed and to mix the seed with moist dirt for better germination.

Brush harrow used to break lumps of dirt and cover the seed. (USDA)

When small grain was ready for harvesting, it was cut with a sickle, not unlike that used in biblical times. The cutting was done by pulling the sickle across the grain with one hand while holding the grain with the other. A person could not cut more than a fraction of an acre a day. Then the grain had to be raked and bound into bundles.

By the middle of the seventeenth century the scythe came into use for grain harvest. Originally, the thick, heavy "bush" scythe had been used only for cutting grass because it probably shattered too much grain. In 1646 a longer, thinner blade, strengthened by a bar of iron on its back side, was introduced. This was such a major improvement that the inventor was given a 14-year patent for the exclusive purpose of manufacturing it.

No other major improvements were made on the scythe until about the time of the Revolution, when a cradle was attached to it. One person with a scythe could do as much as seven with a sickle. A top cradler could cut 2 to 2½ acres of grain a day. Equally important was that the grain was left in a bunch, or swath, making it easier to bind it into bundles. Even such a simple innovation as the scythe was long resisted. In New England the scythe did not replace the sickle until after 1800. Even in an area as progressive as Pennsylvania, the sickle was used as late as 1832.

Cradlers and binders harvesting wheat in Wisconsin. (National Agricultural Library; hereafter NAL)

Once the grain was harvested it was hauled to the farmyard, where it was cured during the winter and was threshed in the spring. Most of the bundles were threshed with a flail, which was nothing more than a heavy stick attached to a pole and swung by the farmer. Then the grain had to be winnowed, which was done by throwing it into the air and letting the chaff blow away, leaving the heavier, cleaner grains. In more prosperous Pennsylvania, the bundles were hauled to the barn and stored for the winter. The grain was flailed on the barn floor and cleaned by winnowing.

Land and Crops

During the 1700s New England farms ranged from less than 25 to more than 300 acres. Near the centers of population a more intensive agriculture was practiced, and the largest farms seldom exceeded 100 acres. In the new areas farther into the frontier, farms ranged from 150 to 300 acres. In spite of the availability of land, it was not possible to farm large areas because of the limited labor force and the lack of technology. By contrast, in 1765 in the Middle Colonies, the average size of 3,293 farms was 135 acres.

Although there was an abundance of land, less progress was made in farming in the United States than in England during the years from 1607 to 1800. Probably the only exception was in the development of grasses, because the more severe winters demanded winter feeding. In 1720 Timothy Hanson developed timothy, which proved to be the most important hay crop of the pre-Revolutionary period.

New England farmers tried nearly every crop they were aware of in an attempt to learn which did the best in the various regions. They combined their efforts from working the land with what they gathered from nature to be as self-sufficient as possible, but some things could not be raised on the farm. The most common items that farmers bartered for were salt, iron for tools, chains for ox harnesses, and nails.

Either flax or wool provided the material needed for clothing. The forest originally provided a source of meat. This was smoked for preservation until such time as icehouses were built and ice was harvested for use as a preservative. The forest also provided maple sugar and wild honey for sweeteners. Nearly all the utensils and dishes were carved out of wood. Most farmers planted apple trees quite soon after settlement, and within a few years apples and cider were available.

Corn made up 60 to 70 percent of the field crops because it grew under the widest climatic conditions and produced the most food per hour of labor. It could be used as food grain for humans or animals. Prior to the development of a good domestic hay, corn provided the best available source of fodder.

Collecting maple syrup in Vermont, 1975. The only basic difference from 1775 was that oxen were used then and the equipment was more primitive. (NAL)

Generally, most farmers "checked" corn so it could be cross-cultivated to control weeds, to provide for more sunlight, and to aid in planting other crops in the same field. ("Checking" meant that fields were marked in both directions, and where the lines were crossed, the seed was planted.) When corn was a hand's length high, the weeds around it were cut and the soil was loosened with a hoe. The weeds between the rows were plowed under. When the soil became exhausted after several crops, two or three fish were placed near the seed at planting.

Corn was surpassed only by wheat as a good cash crop and was sometimes used in payment of taxes. Wheat was not competitive with corn, for it yielded less per hour of labor and per acre because it was not as adaptable to the climate, the soil, and the technology of New England. On the other hand, in the more fertile soil of the Middle Colonies, wheat was more productive and was raised there successfully throughout the eighteenth century. It was not long before wheat was exported from New York and Pennsylvania to New England.

By 1664 the wheat crop of New England was affected by black stem rust, which reduced yields greatly. It was soon obvious that the worst infestations were near barberry bushes. Because berries from the barberry were desired for making jams and jellies, there was reluctance to destroy the bushes. It was not until 1726 that Connecticut became the first colony to eradicate barberry and not until the 1750s that other colonies followed. Most of the legislation did not have adequate teeth for enforcement.

In the mid-eighteenth century, winter wheat was planted instead of spring wheat because it proved more resistant to black stem rust and the Hessian fly. Through experimentation it was learned that late planting, more fertilizer, and better tillage minimized losses to the Hessian fly.

Many New England and Middle Colony farmers shifted from wheat to rye when it proved more resistant to both black stem rust and the Hessian fly. Fortunately, rye was adaptable to the sandy soil. It outyielded wheat and replaced it as the bread grain. Rye also was valued for its straw, which was used for making thatched roofs, for tying corn fodder, for making bee hives, and for weaving breadbaskets. Some rye was exported to Europe.

The next most abundant field crop was oats. It was estimated that oats were grown on about 12 percent of the land.

By the mid-1600s efforts were made to interest farmers in the production of flax for linen. Flax slowly became a staple for its use as a fiber. After a press mill was erected in 1718, production increased because flax was used for linseed oil for paint, medical purposes, and other industrial needs.

In 1719 one of the most important events in New England agriculture took place when Irish immigrants brought Irish potatoes to New Hampshire. Up to that date potatoes were imported for immediate consumption and not for seed. Some farmers were reluctant to adopt them, but in the space of only a few years they became a staple food crop for the European settlers and the Indians throughout the colonies.

It was discovered that potatoes did well in the climate and soil of Maine, giving that colony its greatest claim to agricultural prominence other than forestry. Although the acres planted to potatoes generally were not as great as those planted to grains, the yield per acre was much higher. Potatoes provided a major addition and improvement to the diet.

By 1728 a botanical garden was established in Philadelphia where many imported varieties of plants were cultivated. In 1730 an orchard of fruit trees was established in Maine, and the first commercial nursery was developed on Long Island. Two years later several experimental gardens were established in Georgia. Because crops that were needed in England could be grown there, Parliament granted funds to Georgia for experiments in agriculture. Within a few years Georgians had learned that they could not successfully grow tropical crops.

By 1700 New England farmers complained that their soil was becoming exhausted, and they started moving west and north in search of fresh soil. Those who lived near the centers of population and had a cash market decided to do a better job of farming. Farmers near Long Island Sound, who grew what are now called "truck garden crops," put as many as 10,000 fish on an acre for fertilizer. The value of the fish lasted a considerable length of time. In addition to using fish, farmers hauled horse droppings from the streets to the fields, as well as the manure from the city cow and horse barns. By-products from the potash kilns were also collected to be spread on the farms.

Jared Eliot was disturbed because he realized that unless farmers began better farming practices they would have to continue moving west. In 1747 he published *Essays in Agriculture,* which is considered the first literature published in America on agriculture. For many years people interested in writing about agriculture refrained from doing so because of the suspicion and bias most farmers had against "book farming."

The concept of preserving the soil for the public good had not yet been accepted. The goal was to get the most out of the soil with the least labor. Moving to new land was the obvious choice. Rebuilding land by fallow took from 7 to 15 years, plus it required extra labor. It cost more to manure an acre of land than to buy another acre, so it was not uncommon for a farmer to build a new barn when the old one became full of manure rather than clean it out.

After 1750 some farmers began to seed clover as a step in soil rebuilding. Others, particularly the German farmers in Pennsylvania, began applying limestone, marl, and gypsum, in addition to barnyard manure, to maintain soil fertility. With a demand for dairy products in the areas of German settlements, the practice of irrigating pastures was adopted. Even though yields increased greatly, only a few localities in New England and New York practiced irrigation.

The Need for Markets

Although the term "subsistence farming" has long been used to characterize pioneer farming, it does not accurately describe what took place. From the start it was clear that if the colonies were to survive, agricultural products would have to be produced for the export market. "Self-sufficient farming" comes closer to describing the type of agriculture most farmers were engaged in. Farmers, who were by far the majority of the population, fell into the trade that was most beneficial for them and the country.

Colonial leaders knew that support from English investors would not be long sustained if profits were not forthcoming. Forestry was the first phase of agriculture to be developed, because trees were an immediate source of income. In 1608 the first shipload of lumber for the royal navy was exported. Trees not used for timber were burned to produce potash for export. Exports of potash grew each year, peaking in 1825. On this basis, more aid was provided. Skilled millwrights were imported from Scandinavia to help the Dutch establish windmill-powered saw rigs. Logging and milling became a far larger business than supplying products for the royal navy had ever been.

Before long, sawmills were erected along the Atlantic coast from Maine to Florida. Shipbuilding, lumbering, furniture-making, cooperage, and wagon-building firms were abundant. The forests provided the first marketable products although they were an obstacle to those who wanted to farm. Many people realized that even though it seemed as if there was no end to the trees, the era of lumbering could not continue indefinitely and traditional tilling of the soil would have to take place.

Many of the first settlers did not benefit from forestry other than working as day laborers in the woods. Their chief aim was to become farmers, but first the trees had to be removed.

Of the economic problems farmers faced in addition to securing land and credit, their greatest concern was finding a market. Without a market there was little reason to produce more than was necessary to survive.

Challenges to Production

The first obstacle to profitable agriculture in New England was realized soon after the trees were removed. Then it became known that the land was better suited for timber than for farming. The settlers adopted the Indian practice of burning off the trees, which was the easiest way to dispose of them.

In New England, fishing and shipbuilding greatly overshadowed farming. Both were important to farming, for they took the surplus agricultural production. By 1625 other farmers traded their surplus production to the Indians for furs, which were then exported to England. This was the first instance of commercial agriculture involving other than forest products.

Forests were being removed so fast in Plymouth Colony that in 1626 an ordinance was passed stating that no one could sell or transport timber out of the colony without approval. The law was backed by an adequate fine. In 1633 acts were passed that forbade the setting of fires between September and March and allowed doing so only by permission during other months.

By 1644 the British government granted privileges of making and supplying forest products but with restrictions. By that time several colonies forbade the cutting of any trees unless they were to be used for personal buildings, fences, or firewood. In 1694 a license was required in Massachusetts Bay Colony to cut timber for commercial use.

In the early 1700s more acts of Parliament conforming with the Navigation Acts were passed, not so much as forest conservation acts, but as acts to preserve timber for the British navy and the merchant marine. American farmers, many of whom also logged, resented those restrictions because they limited the farmers' potential sales to the English market.

By 1627 the Pilgrims were successful enough as farmers to buy out complaining English shareholders. The remaining shareholders realized that agriculture was the key to survival of the colony. They allotted 20 acres to each man and 20 acres for each additional family member, up to 100 acres per family. Cows, goats, and hogs were also divided among the farmers. Within two years the Pilgrims were trading farm produce for furs, which were then exported.

Originally, the settlers of Massachusetts Bay Colony had tried to survive by fishing, but by the late 1620s they realized that farming was necessary. They prospered in farming because of good leadership, and they learned from the experience of Plymouth Colony. By 1632 corn was exported to nearby towns. Between 1630 and 1640, over 20,000 English immigrants came to the Massachusetts Bay area. Most of the newcomers were not farmers, so this challenged agriculture to increase production. Fortunately, the balanced system of livestock and crops of Old England provided a good model, and within a few years there was little need for imports to care for the new immigrants.

The Civil War of the 1640s in England provided the next market opportunity for colonial farmers. The British traders were occupied in the domestic market, which provided the colonists with a chance to enter the West Indies market with livestock, barreled beef and pork, plus bacon and hams. Even though hogs were fed little corn, they did quite well on the food that they found in the woods. Market hogs weighed 160 to 170 pounds and produced a better-quality product than cured beef.

The growth of towns provided the major market challenge for New England and Middle Colony farmers. New England farmers could not compete with those of the Middle Colonies and the South in the West Indies markets. But they survived because, as one writer noted, "Good markets make good farmers." Their good market was the rapidly growing population in the nearby towns.

Market days and market squares were often required in New England town charters. These provided an early opportunity to market any surplus production, with farmers selling directly to consumers. By 1670 Springfield and Boston were provided with winter-fattened cattle. The price of animal products in New England increased steadily in the late 1600s and early 1700s as the population grew, even though farmers did not make improvements in their livestock or methods of production. Once transportation facilities improved and western livestock economically reached eastern markets, the New England farmers quickly lost out.

As the towns grew larger they established regulations to the benefit of both the buyer and the seller. Old customs were dropped, and in 1742 Boston erected a building specifically as a place for buyers and sellers to meet. Because of their location close to population centers, Massachusetts farmers learned commercial agriculture at an earlier date than most colonial farmers, except those with foreign markets.

To stop the direct sale of butchered meat products by farmers, Boston butchers formed a guild to supply meat at prices set by the local town officials. To prevent the sale of meat from animals stolen by Indians, regulations were written requiring proof of ownership. Once public slaughter houses were built, the killing of livestock for resale was required to be done under the direction of inspectors. A fee was charged for each animal as valued by the inspectors.

Firewood, vegetables, fat cattle, and dairy products were transported only a limited distance to the local markets by the farmers, or more often by their wives. The volume was sufficiently small so that, except for the cattle, the goods were carried on their backs or in large baskets fastened to the backs of horses. As towns grew and roads improved, farmers traveled up to 50 or 60 miles to market their products. When two-wheel, two-horse carts came into use about 1750, the distance extended even farther. Cattle and hogs might be driven hundreds of miles.

Jared Eliot wrote that the poor practices of the New England farmers were probably as much caused by "ignorance, risk aversion, and hidebound conservatism" as by the lack of fertile soil. The lack of initiative caused by the supposed lack of markets does not entirely explain the condition of haphazard agriculture, for even inland communities provided some market for farmers.

Restrictions to production and to marketing were remembered, but the aid that was received by the British and colonial governments was overlooked. Bounties were never objected to, even when the sole purpose was to secure an adequate supply of a commodity for British needs.

The Impact of the Revolutionary War

Farmers were the great majority of the American population, so it is logical to say that the Revolution was a farmers' cause. Their resentments included the perpetual quitrent, after 1763 the forbidding of settlement west of the Alleghenies, and the British attempts to control export marketing of farm products through British ports. The chief purpose of the export controls was to tax farm products before they entered other markets.

Farmers were often caught in trade conflicts between England and France, which temporarily shut off some markets. On the other hand, being part of the British Empire gave farmers access to the largest market in the world, which included the West Indies, the most profitable market of the eighteenth century. Except for what was smuggled to the Indies, trade was lost briefly during the Revolution. After the war, in spite of British efforts to prevent it, that trade was quickly regained.

During the Revolution, farmers in the path of troops, especially the British, profited by the sale of goods at high prices. Most small-scale producers geared to selling in local markets were not affected. The colonies were cut off from British textiles; therefore, sheep, hemp, and flax growing were encouraged and prospered during the war.

The South suffered directly when tobacco exports were reduced, because of the loss of the British market. The indigo market collapsed when the British subsidy ended. The southern agricultural economy was so seriously affected that slavery started to decline and probably would have continued to do so if it had not been for the "inevitable" invention of the cotton gin.

New England agriculture, which had experienced a decline of its livestock sector since 1750, felt an additional loss when commercial wheat growing ceased and wheat was imported. Only the production of sheep to supply the woolen mills and the raising of horses for export to the West Indies prospered. Prior to the Revolution most of New England had ceased to be a predominately agricultural region. Changes that accompanied the war hastened the demise of farming.

When the Revolutionary War started, farming was much as it had been from the time of settlement. Farmers had learned the agriculture of the Indians well. They had experimented with many crops from other countries, but they still practiced a system of earth butchery that affected the future of farming. They were generally middle-class people who enjoyed comforts unknown to European farmers. To them the Revolution brought economic, scientific, and political changes that were destined to affect agriculture.

Within a month after the Revolution ended, the General Court of Massachusetts forbade the cutting or destroying of white pine trees 24 or more inches in diameter 1 foot from the ground unless licensed by the legislature. That was followed by an act requiring surveyors in every township to prescribe specifications for timber products and their inspection. Such legislation was aimed directly at the preservation of forests, which had been ruthlessly destroyed in the past.

After 1783 agriculture faced a new future, for all traces of feudalism were gone. The government sought programs to put land in the hands of individuals as quickly and as easily as possible. This was done by providing easy terms for the purchase of land.

Conclusion

By the time the American Revolution was over, agriculture was posed for a new era. During the colonial period many of the customs, systems, habits, and laws of Europe remained in force. The shackles were broken, and farmers were at liberty to practice farming without restraints from the past. A new economic, political, and scientific system that would have a direct bearing on agriculture was in the process of being developed.

During the colonial period farmers proved that they were good adapters. They took knowledge from the Indians and applied what was best suited for commercial farming. They applied new knowledge, some of which came from a rapidly changing European agriculture. The new nation basically had an adequate food supply, and agricultural exports were more than 75 percent of all exports.

New England, which seemed least willing to adapt, experienced a rapidly declining agriculture. It was the first section of the nation to develop an industrial economy, relying on food from the virgin land to the west. Overall, the progress in colonial agriculture was at about the same pace as in previous centuries in other countries. The pace changed rapidly after the Revolutionary War.

References

Bailey, L. H., ed. *Cyclopedia of American Agriculture: Farm & Community*. New York: Macmillan Publishing Co., Inc., IV, 1909.

Bidwell, Percy Wells, and John I. Falconer. *History of Agriculture in the Northern United States: 1620–1860.* New York: Peter Smith, Carnegie Institution Publication 358, 1941.

Bizzell, W. B. *The Green Rising: An Historical Survey of Agrarianism, with Special Reference to the Organized Efforts of the Farmers of the United States to Improve Their Economic and Social Status.* New York: Macmillan Publishing Co., Inc., 1926. Reprinted, Wilmington, DE: Scholarly Resources, Inc., 1973.

Carrier, Lyman. *The Beginnings of Agriculture in America.* New York: McGraw-Hill Book Co., Inc., 1923. Reprinted, New York: Johnson Reprint Co., 1968.

Clemens, Rudolf Alexander. *The American Livestock and Meat Industry.* New York: Ronald Press Co., 1923. Reprinted, New York: Johnson Reprint Co., 1966.

Cochrane, Willard W. *The Development of American Agriculture: A Historical Analysis,* 2nd ed. Minneapolis: University of Minnesota Press, 1993.

Coombs, Charles I. *High Timber: The Story of American Forestry.* Cleveland and New York: The World Publishing Co., 1960.

Gates, Paul W., and Robert W. Swenson. *History of Public Land Law Development.* Washington, DC: Public Land Law Review Commission, 1968.

Gray, Lewis Cecil. *History of Agriculture in the Southern United States to 1860.* Washington, DC: Carnegie Institution of Washington, Publication 430, I and II, 1933.

Holly, Marilyn. "Handsome Lake's Teachings: The Shift from Female to Male Agriculture in Iroquois Culture: An Essay in Ethnophilosophy," *Agriculture and Human Values* 7, Nos. 3–4, Summer/Fall 1990, 80–94.

Hurt, R. Douglas. "The First Farmers in Ohio Country," *Agriculture and Human Values* 2, No. 3, Summer 1985, 5–13.

Jensen, Joan M. *Loosening the Bonds: Mid-Atlantic Farm Women, 1750–1850.* New Haven, CT: Yale University Press, 1986.

Jones, E. L. "Creative Disruptions in American Agriculture, 1620–1820," *Agricultural History* 48, No. 4, October 1974, 510–528.

Kulikoff, Allan. *Tobacco and Slaves: The Development of Southern Cultures in the Chesapeake, 1680–1800.* Chapel Hill: University of North Carolina Press, 1986.

Mead, Elwood. "Rise and Future of Irrigation in the United States," *Yearbook of Agriculture, 1899.* Geo. Wm. Hill, ed. Washington, DC: USDA, 1900, 591–612.

Rasmussen, Wayne D., ed. *Agriculture in the United States: A Documentary History.* New York: Random House, I–IV, 1975.

Ross, Earle Dudley, and Louis Bernard Schmidt, eds. *Readings in the Economic History of American Agriculture.* New York: Macmillan Publishing Co., Inc., 1925.

Rothenberg, Winifred Barr. *From Market-Places to a Market Economy: The Transformation of Rural Massachusetts, 1750–1850.* Chicago: University of Chicago Press, 1992.

Russell, Howard S. *A Long, Deep Furrow: Three Centuries of Farming in New England.* Hanover, NH: University Press of New England, 1976.

Sanford, Albert H. *The Story of Agriculture in the United States.* New York: D. C. Heath and Co., Publishers, 1916.

Shannon, Fred A. *American Farmers' Movements.* Princeton, NJ: D. Van Nostrand Co., Inc., 1957.

Sharpless, Rebecca. "Southern Women and the Land," *Agricultural History* 67, No. 2, Spring 1993, 30–42.

Spruill, Julia Cherry. *Women's Life and Work in the Southern Colonies.* New York: Russell & Russell, 1938. Reprinted, Russell & Russell, 1969.

Thompson, James Westfall. *A History of Livestock Raising in the United States, 1670–1860.* Agricultural History Series No. 5. London: Hutchinson & Co., Ltd., November 1942. Reprinted, Wilmington, DE: Scholarly Resources, Inc., 1973.

Whitaker, Arthur P. "The Spanish Contribution to American Agriculture," *Agricultural History* 3, No. 1, January 1929, 1–14.

Wissler, Clark. *The American Indian: An Introduction to the Anthropology of the New World.* Glouster, MA: Peter Smith, 1957.

Wissler, Clark. *Indians of the United States.* Garden City, NY: Doubleday & Co., Inc., 1966.

Unit II

1783 to 1860

Some Events and Technological Innovations That Affected Agriculture

- **1785** Land Ordinance of 1785
- **1785** Philadelphia Society for the Promotion of Agriculture organized
- **1793** Cotton gin invented
- **1803** Louisiana Purchase
- **1819** Farm journals appeared
- **1825** Erie Canal built
- **1831** McCormick reaper invented
- **1837** John Deere plow built
- **1839** Congress made first agricultural appropriation
- **1846** British repealed their Corn Laws
- **1855** Michigan Agricultural College established

Introduction

As soon as the American Revolution ended, the American people were free to change their political and economic thinking. This had a positive impact on those engaged in agriculture. There is some indication that this change may have started as early as 1763. Even though farmers made up more than 90 percent of the population, Congress had a difficult time deciding what to do with the nation's land. It was several years before Congress took a stance regarding the disposal of the public domain. The outcome proved to be wise, for land and individual liberty were the magnets attracting newcomers to our shores.

Without settlers to till it, the land was virtually worthless. It was the immediate obvious resource of the new nation. Exploiting the land for what it had to offer, whether farm crops, timber, or minerals, was the chief means of securing goods for export. Produce from the land was exchanged for commodities needed to improve the level of living and to develop industry.

Most of the new immigrants, however, preferred other ways to make a living and did not settle on the land. That did not relieve the shortage of labor in agriculture, but it helped to build an urban industrial society. This proved to be good for agriculture, for it created a market. Some farmers realized that this gave them an opportunity to make a profit, and they therefore did a better job of farming.

In addition to adopting a generous land policy, the government provided aid to agriculture in other ways, such as collecting information from other countries, conducting studies on ways to improve agriculture, and creating institutions to study agriculture. This work was done even though many farmers were not inclined to heed the advice of "book farmers" and ignored the information available.

A favorable political climate encouraged other sectors of the economy to expand, which in many ways proved beneficial to agriculture as well as to the nation in general. Societies to promote agriculture, improved means of transportation, inventions that enhanced the processing of agricultural products, inventions that enabled farmers to produce more with greater ease, and institutions that aided in providing more, better, and less costly food for consumers—these innovations were all part of a massive movement in which agriculture continually benefited the nation by a steadily increasing supply of exports, products for industry, and food for consumers. This bounteous agriculture served as the foundation for the nation with such ease that most people took it for granted.

Even though production had greatly improved after 1800, farmers were not always amply rewarded for their efforts. During the colonial period, farmers generally were quite indifferent about their methods of farming. A large portion of them acquired more land than they could work properly, with the hopes that if they improved it the resale price would be higher. They were more inclined to speculating than to treating farming as a business to be operated for a profit. This attitude about land and how to farm changed slowly while land was still available on the western frontier.

This is all part of the story of how the nation grew to a dominant position in world agriculture but at the same time provided the economic backbone for the country's rise as an industrial giant among nations. We could always rely on agriculture to open the way—a luxury no other nation has had.

Early Agricultural Policies

One of the first challenges of the government was to determine a long-range policy toward the Indians. The first reference to property rights of those people was written in 1633. It took from then until 1887 to finalize a concept as to how the question should be settled.

The Indians' natural way of life required a vast area of land, which ran contrary to the desires of white settlers. By 1800 Thomas Jefferson realized that the best solution to prevent the extermination of the

Indians was to attempt to teach them a more sedentary agriculture. Efforts to get them to adopt the farming methods of the whites were resisted because those methods were too restrictive. Colonial authorities realized that the Indians had to be protected from the wrongdoings of the whites. By the mid-1700s guardianship was often provided by colonial governments. But after 1783 the new government conducted a policy of progressive extension of authority over land.

An 1817 treaty with the Cherokees provided that the head of each family living east of the Mississippi River was entitled to a life estate in a section of land. Upon his death the children inherited the property in fee simple, which meant they could sell it without restriction.

During Andrew Jackson's administration (1829–1837), a policy of segregation was established, and land was assigned to those willing to accept the social and political system of the whites. In a treaty of March 1854 the idea of individual allotment was determined. This treaty allotted 80 acres per person, 160 acres per couple, and up to 640 acres for a family of from 6 to 10 people. With the Dawes Act of 1887 land allocation and citizenship policies were defined.

The Rise of a Myth

At the time when farmers made up more than 90 percent of the population, it was probably understandable that there was an air of superiority about them. Leaders, particularly George Washington, Thomas Jefferson, and later Abraham Lincoln, saw the exciting changes taking place in European agriculture, in contrast to the stagnation in American farming, and realized the tremendous potential there was for change here.

Washington was intensely interested in agriculture and was responsible for many innovations. He was in constant contact with Europeans and was an avid reader about the changes taking place in their agriculture. He imported animals and plants and had ongoing experiments on his farm, Mount Vernon. But there is little evidence that Washington did much thinking about a philosophy of agriculture or farmers.

Jefferson, on the other hand, had a definite philosophy that was derived to a great extent from his being born and reared on the frontier. The frontier was a great democratizing influence on society, for it was there that age-old traditions were eroded and the past had little bearing on who you were. Jefferson was caught up in that spirit.

Jefferson, like Washington, was an innovator in agriculture. He practiced scientific and experimental farming, using the latest in technology, crops, and methods. At the same time, he was striving to make his farm, Monticello, self-sufficient.

Because Jefferson asserted that farming was the most pleasing phase of his life, he developed his interest into his philosophy. He believed that "Cultivators of the earth are the most valuable citizens. They are the most vigorous, the most independent, the most virtuous. . . ." He felt that agriculture should be a way of life. Historian August Miller wrote:

> Agrarianism was the most significant force in his entire philosophy. As a farmer, Jefferson was anxious to preserve a pattern of civilization which was essentially agricultural. . . . His vision of America was that of a large country in which every citizen would reside on his own farm and live off the products of his own land.

Jefferson initially distrusted urban economy and urban civilization. He feared industrialism would destroy democracy. However, after the War of 1812, he changed his mind. He wrote, "[A]n equilibrium of agriculture, manufactures, and commerce is certainly becoming essential to our independence." He agreed that once farmers produced more than they could sell domestically or abroad, the surplus workers should turn to something else. Apparently he did not state whether they should continue to live on their farms.

During Lincoln's administration some of the most far-reaching legislation on behalf of agriculture was passed. However, Lincoln appeared somewhat doubtful about the virtue of farmers, even though he was born in a three-sided log cabin on the Kentucky frontier. He was a great booster of agricultural fairs, but at one of those fairs he said that he suspected that the reason there were so many attempts at flattering farmers was that there were more farmers to cast votes than any other group. He continued that he was convinced that "They are neither better or [sic] worse than other people."

Acts Prior to 1800

The Revolutionary War did away with all vestiges of English land legislation. All unappropriated lands passed to the newly established federal government. After 1783 the new government faced one of its most challenging issues—a policy on how to allocate those lands. Selling land to pay off the federal debt was an obvious first choice of eastern leaders. Large blocks of land were disposed of under this concept. Westerners wanted land to be dispersed to anyone who wanted to farm. However, it was not until 1821 that this second and most important phase of land legislation was passed. By that time a sectional rift had been created between the East and the West.

From the Revolution to the Civil War, agriculture expanded steadily, and farmers in general fared well. Liberal land policies had a positive impact on agriculture. The rapidly growing population, combined with industrial markets, created a steadily increasing demand for farm products. The adoption of science, technology, transportation, and imported livestock breeds by innovative farmers was responsible for rapid changes in the industry.

The Land Ordinance of 1785 had no restrictions as to how much land could be purchased by a person or a company. Large blocks of land were sold for the minimum cash price of $1 per acre. This created hard feelings because many people felt that the moneyed interests were favored over the settlers.

The Act of 1796 set the minimum price at $2 per acre and the minimum tract at 640 acres. Terms for payment were one year. By 1800 the minimum tract was reduced to 320 acres, which was to be paid for in four annual installments at 6 percent interest. In 1804 an act eliminated interest and reduced the minimum purchase to 160 acres. Later the minimum was 80 acres. Ohio, Indiana, Illinois, and Missouri were the four major states of settlement under the Act of 1800. The government's chief interest was still raising revenue.

Dispersing the Public Domain

From the passage of the first land act in 1785 until the frontier came to an end in the 1890s, land speculation was almost a national craze. Squatting started in colonial times. Settlers learned that they could profit from it, so they continued the practice even though it was in violation of the law. The new Congress recognized the problem of preemption as early as 1790, but it was not until 1803 that the first preemption act was passed in response to pressure from squatters. The act gave them the advantage of purchasing land after the area was opened at the minimum price before speculators and legitimate settlers arrived. Many times troops were used to enforce the laws against squatting. Even though their crops and buildings were destroyed, the squatters returned as soon as the troops left.

After the Revolution, states gave soldiers up to 400 acres and sold land liberally to others, but generally they required settlement to retain title. Purchasing land under the conditions of the early acts proved to be difficult. In 1820 Congress revised the guidelines, enabling settlers to buy as little as 80 acres at the reduced price of $1.25 an acre. These terms were within reach of most settlers. The Act of 1820 marked the shift from raising funds for the government to meeting the needs of the settlers.

Some studies indicate that a 40-acre farm was sufficient for full employment of a family and for its livelihood, in addition to providing a small surplus for the cash market. Farms of 40 and 80 acres were livable units, but they generally were not profitable. By the 1850s it was determined that a midwestern farm probably had to be 160 acres to yield a profit. That included capital gains on the land.

With more liberal legislation, over 75 million acres were auctioned off from 1820 to 1841. The terms were still not generous enough to suit everyone, so many squatters continued to preempt the public domain. This eventually caused Congress to yield, and the Preemption Act of 1841 gave individuals the right to settle on public domain before purchasing it. Homesteaders liked this act because it enabled them to avoid bidding against others with the risk of having to pay more than the minimum price.

Part of the craze for land was caused by the fact that the country was industrializing so quickly that many settlers felt that this would bring them profits. They overlooked the fact that new lands were opened so rapidly that surplus products kept the prices down. For the sake of the farmer, the dispersal of public domain had not worked too well, but the nation's consumers benefited.

Attempts to Create a Department of Agriculture

In 1790 George Washington suggested that Congress do what it could to "encourage agriculture, as well as commerce and manufacturers." Washington felt that there was no better way to help the country than by improving agriculture. He encouraged Congress to do so in several of his messages to that body. In his farewell address in 1796, he recommended a national board of agriculture. Congress discussed the proposal for several years, but in 1801 Thomas Jefferson opposed the idea because it was not within the enumerated powers.

Agricultural societies and journals promoted the cause, but Congress took no action. Not until 1816 did the idea of a national board for agriculture reappear, and again it was sidetracked. In the meantime, members of consuls and naval officers collected seeds and animals from throughout the world and brought them to the United States. It was not until 1825 that Congress created a standing committee on agriculture. Some committee members felt that agriculture was as important as commerce and manufacturing. Prevailing sentiment, however, was against the government being involved in agriculture. This attitude kept the collection and distribution of agricultural statistics out of the hands of the committee.

When the agricultural statistics program was established in 1839, it was assigned to the Patent Office. The Patent Office commissioner's strong interest in agriculture prompted him to secure $1,000 from Congress for the purpose of collecting data, seeds, and plants. This was the first money allocated for agriculture and the beginning of what became a department of agriculture.

The commissioner's annual report on his activities was dominated by agricultural information. This led to the inclusion of agricultural data in the Census of 1840, a practice which continued thereafter. In the 1842 report, foreign agricultural markets and commodities were mentioned for the first time. This opened the potential for new markets.

The navy conducted a three-year exploration expedition, from 1838 to 1840, and collected seeds and plants, which became the core of the Botanical Garden. An international exchange system established in 1848 became the chief means of introducing new seeds into the country. In 1849 the Patent Office was transferred to the Department of the Interior. From that date on, the annual agricultural report was issued separately.

In 1852 the Patent Office employed a specialist to conduct experiments in agriculture. That was followed by a garden plot to investigate the cultivation of certain crops. At the same time, the Committee on Agriculture came under pressure because many people felt that it did not have a purpose. This led to its eventual abolishment.

In 1841 a national agricultural convention was held in Washington, D.C., with the aim of founding a national organization. By 1843 all was forgotten. However, in 1851 the Board of Agriculture of Massachusetts took action. Delegates from 23 states assembled in Washington, D.C., and in 1852 established the United States Agricultural Society. It met annually in the Smithsonian, where it conducted its national agricultural fairs. The Society's job was to keep all agricultural issues alive, especially any action regarding greater recognition by the government. However, every time the subject of agriculture was brought before Congress, it was sidetracked.

The Society was the major force behind the Land Grant Act for the support of land-grant colleges and also for the creation of a department of agriculture. Those two goals were assured with the election of 1860, when the Republican Party was able to deliver on its campaign promises. The United States Agricultural Society had accomplished its goals, and in 1860 it was disbanded. George Washington's wish for a department for agriculture finally had been fulfilled.

Farming in a New Setting

Farmers have long prided themselves on their independence and reluctantly organized when it was for their mutual benefit. Most of them never did well economically except for gains from land appreciation. Agricultural historian Danhof wrote of a farmer who observed that there were four basic classes of farmers: (1) those who were always poor; (2) those who barely made a living throughout their lifetime of farming; (3) those who made a comfortable living and were always improving their competence; and (4) those who became wealthy. This farmer added that the numbers declined from classes 1 to 4. He noted, "The man himself, and what he is made of, determine to what class he will belong."

Fortunately, there were always a few visionaries who realized that agriculture could not remain static. Ironically, many of them were not directly involved in agriculture, or they had other major interests. In any case, they were not practical working farmers. They were progressive people who realized that agriculture had to do well if commerce, industry, and the nation were to profit.

After the Revolution, several events combined to help improve agriculture: national leaders recognized that conditions in agriculture had to improve; agricultural societies were formed, but unfortunately they did not attract practical farmers; and prizes were offered to stimulate improvements in agriculture.

The Philadelphia Society

The Philadelphia Society for Promoting Agriculture was established in March 1785. Its charter members were mostly professional people with close ties to farming. Only a few of the original members were farmers. The members felt that an organization was necessary to develop the resources of the nation. This was not a concern of practical farmers, who generally rejected what the Society had to offer.

The Society's chief aim was to improve agricultural practices. Society members preferred to adopt ideas from English farmers because they probably were the most advanced of the day. But there was a problem: most English farmers used intensive agriculture. This was not appropriate in the United States, where most farmers were engaged in extensive agriculture. Other countries had to be investigated to look for newer practices.

The Philadelphia Society devoted most of its studies to regional agriculture rather than taking a broader national approach. By 1788 a split developed in the Society that was caused more by social and moral issues than by agricultural ones. But this did not sidetrack the Society from focusing on the major problems: overcoming soil exhaustion; increasing crop yields; improving livestock; and developing more efficient farmyards.

The Society also systematically conducted experiments to improve technology in agriculture. Livestock and poultry farming benefited because nearly all the pioneers in veterinary medicine were members of the Society. The long-term benefit to American agriculture was derived from the publication, between 1808 and 1826, of five volumes of the activities and the findings of the Society.

The Philadelphia Society laid the foundation for a large number of societies created in the late 1700s and early 1800s. There was a

The Agricultural Press

Pioneer farmers were often very superstitious, which tended to make them question any innovation. This sometimes made them conservative to the point that it retarded progress. They did not understand basic principles of animal and plant breeding or nutrition. Nor did they comprehend what tillage did to aid plant growth or what they needed to do to maintain soil fertility. Knowledge was so limited that agriculture had actually regressed from its earlier levels.

Little was written about agriculture, probably because of the known resistance of farmers to "book farming." It was not until the 1740s that the first essays appeared on American agriculture. They concentrated on what farmers could do to improve their field practices. In 1813 the *Arator*, the first book on southern agriculture, was published. It dealt with the value of rotation and soil building.

On April 2, 1819, *The American Farmer*, the first farm journal, was published in Baltimore by John Skinner, who is generally credited as being the father of U.S. agricultural journalism. It carried a wide range of articles in the hopes of catching the interest of as many farmers as possible. *The American Farmer* continued publication until 1897.

Within a few months the *Plough Boy* was founded in Albany, but it survived only until 1823. The *New England Farmer* proved to be a more influential and longer-lasting journal, existing from 1822 to 1846. Probably because of its New England location, its major emphasis was on dairying.

In the 1830s about 30 journals had an estimated 100,000 readers. By the 1860s the number of journals had increased to about 60, with a combined circulation of more than 250,000. Over 400 agricultural journals had been founded by then, but most failed in less than three years.

Despite the efforts of the journalists, it is estimated that as late as 1850 fewer than 1 farmer in 2,000 subscribed to an agricultural paper. Many of the journals were poorly adapted to the needs of the average farmer, but they exposed the farmers to what was taking place elsewhere in the nation and in Europe. After agricultural colleges were developed in the 1860s, a wider range of information was available for the journals.

The journals did much to minimize the superstitions of farmers, many of whom still farmed "by the moon" or followed similar fallacies that often were contrary to good practices. Most of the editors campaigned against moon culture. Not surprisingly, journals called attention to new farm machinery. They were careful as to how they presented their information, for many farmers were hostile to innovation and to change.

Labor-saving ideas and agricultural education were favorite topics of the editors. Plowing matches, fairs, and reaper trials were widely promoted because they helped to make farmers aware of change. Agricultural editors were determined to upgrade agriculture because they felt strongly that it was the backbone of the nation. The challenge was especially difficult in the years before the Civil War, for so few farmers subscribed to and read farm papers.

Agricultural Fairs

The foundation for what eventually became the agricultural fair was laid in the 1600s, when weekly markets and semiannual fairs were held under direction of local authorities. Farmers brought their livestock and produce to the village in the hopes of gaining a premium or of at least exchanging them for goods they needed. These fairs were patterned after those in Europe, where they played an important role in the life of the people.

The early agricultural societies did not appeal to the average farmer because they tended to be elitist oriented. Most average farmers were not market oriented and were satisfied chiefly to gain a way of life and survival. To make agriculture more popular for the rank-and-file farmer, the farm organization had to be more democratic and have an element of competition. Premiums had to be offered for the best exhibit or for events, such as sheepshearing contests.

The concept of agricultural fairs commenced in 1807 with the creation of the Berkshire Agricultural Society at Pittsfield, Massachusetts. Elkanah Watson was the recognized father of the event. In 1809 the Columbian Agricultural Society of the District of Columbia tried to promote domestic and rural economy by having a series of six semiannual affairs, which were a combination sale and show.

In 1810 Watson convinced 25 others to join him in holding a livestock show, which served as the model for most of the county fairs that were founded between 1815 and 1840. Watson understood what it took to motivate rural people. Prizes were given for the best animals, which created a competitive spirit among farmers. Displays of women's domestic manufactured goods also competed for prizes. Rural clergy were involved, and there were events for rural youth.

Watson wanted a fair in every county. The concept spread rapidly, and by 1819 it was estimated that 100 fairs were held in New England, in New York, as far south as South Carolina, and as far west as Illinois. After 1825, when state funds were diverted away from the fairs, they died out until revived in the 1840s.

Farmers initially hoped that the fairs would give them ideas about how to ease their economic plight, but instead they learned about ways to improve production. The animal exhibitions were popular, as were the plowing matches. But farmers realized that in addition to the educational aspects of the fairs, they also benefited from the social and the recreational opportunities.

The aim of the fairs changed to collecting information on production, spreading knowledge on eradication of crop pests, and encouraging production of new crops. Public concern was widespread about the need for improvement in agriculture, so it was possible to get statewide organizations founded on the model of the Berkshire plan. Science, transportation, and communication all had innovations that stimulated agriculture. Industrialization in the East created a demand for more production. Traditional self-sufficiency gave way to commercial agriculture.

After the revival of fairs in the 1840s, most state legislatures promoted them and offered financial aid. New York led the revival in 1841 with what might be called the first state fair. Other states followed. At the same time, the federal government provided funds for the purpose of collecting statistics and information about rural conditions. For several years the allocations increased annually due to the success in securing information.

By 1852 there were about 300 local and county agricultural societies in 31 states for the purpose of promoting fairs. The number increased to 1,367 by 1868. The period from 1850 to the 1870s was sometimes called the "golden age of the agricultural fair." Like the farm journals, fairs were major contributors to the great technological changes taking place in agriculture.

Early Agricultural Education

Despite the reluctance of farmers to become involved with any type of "book farming," there were always a few who did what they could to foster knowledge of an improved agriculture. As early as 1749 Benjamin Franklin advocated the teaching of agriculture in every town. No doubt the idea was too revolutionary, considering the fact that most towns were still dominated by farmers. Probably the first formal instruction in agriculture was offered in 1792 at Columbia University, in New York City, where a professorship of natural history, chemistry, and agriculture was established. It was supported in part by funds from the New York legislature.

Between 1821 and 1823, the first school for agriculture opened at Gardiner, Maine. This was a combination agricultural, industrial, and technical school providing the elective system, courses in natural philosophy, winter short courses, and an experimental farm. After 1831 more courses were added.

In 1824 Stephen Van Rensselaer, who had inherited over 700,000 acres, established the Rensselaer Institute at Troy, New York. The objective of the Institute was teaching and applying science to practical purposes in life, with special emphasis on farming and mechanical pursuits. The school has continued to this day.

Massachusetts maintained its leadership as a progressive state when several schools for agriculture were established after 1824. In 1843 Amherst offered a course in agricultural chemistry, and by 1852 that work led to the creation of a separate department for agriculture in the state.

The push to develop universities to train agricultural and other scientists led to the establishment of Cornell University at Ithaca, New York. Harvard also expanded its curriculum to include a department of chemistry headed by Eben Harsford, who had studied under the famed German chemist, Justus von Liebig. It was the state of Michigan, however, that established the first educational institution of its kind in the nation, when on February 12, 1855, it created a college for the purpose of improving and teaching the science and practice of agriculture. This became a model for similar colleges that came about after the passage of the Land Grant College Act of July 2, 1862. The passage of that act, along with complementary legislation, provided for the scholarly activity necessary to open the way for the rise of our nation's agriculture to a worldwide position of dominance.

Scientific Agriculture

The three presidents discussed earlier as being intensely interested in agriculture were also among the leaders in developing scientific agriculture. Washington was interested in scientific research, conservation, diversification, and mechanization. He said that the more involved he became in farming, the more he liked it. Although he owned nearly

70,000 acres at one time or another, he preferred intensive farming because he felt that the nation would profit more from it.

Washington was disturbed by what he called the shiftless methods of most farmers, in which their wheat yields dropped from 30 bushels per acre when the soil was fertile to 8 to 10 bushels once the land was exhausted. This made him a constant experimenter with crops, livestock, and methods of management. He established rotations of as long as seven years, one of which included buckwheat as a manure plowdown in the third year. Others had clover and grasses in the last three years of the seven-year plan. To make such a plan feasible, livestock production had to be maximized.

Initially, Washington raised tobacco, but he curtailed production when he realized that it was too hard on the soil. His dislike of slavery probably made that decision easier. Working with slaves was a constant source of irritation for him. Probably his three major contributions to agriculture were his introduction of mules, his introduction of Merino sheep, and his introduction of legumes to the rotation.

Jefferson was one of the first to use animal and vegetable manures, along with commercial fertilizer and a rotation, to help the soil regain its fertility. He also was a pioneer in introducing foreign plants. He grew 32 different vegetables, in addition to other plants, shrubs, and trees, to acclimatize them to the new setting.

Jefferson was responsible for developing a seed drill and a hemp brake. He made improvements on a threshing machine, and his most lasting contribution was designing the moldboard plow so it gave the least resistance.

Lincoln spent his early years on four farms, none of which were substantial farming operations. He learned what pioneer exploitation of the land was in contrast to a more settled cultivation. He observed what farmers lacked and understood frontier democracy. He realized the haphazard methods of most farmers were not the most efficient way. As early as 1859 he expressed the hope that steam power would be successful on the farm so that farmers could do a more thorough cultivation. Like Washington and Jefferson, he felt that education was a key to better farming.

Soils By the last decades of the 1700s farmers were aware of the declining fertility of the soil. However, they did not respond quickly to change, because they lacked knowledge of what caused the problem. Lime was used as a soil additive as early as the 1680s, but it was not until after the Revolution that its use became widespread. The commercial lime industry was not developed until the 1840s. In 1784 a Virginia farmer started a 19-year experiment using gypsum on several grains and grasses, with positive results. The use of gypsum encouraged the growing of clover and grasses, which made the production of more livestock economical.

In 1821 Edmund Ruffin demonstrated the value of crop rotation and legumes to maintain soil fertility. He applied decaying seashells and clay marl, which increased productivity of the acidic soils on his farm. Ruffin published the results of his experiments in a book reputed to be the "most thorough piece of work on a special agricultural subject ever published in English." It was historical, scientific, and practical, attracting wide attention. Unfortunately, most farmers preferred to ignore his findings and continued to let the land revert to weeds or brush, or they moved west.

Even though farmers were reluctant to change, once they realized that they had to do so to survive, they adapted. Probably no sector saw such a dramatic shift as New England, where wheat was dropped in favor of sheep. The growth in the woolen industry provided profit opportunities, and the agricultural sector experienced a rapid upgrading.

By the 1840s the cause of soil exhaustion was well known. The problem was less obvious in New England than in the South, because soil in New England was poor from the start. By then some farmers had started a cycle of settlement, exploitation, ruin, and a move west to new ground. Others, especially in New England, sought alternatives: move to town and enter a trade; raise sheep instead of wheat; go into dairying or truck gardening; send the children to work in the factories; or move west. Most of them learned that the best opportunities were off the farm.

The first known tiling was conducted by a New York farmer who imported a pattern of drain tile in 1821 but did not begin to drain his farm until 1835. In 1848 a tile-making machine was imported, which greatly reduced the cost of producing tile. Within the next few years the New Yorker had laid 47 miles of tile.

Justus von Liebig's work, *Organic Chemistry in Its Applications to Agriculture and Physiology*, was quickly reprinted and widely accepted. This was an indication that many Americans understood something about agricultural chemistry. It marked a beginning in scientific agriculture by providing answers to many of the problems, but more importantly it showed the need for scientific education. A new interest in soil testing opened an awareness for agricultural experiment stations.

Americans realized that they were behind Europe in scientific agriculture and invited a Scottish agricultural chemist to the United States in the 1850s to study the soil. Some farmers used crop rotation and fertilizer as well as other innovations and realized positive results. Initially, it was believed that reduced productivity was caused by chemical deficiency, but it was soon learned that bacteria, biological conditions, and erosion were also important. Heavier rainfall and hilly land made soil erosion a problem that had not been experienced in Europe, where rains were less intense. Constant hoeing exposed the organic matter and reduced soil tilth as well as fertility.

Gypsum, marl, lime, and guano were among the additives that could be purchased to maintain soil productivity. Southerners were more aggressive in purchasing such products because of the greater profit potential of their cash crops. Northerners found it was simpler to move west. In 1843 the first commercial shipment of guano arrived from Peru. By 1849 the first commercially manufactured fertilizers with chemical analyses were produced. By 1860 seven factories nationwide were in production, but improved transportation was necessary before widespread use of commercial fertilizers was feasible.

Some of the most innovative activity in rebuilding and maintaining soil fertility took place nearest the urban centers once farmers realized the profits from being able to supply those markets. This was particularly true with specialty crops. Manure was hauled from the towns to the fields. Firewood and farm crops were readily traded for manure and other wastes from the towns. Composts of earth, fish, peat, salt mud (shore silt), broken shells, fish heads, and seaweed, as well as slaughterhouse and tanner's waste and offal, were valued as manure.

Probably one of the greatest advances in working the soil was the application of horse power. Land could be tilled better than at any time in history on a larger scale without backbreaking human effort. The use of horse power by the 1850s opened the way for a new system of farming and enabled the North to increase its agricultural production with reduced labor during the critical Civil War period.

Livestock

George Washington was not only one of the largest landowners in his time, but he also was one of the best farmers. Through his connection with the king of Spain, he received some high-quality mules as a gift. His work with mules laid the foundation for the increase in the use of those animals, which became popular in the South.

In the early 1790s three Merino sheep were imported from Spain, but nothing came of them because they were eaten rather than used for breeding. It was not until 1801, when E. I. duPont imported Dom Pedro, an outstanding Merino ram, that the sheep industry got a serious start. The Merino was the right sheep for the time.

Merinos caused a significant change in the nation because their fine wool aided the establishment of the woolen textile industry and reduced our dependence on Europe for cloth imports. Vermont became the leading sheep state, while textile manufacturing grew rapidly in Massachusetts and Rhode Island. The War of 1812 skyrocketed demand, and raw wool prices rose sharply, encouraging major expansion in sheep production. The foundation of the textile industry was assured.

The year 1783 probably marked the first importation of purebred cattle, when Shorthorns were shipped from England to Maryland, Virginia, and Kentucky, but farmers were slow to upgrade. Prior to 1817 dual-purpose Devons were imported. Soon more Shorthorn cattle and Berkshire hogs arrived. Henry Clay introduced Herefords at about the same time.

In the 1850s Ayrshires, the first single-purpose cattle, came. They gave far more milk than the dual-purpose animals. Their production was soon challenged by the Jerseys. The Dutch had introduced Holsteins, but it was not until the 1850s that they were reintroduced into Massachusetts. Further importation was delayed after they were attacked by a disease that was believed to be unique to the breed.

From 1830 to the 1860s considerable upgrading in the breeding and care of sheep, horses, hogs, and beef took place. Americans were responsible for breeding Poland China, Duroc-Jersey, and Chester White hogs. As late as 1825 hogs still roamed at large as scavengers on the streets of New York and Philadelphia. After the 1840s farmers in the Northeast could no longer compete with western farmers and withdrew from hog production. Areas began to specialize according to which animals did the best or had the strongest markets. Farmers learned that they could gain more from their farms by having more and better animals.

In general, dairy herds were improved more slowly than other livestock. It was not until well into the 1800s that dairying became a specialty and the industry progressed. Traditionally cows calved in the spring, milked heavy in the summer, and went dry in the early winter. It is estimated that in the 1830s a cow of average quality produced 2,200 to 2,600 pounds of milk annually. In the South, most of the milk was churned into butter, and buttermilk was drunk.

Milk cows were kept in the cities until the advent of rail transportation in the 1840s and 1850s, when dairy products could be shipped in. Until then, virtually all farmers had dairy cattle, but when the industry started to specialize, many farmers discontinued milking.

New England, central New York, and northern Ohio became the early dairy centers. By the 1850s they switched to single-purpose breeds. Holsteins became the leading dairy breed. After a cheese factory was opened in Rome, Orange County, New York, in 1851, followed by the first creamery for butter production in 1856, there was more incentive to increase in the dairy sector. By 1869 the nation had 1,000 cheese plants.

Early Protests

The first farmer protest activity took place when the price of tobacco collapsed. Because the British failed to support them, tobacco growers organized to bargain collectively. When that failed, they destroyed their crops.

No other significant protests occurred until after the Revolution. Prices for farm products rose dramatically, and farmers did quite well, especially those who were able to sell in hard currency to the British. Then the wholesale price index dropped from 225 in 1780 to 90 in 1786. Farmers could not meet obligations they had contracted during the years of high prices. Many lost their farms, and some were sentenced to debtors' prisons. The result was Shays' Rebellion of 1786. It was repressed by Massachusetts, but eventually the legislature enacted many of its demands.

In the Whiskey Rebellion of 1794, farmers revolted against the tax policies of the new government. They disliked federal taxes, and in particular they felt that they were being discriminated against by the 10¢ per gallon tax on whiskey, in addition to a tax on stills. On this occasion the newer and stronger federal government used force to suppress the activity.

Sectional differences caused Fries' Rebellion in eastern Pennsylvania in 1799. This was prompted by a federal tax on land, dwelling houses, and slaves that farmers considered unfair. The rebellion ended when Fries was captured by federal troops.

Westward Expansion

One of the unique phases of the nation's history had been the way in which farmers' thinking was affected by the massive land frontier. That frontier confronted or comforted farmers from the 1600s until the 1890s, when the bulk of the free land was settled. According to historian Frederick Jackson Turner, the presence of available land affected the thinking of most Americans, not just farmers. Even if some question the theory, there is no doubt that when the frontier ended, farmers were concerned about their future.

Settling on the Frontier

When the first farmers left the original settlements, they usually remained in immediate contact with the older communities for two reasons: they needed protection from the Indians, and they needed a market for their surplus production. These farmers were interested in providing for their families and in gaining virgin land. In this respect the movement was so gradual that it was not thought of as heading into a new frontier.

But a few ventured farther west. As early as 1746 flour and other goods were shipped from Indiana down the Wabash, Ohio, and Mississippi rivers, but except in the South, there was little need to go far from established settlements. However, after the Proclamation of 1763, when the British forbade settlement beyond their established line, farmers felt that their rights were being violated.

By the 1770s settlers moved west of the mountains, especially to Kentucky and to Pennsylvania. Initially they attempted to maintain small settlements, where they built log cabins facing each other in the form of a rectangle so that the back sides made a solid wall. For them, moving west to gain a farm meant independence. It gave them a feeling of equality and was an avenue of opportunity. In 1775 Kentucky had sufficient settlers so that it had to pass two laws involving agriculture—one to preserve the quality of horses and another to preserve the range for livestock.

By 1800 Pittsburgh was a recognized farm center for exporting corn, flour, and salt to the South and for importing flour and smoked and salt pork from the West. From the start, river traffic was important to Pittsburgh, and after 1811, when the first steamboat arrived on the Mississippi, the western flow of people and goods increased rapidly. However, high steamboat freight rates kept flatboats, barges, keelboats, and rafts busy carrying most of the bulk goods from Pittsburgh to Mississippi towns.

After the War of 1812, hard times caused many people to go west to farm or to establish businesses in trade centers. The soil in the Kentucky and Ohio region was heavy and difficult to work, but corn yields of 60 bushels an acre, together with yields for oats of 50 bushels and those for barley of 40 bushels, attracted many newcomers.

The war also brought an increased demand from the South for northern farm products. This was caused in part by the westward movement of southern farmers with their slaves and by greater specialization in cotton. This demand and improved river transportation encouraged settlers to move into the Mississippi Valley.

During the westward movement some people worked haphazardly, developing their farms just enough to survive. When more prosperous settlers came, the restless ones were quick to sell out. They then moved to the next area and started all over again. Part of the reason for this willingness to move was that the task of developing the midwestern prairie was not as difficult as had been the task of developing the South or the wooded areas of the East.

The price of land in the fertile valleys rose rapidly and soon was beyond the point of profitability. With the Panic of 1837 the price of land dropped and many farms were lost. At the same time, factories in the East terminated many employees, causing large numbers of them to move west and settle primarily on the prairie east of the Mississippi River.

The tough prairie sod was broken and planted to corn by dropping seed in the furrow or in openings made with a mattock. After the first harvest, the land was cross-plowed and planted to wheat. Sometimes the wheat was planted among the standing cornstalks. In any case, the drudgery was not as severe as it had been on previous frontiers, because much of the equipment that once had to be made on the farm could now be purchased. The new factory equipment was much sturdier, worked better, and did not break down as often. However, husking corn and making hay were still strenuous tasks.

The preference for settling on individual farmsteads instead of in small communities as in Europe kept farm families isolated. There is little doubt that this was a negative factor for the families, at least from a psychological standpoint. Hunting and fishing often were given as the reason for isolated farmsteads. Some said that they did not want the poultry and the livestock to mingle, but others simply did not want to have close neighbors.

By the 1840s soil exhaustion in the East had reduced capital returns to such a low level that farmers were eager to sell and to move. At this point a large portion chose to go to the city, while others hoped to make enough from the sale of their farms to purchase better equipment and buy cheaper land on the frontier. Other eastern farmers realized that consolidating their farms with those of farmers who wanted to sell would enable them to gain from the economy of scale. By adopting the newest techniques, by farming to meet specialized demands, and by benefiting from lower transportation costs than the western farmers, progressive easterners could profit.

Western New York, western Pennsylvania, and Ohio were the leading wheat-producing areas by 1840. The opening of the Erie Canal in 1825 and the Ohio Canal in 1832 made it possible for farmers from those areas to compete in the eastern markets. Indiana, Illinois, Michigan, Wisconsin, Missouri, and Iowa were being opened but were not yet producing for the remote commercial markets because of the lack of transportation.

The Preemption Act of 1841, discussed earlier, was the key to a massive westward migration accompanied by a huge expansion in production. At first many settlers avoided the open prairie because they felt that the wooded areas were more fertile. But once the 6- and 12-ox teams had pulled the plow through the tough prairie sod, which had no stones or tree roots, and the first crops were produced, the reluctance to settle there diminished. By the time the third crop was produced on the mellowed prairie sod, visions of a system of commercial agriculture became clear.

During the years that the midwestern prairies were being settled, activity was taking place along the Pacific Coast. By 1825 farming was being done along the Columbia River near Fort Vancouver in the extreme northwest. Farms of several thousand acres raised grains, fruits, and vegetables. Cattle and horses were imported from California, and hogs from Hawaii. The food produced on these farms was marketed to local trading posts, to whaling vessels, and to Russian posts in Alaska.

The first farms were operated by the Hudson's Bay Company, but later employees left that firm, and others migrated into the region to farm. The Panic of 1837 caused a movement of settlers to Oregon, which expanded to a great migration in 1843. The 1849 gold rush in California stimulated agricultural markets for Oregon farmers.

The first non-Spanish or missionary farmers and ranchers to California arrived about 1835. John Sutter became the best known of these early settlers. Those who did not join the rush for gold made their fortunes raising food for the miners. As in the Midwest, the settlers did not appreciate the fertility of the treeless, barren, parched, and cracked earth. But after they worked it and came to understand it, grain production grew rapidly where horses and cattle previously had roamed. In a few years the value of wheat was nearly equal to that of the gold mined.

Women on the Frontier

As mentioned earlier, every member of the family had to work on the frontier farm to make it survive. In larger families both boys and girls commonly were bound out to work for other farmers or in factories. Boys were contracted until age 21, at which time they frequently received goods and cash. Girls usually were released from their contract at age 18 and generally received a supply of bedding. At that time it was assumed a woman would marry and become mother, housewife, and farm worker. How much actual farm work she did depended on various circumstances, including ethnic background.

Until the past few decades little was written specifically about the role of women in settling the frontier. Assorted studies of the post–Civil War period indicate that women homesteaders were registered on from 8 to 25 percent of the original entries. Many of them filed for their land adjacent to that of their future husbands, and once the legal terms were met, the couples married. From personal interviews this writer learned that in some cases houses were built over the boundary line between quarter sections to satisfy the requirements of homestead regulations.

Unmarried women were enticed to the frontier by school teachers' wages, which often were higher than in settled areas. Single women also went west to be laborers, missionaries, dressmakers, or seamstresses, or they went to homestead. Because men outnumbered women on the frontier, women often went west to find a mate. Some went to escape unpleasant marriages.

Married women went west to please their husbands who wanted to go. Others encouraged the move in the hopes of improving their income. They came with the emotion of optimism and not with reluctance, as was so often portrayed. They sensed the push and pull of others around them. Some thought a change of climate might be good for their health. During the 1840s and 1850s, many came to the frontier because it was free from slavery. Others came because they wanted to escape paying accumulated debt or to avoid paying taxes on their rundown eastern farms.

It appears that women generally were as optimistic as men about the spirit of the frontier and the hope it held for them. Pleading and convincing letters from those already on the Iowa frontier encouraged others to come there. An effective propaganda campaign designed to lure people to the cheap fertile land played a major role. Women's historian Glenda Riley added that frontier farm wives realized that they were "fully immersed in a system of economic partnership" with their husbands. But they did not consider themselves to be "obedient, passive, docile, and quiet" wives. The granddaughter of a pioneer farm couple related to this author, "Grandfather always ruled the family and farm until Grandmother put her foot down. He never stepped over the line."

Probably the greatest change on the midwestern farm frontier in regard to women related to education. By the 1840s it was expected that girls would be as well educated as boys. Whether that was a more liberal frontier attitude or was a reflection of the times is uncertain.

Forests

Forests originally covered about half the area of the 50 states. In 1990, about 32 percent was in forests. Pioneers depended heavily on the forests for their personal needs and for the general economy. From the start, forests had to be cleared for agriculture and to provide fuel and building material. Many farmers avoided nonforest land because they felt it might be inferior to other soil types.

Forests also were the source of the fencing material so essential to farming. At one time more forest products were required for the construction of rail fences than for any other use. Until the 1840s the amount of wood used in fencing exceeded the total use of lumber. To fence a square 40-acre field with zigzag fence required about 8,000 fence rails. A person could split 50 to 100 rails a day. Add to that the requirement of about 4.5 cords of wood for fuel per capita annually during colonial times.

Rivers often ran through forests and were important to the food supply and for transportation. Using the Indian method of clearing the forests for farming was less costly than breaking the prairie, in terms of horse or ox power required. But the eventual agricultural use of the land yielded far more than the forests could. Although the forests were a major source of exports in the early years and the basis of an extensive industry, forest products of pioneer days did not compare with the demands for personal use of the land.

The Commercial Livestock Industry

Prior to mechanized agriculture, farmers did not have the ability to produce all the food crops necessary to survive. This made it necessary to rely on livestock to supplement the diet. It also provided the means for additional income.

Sheep Production

During the early 1800s, northern New England changed from self-sufficient farming to commercial agriculture. Between 1815 and 1840, manufacturing in the inland towns of New England expanded with numerous woolen mills in Maine, New Hampshire, and Vermont. This increased the demand for wool, and the price rose accordingly. The sheep growers responded and discontinued most other agricultural production. Hay became the leading crop. This proved to be good for farmers who were not doing well, for they sold their land to sheep growers. They then either took jobs in the mill towns or went west to start over in farming.

After 1840 sheep numbers began to decline because the protective tariff was reduced, causing a drop in the price of wool. At the same time, there was a slow but steady increase in wool production in the West. The newly developed canal system enabled competition with the eastern growers. The market was saturated and the prices continued to drop.

New England farmers had an annual sheep cost of $2.00 per head compared to $0.25 to $1.00 for the western raisers. By the 1850s the cost of maintaining a sheep was more than the return from wool. Except for a spurt in demand during the Civil War, the New England sheep industry was doomed. By then the demand for dairy products had increased, and farmers turned to dairying, which could more effectively use family labor. However, the sheep industry continued for a few more decades in the hill country, which had no other use at that time.

Expansion of the Livestock Industry

After the Battle of Fallen Timbers in August 1794, the Ohio country was cleared of hostile threats, and livestock numbers increased rapidly. Initially, cattle were driven from the East to Ohio and to Kentucky to be fattened on the rich grass of that region. Kentucky livestock people quickly recognized the fertility there for cattle and became leaders in upgrading the breeds. These improved cattle found a ready market at the slaughterhouses of the larger eastern towns.

As early as 1756 Boston had established a centralized livestock market and packing center. After the Revolution, as many as 5,000 cattle were sold and slaughtered there weekly. Originally, this market helped New England farmers, but when they could no longer compete with western cattle, they switched to sheep production.

The Whiskey Rebellion of August 1794 ended the farmers' chances of selling corn in a liquid form and forced them to feed their corn to hogs. The finished pork products were then floated downstream to New Orleans. Thus, during the 1815–1830 era, when Kentucky and Ohio became the major beef-and pork-producing region, New England turned to sheep production.

Ultimately, cattle were driven up to 1,000 miles to reach eastern markets. They averaged about 33 miles a day and generally rested at "drove stands," which were inns with accompanying cattle pens. Drovers originally went only as far as Philadelphia, but by 1804 they continued to Baltimore and by 1817 to New York. Droving from Ohio and Kentucky continued until the 1850s, when packing facilities and improved transportation ended the need for such activity.

At the same time, settlers in the South moved into the areas of Alabama and Mississippi that were filled with huge herds of wild Spanish cattle, hogs, and horses. Plant life was ample to accommodate the large numbers, but neither the Spanish nor the French had done much to upgrade the livestock breeds or to improve the area. Once the cotton economy grew, plantations entered the area and forced the livestock people farther west.

Even though the Carolinas, Georgia, Alabama, and Mississippi were all good cattle country, they became importers of animal products once cotton production became dominant. By the 1840s, to fill the need for beef in the Cotton Belt, cattle were driven east from Texas. For decades New Orleans was the chief market for Texas cattle. Because of the great numbers of cattle in Texas, they also were driven to Missouri, to Ohio, and eventually all the way to New York.

Indiana was not good cattle country, but by 1830 livestock producers had found Illinois and Missouri very favorable for animal production. Their largest market was downstream in New Orleans. From the start, the supply of meat on the western frontier probably was greater than in the original settlements.

Michigan, Iowa, Nebraska, and Kansas rounded out the cattle and hog frontier from 1830 to the Civil War. Although Michigan was farther to the east, its settlement was delayed because of problems with the Indians. It was not until after 1830 that any real activity took place there, and then it was restricted to the southern tier of counties because of swamps and forests to the north.

The Pacific Coast was not devoid of livestock for long. Shortly after the mid-1820s, animals were driven overland from St. Louis to Oregon. Later, others were driven north from California. The Willamette Valley was ideal cattle country, and after 1842, when the Oregon Trail was opened, large numbers of livestock reached the area. The California gold rush spurted the Texas cattle industry, which continued strong until the Civil War, when the demand became even greater.

The movement of the livestock frontier to the west resulted in the development of the modern meat packing business in the United States. By 1850 Cincinnati was known as Porkopolis. The principles and methods for the industry were developed there, and the entire nation was aware of what was happening. As the industry moved west, other great meat packing centers were established in St. Louis, Kansas City, Omaha, and Chicago.

Improving Livestock

As with most other innovations in agriculture, improving livestock came about slowly because many farmers resisted. Scrub bulls and boars were allowed to run loose, and any attempt to restrict the practice brought an outcry from small farmers saying that such laws only protected the rich. The imported English breeds, especially in the South, degenerated rapidly when forced to compete with native cattle that were accustomed to scrounging off the countryside as they had done since Spain controlled the area.

Some upgraded British cattle were imported shortly after the Revolution. They soon were dispersed to the South and to the West, and it is doubtful that much benefit came from them. In 1817 eight Shorthorns and four improved Longhorns were imported into Kentucky. The Shorthorns were a success and multiplied rapidly, while the Longhorns lost popularity in the area. During the next decade serious efforts were made to upgrade both the Shorthorns and the Herefords. Shorthorns soon became the leading breed, because they were better dual-purpose animals.

Packers acknowledged a general upgrading of quality in beef cattle but not as much as in sheep and hogs. The success of improved breeds in sheep came with the importation of Merinos. Then, the development of Poland China hogs, combined with high beef prices in the 1830s and in the 1850s, caused renewed interest in improved livestock breeding. Additional income from livestock during those periods, plus competition at fairs and encouragement from the agricultural press, convinced farmers to improve both beef and dairy breeds. The farmers realized that the best way of doing so was by importing better-quality European breeds.

By the 1860s sufficient progress was made so that specialty dairy and beef belts were established. The dairy centers were near the larger cities along the Atlantic Coast, and the beef belt extended west from Philadelphia, generally between the 36th and 43rd parallels.

Between 1840 and 1860, horse and ox breeders reached a turning point. Previously, oxen were preferred as draft animals, but as machinery requiring faster motion to operate properly came into use, horses were needed. Because horses worked faster and were easier to teach and to manage, they became more valuable for use in the city and on the roads. On farms that hired labor, horses were desired because they kept the rising cost of labor down. By the census of 1850 it was clear that the number of oxen was declining in comparison to the number of horses.

The Cotton Economy

A turning point was reached in southern agriculture in 1775, when long-established trade and credit relationships with England ended. In pre-Revolutionary days, cotton was not much more than a garden shrub, with only a small amount processed. Tobacco, rice, and indigo were the dominant crops, but prices for those crops dropped sharply with the loss of British markets. The rural economy was in for a change. Slave trade was abolished over the next 20 years, and colonial land policy was at an end.

From 1775 to 1782, exports to England dropped to less than 4 percent of their previous volume. Virginia and Maryland shifted much of their tobacco land to wheat, while North Carolina increased its wheat production for export to the West Indies.

Sometime between 1778 and 1786, sea-island cotton, accidentally or illegally introduced to Georgia, proved superior to the cotton then grown and exported. There was a surprising reversal of trade, and exports of the new variety exploded. This new short-staple cotton filled the demand for an industry that had experienced major changes in technology in recent years. Now the South faced a new problem: it could not meet the demand because of a bottleneck in removing the cotton seeds. A crude machine called the "churka" was capable of cleaning the short staple at a rate of less than 1 pound per day, or only about 4 pounds per week if operated evenings and during other free time.

In 1793 the cotton gin was invented. The first hand-powered gins could clean 50 pounds of cotton per day, and power-operated machines could clean 1,000 pounds per day. Until then, the cost of cleaning out the seeds had made cotton fiber more expensive than wool or linen. Now the picture reversed, and prices for raw cotton rose sharply to meet the demand. From 1794 to 1804 production increased eightfold, and that was just the beginning.

The production area expanded rapidly to the Mississippi Valley, which meant that labor needed to work the new lands continued to be in short supply. The Constitution forbade importation of slaves after 1807, and low-cost or free labor was not to be found. Ironically, other than the adoption of new varieties, little was done to improve the methods of production to reduce the demand for labor. Southerners believed that cotton production and slave labor were tied together.

In any case, the number of blacks increased from about 750,000 in 1790, nearly all of whom were in the South, to 4.4 million in 1860, 90 percent of whom were in the South. Most of them were slaves. Buying slaves in advance proved very costly, causing the system to become highly capitalized. Overcapitalization of labor became a serious problem as the cost of slaves rose while the price of cotton remained steady. In some respects, only the natural increase in the slave population, which added to the assets of the planters, kept the system economically viable.

The failure to maintain soil fertility, to grow alternate crops in rotation, and to adopt new technology to gain from the economy of scale further added to the burden of the planters. The servile labor system also harmed the South because it effectively blocked other enterprises from entering the region. Small farms intermingled among the plantations produced a variety of crops but not enough to make the South independent of food exports. As in other sections of the nation, some of the small farmers were progressive and others were lazy and shiftless, doing little to aid society. Small towns that were few and far between made up much of the rest of southern society.

The Mormons in Agriculture

In 1846 the Mormons left Illinois for Salt Lake City, Utah, in a caravan of 16,000 people, 2,000 wagons, and 30,000 head of livestock. They placed supreme importance on agriculture, so as soon as they arrived in July 1847, they began constructing an irrigation system. The following year they divided the agricultural land into strips of 5-, 10-, 20-, 40-, and 80-acre plots along the irrigation canals. The five-acre strips were the choicest land and given to those who conducted a trade in the city. Larger strips were given to full-time farmers who lived on the land. Each man in the community was assessed a fixed number of days annually to build canals and to keep them in repair.

The California gold rush of 1849 proved to be a boon for the new settlement. It gave the Mormons an opportunity to buy provisions from those traveling through and allowed them to sell their surplus crops and horses at high prices. By 1850 the remaining 11,380 settlers produced crops that had no comparison with those of other early farmers in that area. They had ample grains and hay; a surplus of meat; butter, cheese, and garden products; plus the surviving 1,304 horses and mules and 5,552 head of cattle.

Crop prices doubled and provided funds to develop manufacturing, which by 1853 had passed the primitive stage. Agriculture boomed until 1860, as gold seekers provided a good market. Sugar processing was attempted but failed for lack of capital. Flax, silk, and cotton were all grown in an effort to secure fiber for clothing; none succeeded. Sheep, however, were profitably raised in all parts of the territory, and by the 1860s wool was in sufficient supply to create a thriving woolen mill industry. Lumbering and leather tanning prospered until railroads arrived in 1870 delivering cheaper and better-quality goods.

Rivers and Canals

About 1800 one ton of goods could be shipped the 3,000 miles from Europe to the United States for about $9, the same charge as for hauling one ton 30 miles overland. A bushel of wheat was worth 50¢ at a Great Lakes port, but the cost to get it to New York City was 75¢. In 1814 the cost of delivering a ton of goods 100 miles overland was $32, and the cost of delivering wheat was still more than its value at the point of origin. It took 100 days and cost $100 per ton to float goods from Cincinnati to New Orleans. The return trip was five or six times more costly. At the same period it was estimated that it did not pay to transport corn inland more than 20 miles.

In some respects this did not bother farmers too much because they had so little to sell. In 1820 they sold only about 20 percent of their production, mostly to nearby urban markets, but to make progress, farmers had to sell more. The Erie Canal started a boom in transportation, which caused several thousand miles of canals to be built. Except for the Erie, most of the canals had little long-range impact on marketing agricultural products, for by the 1830s railroads entered the West.

The Erie Canal opened markets in the East and in Europe for Ohio, Indiana, Michigan, and Illinois farmers. By 1860 railroads had penetrated those states and beyond. Rail rates were higher than river and canal rates, but the rails were more reliable and could go nearly anywhere. By 1840 the development of canals and railroads enabled Ohio to replace Pennsylvania as the leading wheat state.

By 1860 there were 3,566 steamboats on western rivers, but, like wagon trains, they disappeared as soon as the railroads came. Canals had reduced the cost of transportation to less than one-fourth of the overland cost, but railroads were still desperately needed to make the low-priced western land profitable to farm. By 1870 urban markets took 40 percent of farm production.

Highways and Railroads

Because the distance between most communities in the 1700s was extensive, building roads between them was too costly for the communities to bear. Most of the early roads were built by citizens who received tax credits for their efforts. Connecting links between most towns were no more than dirt trails through forests and prairies. About the only regulation on road building was that stumps left in roads should be no more than one foot high. Logs were placed in wet spots to make corduroy roads. Only the most important roads had drainage ditches or a rock base with gravel coating.

The Lancaster Turnpike from Philadelphia, built in the late 1790s, was the first macadamized improved highway in the nation. This fostered the craze for turnpikes, and many private companies started to build them. In 1806 the federal government funded the National Road, which started at Cumberland, Maryland, and about 1850 reached the Mississippi River. Between the 1790s and 1825, much of southern New England was covered with a turnpike road system. New York and Pennsylvania built theirs prior to and immediately after the War of 1812, and the southern states followed with turnpike construction in the next two decades.

Turnpikes were a boon to the average traveler, but they were too costly for those who had heavy or bulky goods to move over long distances. Unfortunately, those were the items that produced the best revenue. This prevented turnpikes from being profitable.

Much of the country was so sparsely settled that roads could not be justified. Even stagecoaches and freighters traveled over simple marked trails. By 1820 stagecoaches appeared as far west as St. Louis, and their routes spread out from there as new settlements appeared. Mail contracts supplemented their passenger income.

The Conestoga wagon was developed by the Pennsylvania Dutch for hauling goods. First used during the Revolution, it was capable of a two- to three-ton load. It had a durable, flared box covered with bows to support a linen tarp and was pulled by five or more horses. A slightly lighter version of the Conestoga wagon was used west of the Mississippi.

Under ideal conditions wagon trains could travel from St. Louis to the West Coast in 25 days. The owners of individual wagons usually pooled them in caravans. The initial major task of the wagon trains was to supply the 100 forts scattered throughout the West, then the mining camps and other communities. Buffalo hides made up the major portion of their limited return haul east.

In 1848 the first plank roads, called "farmers' railroads," were built in a spokelike fashion out of Chicago. The railroad had already arrived, and the purpose of these roads was to move farm produce to the railhead. These roads enabled a horse to pull two to three times the normal load at a faster pace than on dirt. Farm land along the roads increased in value, and consumers gained because goods were delivered at a lower cost.

Except for the rivers, a few canals, and the National Road, most systems to transport goods overland were of little permanent value. However, when railroads were first constructed they were looked upon as a substitute for canals. Some canals were built where railroads already existed. Few dared to think that railroads would ever be used to transport the massive volume of grain, livestock, timber, coal, and equipment over long distances. But between 1830 and 1860 the railroad system startlingly changed economic life and greatly improved social life.

After railroads were completed to major mid-western towns, common freight, such as barreled flour, was hauled to the Atlantic seaboard for less than it could be shipped 30 miles in pre-railroad days. An 1839 study indicated that rail rates were competitive with canals and steamboats on the Great Lakes and were 1/5 to 1/10 the rates on macadam roads and turnpikes. Only steamboats on the Ohio and the Mississippi rivers were cheaper.

A railroad craze struck the nation, and only the shortage of capital and labor kept speculators from building new roads as fast as they had hoped. The government also profited from the drive to secure land, for public domain that was accessible to the railroad sold for as much as $10.00 per acre in contrast to the established minimum price of $1.25. Fortunately, the world price of wheat skyrocketed in the mid-1850s, just as the railroads entered the Midwest, and the newly established farmers capitalized on the opportunity. By 1858, only a decade after railroads had entered Chicago, nearly 75 percent of the goods exported from there left by rail.

In some respects the South was more desperate for railroads than any other section of the nation, for it relied on the export of tobacco, rice, and later cotton. Equally as critical, southerners needed to import most of their food supplies and manufactured goods. The South hoped that railroads would make it more competitive with the North in securing population, wealth, and strength. Every rail line, short or long, was seen as another connecting link in a vital chain of railroads to any part of the nation.

Mechanization

In the closing decades of the eighteenth century a few innovations made life a bit easier for the farmer. In 1776 the addition of the cradle to the scythe speeded the process of harvesting small grain. Jefferson's discovery of the proper mathematical principle of the plow became

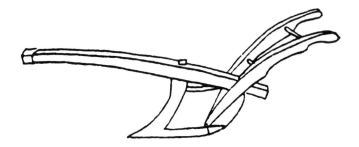

Charles Newbold's plow. (USDA)

public in 1793. The plow was improved when Charles Newbold made and patented the first cast-iron plow in 1797. The new plow was further upgraded in 1814, when Jethro Wood patented an iron plow. The fanning mill, which provided a means of mechanically cleaning grain, was also developed in 1797.

From 1830 to 1860 a series of inventions caused a revolution in agriculture. As population and industry grew, the demand for agricultural products increased rapidly. Machines were developed to ease the farmers' task of opening new land to meet that demand. The cost of farm products decreased as the cost of production declined with the adoption of labor-saving equipment. Mechanization put a premium on large-scale farming because it enabled economies of scale. The American nation and the world's consumers have been the constant beneficiaries of the movement that started in those decades.

The major emphasis of the new technologies encouraged a shift from human power to animal power. This was particularly so in the North, where the labor shortage persisted. In the process, northern farmers moved from self-sufficient to commercial farming. Ironically, the South, which already had commercial agriculture, could have gained as much or more by taking advantage of mechanization but chose not to do so.

Each invention seemed to open the way for another. One of the best examples was the development of the reaper in 1831. Only three years later, the mower and the threshing machine came into being. Then in

1837 the steel plow was produced, and during the 1840s the grain drill was perfected. Those machines were a major factor in the reduction of the time required to produce an acre of wheat from about 37 hours in 1830 to 11.5 hours in the 1840s.

The Sickle and the Cradle

The sickle, which dates almost to biblical times, was still in use at the time of the Revolution. The scythe without a cradle was used for a period before the adoption of the cradle in 1776. However, it was not immediately adopted by all farmers in spite of its advantages. But by the early 1800s most farmers used the cradle for small grain and used the sickle only for corn.

With a sickle a person could cut a half acre per day, with a scythe about three-fourths acre, and with a cradle about two acres. The big advantage of the cradle was that the grain was bunched, which greatly speeded the work of the one who gathered, bound, and shocked the grain. Using a cradle reduced the time to cradle, bind, and shock a 20-bushel-per-acre wheat crop to one-fourth to one-half of what it took to do the same task with a sickle. Two men could cradle, bind, and shock two acres of wheat in a 10-hour day.

The amazing thing about the cradle is how long it took farmers to adopt it. Leo Rogin wrote that as late as 1837 the sickle was still more widely used in New England than the cradle scythe. The reasons most often given were that it was easier to cut downed grain with a sickle than with a cradle, it took too much time to train an operator to do a good job with a cradle, and ripe grain shattered too much when a cradle was used.

The Mower and the Rake

Using the cradle scythe was the common method of harvesting hay until the grain reaper was developed. Several attempts were made after

1800 to perfect a horse-drawn mower, but it was not until 1831 that a machine with a scalloped sickle and a grain divider was patented. Both ideas were adopted by later inventors. Each improvement of the reaper aided the progress of the mower, which was like the reaper without a reel and a platform.

It was not until 1847 that the first practical mower was patented. By 1856 the two-wheel mower with a flexible cutter bar totally distinct from the reaper was marketed. Within the next few years refinements were added, and by 1865 nearly all the hay in the nation was cut by two-wheel cutter-bar mowers.

Between 1818 and 1820 the horse rake was introduced. Prior to that time both grain fields and hay fields were raked with hand rakes. By doing two to three acres per hour, a standard early wooden horse rake could do the work of six men. The reliance on hay made both the mower and the rake popular with farmers everywhere, but the extreme shortage of farm laborers in the West made them particularly desired there and explains their relatively rapid adoption. By the end of the Civil War, mowers and rakes (dump or revolving) were standard equipment on northern farms.

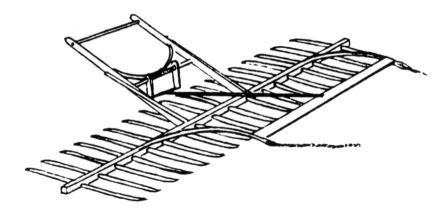

Revolving rake, 1820s. (USDA)

The Reaper

One of the best-known stories in American agriculture is that of the development of the reaper. Robert McCormick had started working on a reaper by 1809. He had one in operation by 1816, but he continued to work on his idea until 1831. That year he made his final demonstration, which he considered a failure, and quit. Fortunately, his son Cyrus continued his father's work, adding a gathering reel and improving the divider. His 1834 demonstration proved successful. A patent was granted in June 1834.

In 1833 Obed Hussey patented a machine that contained a hinged frame and slotted guard teeth so the scalloped knife could vibrate through the opening to cut the grain. Because of the demonstration of 1831, McCormick's machine received priority claim at being the first successful reaper. This marked the beginning of horse-drawn harvesting machinery in the nation.

The reaper, with a 4½-foot sickle bar powered by a bull wheel, required one person to drive the horses, another to rake the grain from the platform, and five to seven people to move the grain out of the way of the horses for the next round. This machine could cut 12 to 15 acres a day. Crude and labor intense as that may seem, the first reaper could do the work of six to seven workers with cradle scythes.

The machines sold slowly at first because of the cost and labor requirements, but the demand was encouraging enough to cause McCormick to move closer to the grain production area. In 1847 he started a reaper factory in Chicago. In 1850 the self-rake was attached, but the $50 additional cost kept this feature from being popular. That year McCormick produced 4,500 reapers, and Hussey produced 500.

In 1855 McCormick adopted the Hussey sickle and sickle guard and produced his first really successful reapers. During the Civil War the labor shortage was so intense that by 1864 two out of three reapers sold were self-rakes. By that time the nearly 100,000 reapers in operation cut 70 percent of the grain, replacing more than 1 million harvest hands.

A self-rake reaper in clover, Michigan, 1890. (USDA)

As dramatic as the adoption of the reaper was, most farmers in New England and in the South still used the cradle. As late as 1919 the cradle was in use on small farms in New England, Virginia, West Virginia, Kentucky, Tennessee, and North Carolina. This writer is aware of the cradle being used in Tennessee in 1958.

The Header

The header was used less widely than the reaper but was important in the progress of harvesting implements. A forerunner of the header probably was invented in New Jersey in 1825. Using a horse to pull the machine and two men to remove the grain, 10 to 15 acres could be cut daily. This machine apparently did not come into general use.

In 1844 George Esterly, of Wisconsin, patented a header that had a reel with beaters, an adjustable permanent knife, and a box to receive the grain. It reportedly was modeled after a Roman header. It was very labor saving and popular, but output was limited because of the side draft problem.

The Haines Illinois Harvester, patented in 1849, also was very popular. It was a pusher model, so the side draft problem was eliminated and the cutter bar was longer. The grain was elevated directly to wagons pulled alongside. Because binding and shocking were eliminated, harvesting was faster and done with less labor.

A four-horse pusher header, North Dakota, 1910. (C. H. Frey)

The greatest drawback to the use of the header was that wheat did not mature evenly, and the grain spoiled if it was not threshed immediately. Sometimes field losses were heavy because harvest had to be delayed until the grain was uniformly ripe. This was not a problem in the Pacific Coast regions, and it was in California that the header was the most popular. A 12-foot header, pushed by six horses and accompanied by three wagons with two horses each and two men at the stack, could cut and deliver 20 acres of grain per day to the threshing site.

The header also made inroads in the Great Plains states of Kansas, Nebraska, and the Dakotas. However, by the time those areas were developed, the combine was on its way.

The Binder

As early as 1850 the idea was conceived of placing workers on the reaper platform to catch the grain and bind it before it hit the ground. In 1858 the first Marsh brothers harvester, with two men standing on the platform binding the grain, was in operation. The ability of the binders was the limiting factor, but the machine used three fewer workers than were required to do the job using a reaper. Next came the drive to build a self-binder. Some of the first work was done in the 1850s, but it was not until 1872 that the first successful wire-tie binder was marketed.

The Fanning Mill and the Thresher

When flail threshing was practiced, farmers complained that as much grain was thrown away with the straw as was threshed out. The loss was reduced if the straw was fed to animals, for they salvaged the discarded grain. Flailing was the most common way of threshing on small farms. An average worker could flail about eight bushels a day. Horses and oxen were used instead of flailing by having them trample over the grain that was in a confined circular area. This was very hard on horses, and many were injured and often ruined in the process. By using animals to thresh instead of men to flail, daily production per worker increased to about 40 bushels.

Winnowing was improved about the time of the Revolution by using winnowing sheets to create a breeze to speed up the process. This was hard work, and on the average, slightly over one bushel per hour could be cleaned in this manner.

The first winnowing fan probably was introduced about 1710, but producing wind artificially was looked upon as slightly irreligious, so fans were adopted very slowly. It was not until the 1840s, when larger acreages of grain were produced on western farms, that the concept of the fanning mill was adopted. The early mills could clean grain at the rate of from 10 to 13 bushels per worker hour.

At the same time the above progress was being made, work was being done to develop a threshing machine. Early machines were made in Europe. The first American thresher did not appear until the 1790s. In the early 1800s, six or seven threshers were in use here. These small machines, powered by from two to four horses on treadmills, required two men and two boys to thresh 10 to 18 bushels of wheat per hour in a 10-hour day. The early machines were complicated, often broke down, were costly, and required more horses than generally were available, so farmers continued to flail.

Treadmills continued to be used to operate small threshers, but after 1825 horse-powered sweeps propelled the larger ones. In the late 1840s portable wheel-mounted steam engines, which were pulled around by horses, powered the threshing machines via the use of belts. In the 1840s the separator was added to separate the grain and the straw, but it was not until the late 1850s that the cleaner was added, making the thresher a three-phase machine: a thresher, a separator, and a cleaner. However, the grain still had to be run through a fanning mill for additional cleaning before marketing. The Pitts and Case threshers, the recognized leaders, could thresh 25 bushels of wheat and up to 60 bushels of oats per hour, which was the peak rate until the 1870s. However, smaller farmers, and most of those on the frontier, continued using the flail well into the 1880s.

After 1840 the growing domestic market, an expanding rail network, the development of steamships for shipping grain throughout the world, and the repeal of the Corn Laws in England all served to create a strong demand for grain. That demand, coupled with a short labor supply, convinced farmers to purchase self-binders so they could grow and harvest more grain. This created a demand for threshing machines. All these steps contributed to major agricultural prosperity and an explosion in production.

The Plow

Plowing in the present sense means turning the soil completely over, but in the centuries prior to the moldboard plow it simply meant opening the soil. This was a difficult and time-consuming task. After Jethro Wood added interchangeable parts to his plow in 1819, the next major breakthrough in plow design came in 1833, when John Lane made a moldboard plow out of material from a crosscut saw. This was changed slightly in 1837 by John Deere.

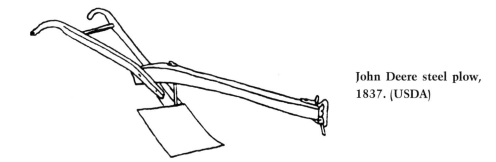

John Deere steel plow, 1837. (USDA)

At the same time, Daniel Webster built a heavy plow with straps of iron on the moldboard, using an iron share and a colter. This plow required six yoke of oxen and from six to eight men, but it could plow 12 inches deep and cut tree roots. However, it was not what the average farmer needed, and it was not capable of turning the sticky, loam prairie sod. John Deere's steel plow, which scoured in the prairie soil and could be operated by one worker using two horses, was what was needed. In 1858 the Deere factory made 13,000 of the 150,000 plows manufactured that year.

The early Deere model plows had two defects: they had blow holes on the surface, and they were brittle. Those problems were minimized in 1869 by James Oliver, who developed the chilled-steel moldboard,

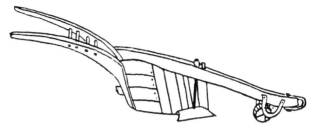

Daniel Webster's plow. (USDA)

which scoured well in heavy soil. Shortly thereafter John Lane, Jr., produced a moldboard that had a steel surface only and a soft center. This moldboard was lighter and less brittle.

In 1864 the plow was improved by adding wheels for riding (the sulky). The double-gang riding plow was introduced in 1879, and in 1881 large plows were developed for steam power. By 1915 unit-frame tractor plows were marketed. From then on, plows were refined and enlarged to match the available power units.

Two-bottom gang plow pulled by five horses. It could do 5.5 acres per day. (Larry Dahl)

Seeders and Drills

Hand broadcasting was the common method of planting small grains and hay seed. To keep birds from eating the seeds and to achieve better germination, the field had to be harrowed at least twice to work the seeds into the soil. Hand seeding required an average of about 1 hour and 20 minutes per acre. In 1857 the hand rotation sower, which seeded up to six acres an hour, was patented. This seeder was almost identical to the small lawn seeder used today. A further refinement was the broadcast endgate seeder, which was capable of about five acres per hour but wasted seed.

In 1841 the first American patent was issued for a grain drill. For the first time, grain was placed into the ground at desired depths and was covered for better germination. By the 1850s the drill was adopted in the more progressive farm areas. The early two-horse drills did 8 to 10 acres a day. By the 1890s the 12-foot, four-horse drill averaged 16 acres a day. The drill moved west rapidly because of the larger-size farms and more severe labor shortage. In the older established areas, with their rocks and stumps, the drill could not be used, and hand broadcasting continued.

Late 1800s version of drill capable of seeding 20 to 28 acres per day. (June Tweten)

The Corn Planter

Corn is indigenous to the Western Hemisphere, so it was only natural that the corn planter was an American invention. Several patents were issued for planters from 1799 to 1836, and by 1869 over 100 patents had been granted. During the 1850s, planters came into use, and by 1860 much of the western corn crop was planted by either one- or two-row hill-drop or drill planters. The two-row check planters, which operated on wheels, reportedly could plant 10 acres a day. By 1857 a horse-drawn planter with a seat for the driver and a row marker that could be used on both sides of the planter was marketed.

Expanding Demand

Commercial Agriculture

Farmers had little opportunity to become involved with commercial/capitalistic activity until transportation made distant markets more readily available. From 1820 to 1860, farmers increased the sale of their production from 30 to 60 percent. Once the market was available, some farmers were eager to participate, but others remained slow to change. Many of them had little business experience, which sometimes caused them to fall prey to swindlers. Others were suspicious of the business community, which prevented them from taking advantage of market opportunities.

At the same time, most farm boys wanted to become farmers and landowners. This was due partly to their lack of skills in other vocations and to limited opportunities off the farm. However, in the 1800s, when industrial opportunities with better-paying jobs increased rapidly, they still preferred to remain in agriculture.

Land was abundant, but it took at least $500 to own enough equipment and livestock to become a renter on a two-thirds–one-third share agreement. It was also necessary to have other credit available to begin farming. To purchase a 40-acre farm and start operations, at least

$1,400 was required, and if the farm had to be fenced, another $1,400 was needed.

Lack of working capital handicapped the transition to commercial agriculture. Part of the problem lay with the farmer, who did not always reinvest income in an enterprise that offered the greatest return. To become a commercial farmer, it was necessary to invest in more equipment and labor than farmers were accustomed to doing. Most farmers were not good record keepers and were not willing to become involved in ventures that required capital and necessitated borrowing from bankers, whom many distrusted. The more farmers relied on mechanization to ease the burden, the more capital that was required and the more dependent they became on the market.

Urban centers continued to grow, and by 1860 they contained 60 percent of the population in the North. Over 50 percent of foodstuffs produced entered the trade channels. Farmers applied better technology and systems to improve productivity to meet the market demand. They soon realized that the shift to commercial production was logical. Because they could produce more than they consumed, they had the opportunity to sell their surplus for cash. By 1860 virtually no food or fiber processing for the market was carried on in farm homes. The transition to commercial agriculture had arrived. This was a key to the agricultural and industrial revolutions that took place during that era.

The Food Chain

It has been determined from archaeological digs that the commercial food chain was well established by the late 1700s. A well-structured food distribution network served large and small towns and isolated fur-trade posts. In the 1700s, consumerism was present, and there were few self-sufficient households in either urban or rural areas. Convenience foods were available at all levels of society. Prepared foods were on sale at bakeries, from street vendors, in taverns, and elsewhere and in some cases probably were less costly than if they were prepared at home.

Food chain in the United States—1800

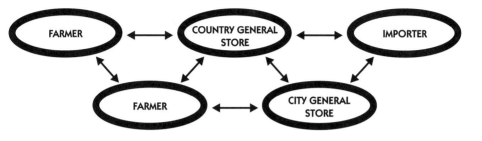

(Lifeline of America)

The food chain was still very simple in the early 1800s, for most of the food produced on farms went to the country general store. Part of that volume was resold to retail customers, and the remainder was delivered to general stores in urban centers. Many city dwellers had gardens, cows, and chickens and were not as reliant on the grocer as today. Much of the produce that farmers sold was bartered at the store for goods they needed. The amount of the total production that farmers sold or bartered was dependent on how near they were to an urban market.

About 1820, as farmers moved to the frontier in increasing numbers, the food distribution system underwent a change. Merchants followed the farmers to supply them with manufactured goods that they needed but also to purchase farm produce to ship to the city. Railroads made this movement possible and profitable. In 1840 there was a merchant enterprise for every 300 persons. In addition to buying goods from farmers and selling to them, the local merchant also often provided a gristmill, a sawmill, an inn, and a livery stable.

Initially, the farm produce delivered to the country store was of poor quality, but because little else was available, it was accepted. However, as transportation improved and both urban and world markets grew, discounts were imposed on poor-quality goods. This encouraged farmers to seek ways to improve the quality of the goods they sold. They were aided by the government, which informed farmers about what consumers expected.

In 1830, weekly market reports were given on the volume and the kinds of livestock sold and their prices. The earliest reports were published in Boston newspapers. Prior to that, market reports were chiefly rumors and were unreliable. At the same time, specialized middlemen became involved in marketing farm products. Cattle and hog drovers were probably the first of such people, followed by wool buyers.

The industrial revolution continued to expand, and by 1850 the food chain reacted to the increased demands. Many farmers still sold through the country stores, which acted as collection points. Soon a far larger portion of sales were consigned to urban central markets, which sold to wholesalers and they to retailers. A portion of the vegetables, poultry, livestock, and grain went to processors, which were just starting to develop in the 1850s. After the goods were processed, they were placed back into the food chain.

In 1859 the Great American Tea Company became the first food store chain. The idea was so successful that by 1865 the company added groceries to its tea sales and had 25 red-front stores. When the company

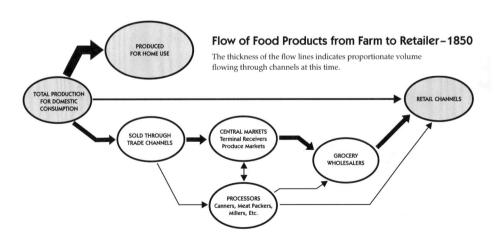

(Lifeline of America)

started moving westward, the name was changed to The Great Atlantic & Pacific Tea Company. By 1880 there were 100 A & P stores, and by 1900 there were 200. The company experienced a virtual explosion after that, and in the 1920s there were 11,000 A & P stores. Other chains also came into existence during those years.

Food Processing

The first known successful cannery in the nation was established in 1819. It processed pickles, ketchup, sauces, and jellies, most of which were exported. In 1820 seafood was canned in Baltimore. The canning industry was based on knowledge from Europeans, who preserved food by sterilizing it in airtight containers. The Europeans also used tin cans. Canned goods initially were not popular in the United States because of cost and the high rate of spoilage.

The introduction of calcium chloride to the processing water sharply reduced the cooking time and enabled canneries to increase their daily production eightfold. This was followed by the pressure cooker, which reduced canning time even more and also provided for greater accuracy in the process.

At first, fruits were eaten fresh and used for beverages, the most common of which was apple cider. Drinking cider was a carryover from Europe but was also done out of necessity because of the lack of pure drinking water. In the early days any fruit that remained after cider making was dried for later use, and the rest was left for animals to salvage.

Little was done in the fruit processing industry as long as the population was limited and the natural supply of wild fruit was ample to provide for the needs of all. However, by 1830 the supply in the wild began to decline, and domestic production had to increase to meet the demands.

In 1846 Gail Borden started experimenting with preserving milk so that it could be transported long distances. By 1856 he had produced powdered milk, and by 1861 condensed milk. The demand was immediate, and by 1879 production of condensed milk was over 13 million pounds. By 1889 it was 38 million pounds. Borden's innovation coincided with the development of cheese factories and creameries to stimulate the farmers toward commercial dairying.

In 1858 Mason jars, sealed by rubber gaskets and screw tops, were invented. Because the early jars were still hand blown, they were expensive. It was not until 1903 that mechanical glass blowing greatly reduced the cost of glass jars and popularized home canning.

Virtually all pioneer farmers kept chickens, but a small flock was also commonplace along with the family cow for a large portion of people living in towns. As commercial agriculture increased, the number of farmers with flocks rapidly decreased. Likewise, as towns grew, urban flocks also became less common. To preserve eggs for shipment to distant markets, they were immersed in lime water or oil to seal the porous shells. It was not until the 1870s that icehouses were used to preserve perishables. After 1880 commercial egg drying and refrigerated cold storage further enhanced the life of eggs. Eggs were stored and shipped to market by packing them in straw in barrels. In the 1880s the 30-dozen egg case with fillers was used, which gave economy in storage and in shipping.

Changing Work Patterns

The labor shortage during the colonial period made it essential that women be involved in all phases of farm work. That made women more autonomous than their European counterparts. Some ethnic groups expected women to work in the fields, while others did not. In the back country, where the most primitive conditions existed, frequently there was no gender division of labor until the community became more established. Women usually were more involved with the poultry and the dairying operations, which were generally thought of as being sidelines to crops, hogs, or beef. This was particularly true after the 1700s from New England to the Great Plains.

Starting about 1800, household industries began to decline almost in proportion to the growth of the industrial revolution. Farmers were taken out of their self-sufficiency mode, and, among other changes, women no longer made wool and flax into cloth. Women became more involved with men in enterprises that produced for the cash market.

Work such as harvesting, erecting buildings, and butchering was frequently shared with other farms, especially in emergencies. Generally, men applied a stricter dollar value to the exchanged time than did women. Women were more haphazard about placing a dollar value on their time, except when they did market-oriented work, such as buttermaking or dressmaking.

Over a period of several decades, wives and daughters found a market for their labor when they were no longer producing marketable products at home. As commercial dairying became more prevalent, many women found jobs on dairy farms. Some worked as dairymaids for as long as 20 years on the larger dairy farms. Young women often preferred working on farms where room and board were provided rather than in the shoe and textile factories, where they had to pay living costs. Such farm work also enabled them to live closer to home.

Fortunately, the industrial revolution created a sharp increase in the demand for women farm workers, which caused a rapid rise in their wages. Indenturing was still practiced, but most employers preferred hiring young women because under indenturing the workers had to be released at age 18.

Farmers also found that working in shoe factories was more profitable than making shoes at home. This caused some of them to start working off the farm. In nearly all cases, every member of the family who worked off the farm was expected to contribute toward its support.

The Rise of Central Livestock Markets

Once cattle production commenced in the Ohio Valley and transportation improved to make the transition possible, it became obvious that farm products should be processed near the chief point of production and only the finished products sent to the consumer. The first major effort in that direction took place in 1811, when farmer-packers did business in Cincinnati. In 1818 that led to the first commercial livestock packing establishment. By 1830 Cincinnati was recognized as the major center of the packing industry for the nation. During the 1832–33 packing season, an estimated 85,000 hogs were slaughtered there.

Until the adoption of ice in the packing industry, the packing season was restricted to November through mid-March. Bad weather often stopped drovers from delivering livestock during those months, and financial panics or even a thaw or heavy rain was enough to stop the packers. Improvements in transportation, gathering of livestock, financing, sanitation, further processing, and marketing systems were needed for the packing industry to grow as required by the increasing demands.

Cincinnati's supremacy in the packing industry was short lived. In 1848 Cincinnati peaked in the slaughter of hogs and then reached a plateau as centers farther west developed. After 1850 Chicago challenged the eastern packing centers. The number of cattle and hogs delivered there became so great that a centralized stockyard was needed.

In 1848 the core of a union stockyard was opened. Then the various stockyards in Chicago were consolidated by the Illinois legislature. The nine railroads that converged on the yard became the major stockholders in the formal creation of the Union Stock Yards Company, which opened in January 1865. The yard had a holding capacity of 120,000 head of livestock. It remained a dominant center in the livestock industry until the 1930s.

Several innovations took place in the slaughtering industry during the next few decades that made it a major force in the food chain. Initially, the by-products of slaughtering in Cincinnati were discharged into the Ohio River and were carried away by the swift stream. This was not possible in Chicago, so the offal had to be delivered to the country, where it was fed to hogs. But as volume grew, this was no longer feasible. Consequently, uses were found for what had previously been discarded.

By the 1840s lard became the first profitable by-product of slaughtering hogs. Lard oil became a substitute for whale oil. In the 1850s, soap and glue were profitable by-products, followed by bristle dressing for brushes, hair for mattresses, glycerin, gelatin, and leather products. Because of the by-products industry, slaughtering and packing by large firms was much more efficient and profitable than butchering livestock on the farm or by small local establishments.

The first shipment of meat in iced cars took place in 1851. In 1857 the summer packing of hogs started when ice was used in the storage of meat and in rail cars. People preferred meat that came iced from the packer rather than that which was winter cured. This meant that packing plants could operate year round and the cost of acquiring livestock could be reduced. Securing an adequate supply of cheap ice then became essential.

Because of the huge losses of livestock in the South during the Civil War, a new supply of animals had to be found. Fortunately, the large Texas herds became the basis for the new cow country of the Great Plains.

The Growth of Foreign Markets

As early as 1545, Spanish officials in the New World promoted the production of hemp and flax. In 1780 Spain again encouraged hemp production in the North American colonies and made a special attempt to raise the plant at the California missions. In the 1790s the demand for hemp to be used in rigging for Spanish vessels brought more encouragement by local officials. In 1807 a Spanish subsidy served to increase hemp production sharply.

Exports were a key to the growth and progress of our agriculture. Tobacco, fish, furs, timber, grain, and rum were our first major exports, and, until recent decades, agricultural products have dominated our exports. After the War of 1812, the foreign market for all exports expanded, with agricultural products making up 60 to 80 percent of the total. From 1820 to 1860, cotton alone made up over 50 percent of all exports. Wheat was exported in steadily increasing amounts to England after the mid-1700s, and after the Revolution it was exported regularly to the continent.

From the 1790s to 1807, the Napoleonic Wars created a strong demand for our products. However, from then until the end of the War of 1812 there was a sharply curtailed growth in our shipping and commerce. Expansion was slow until the 1840s, when the Irish potato famine struck. Repeal of the English Corn Laws in 1846, European revolutions in 1848, the Crimean War of the 1850s, and scattered shortages elsewhere in Europe created new demand for our products. The California gold rush of 1849, plus the above demand, served to set off a wave of increasing prices for agricultural products.

The abundance of good land gave U.S. farm products a comparative advantage in international trade. That was fortunate, for the nation had a constant deficit, which was made up only partially by invisible items of trade, particularly shipping and cash payments. Agricultural trade was essential to the development of the other sectors of our economy, and it led to specialization in the industry by sections of the country.

Summary

During the period from 1783 to 1860, our infant nation relied on agriculture to serve as the economic foundation for its growth. Agriculture performed well as farmers moved west in a steady flow, opening new lands and meeting the demands of a growing urban population, the demands of the rising industrial sector, and the continued needs of the world market.

Fortunately, governmental policies were favorable to the agrarian needs and fostered an advantageous political climate for farmers. The abundance of relatively free land was the cornerstone of the governmental policy and the lure that attracted many immigrants into farming.

The miraculous growth of a transportation system, consisting of the Great Lakes, canals, and, most importantly, the railroads, combined to stimulate all sectors of the economy once they were assured of securing the production of western agricultural lands.

The creativity of our early leaders in establishing agricultural societies, fairs, a farm press, and educational institutions served to persuade the traditionally reluctant farmers to change their ways. The switch from self-sufficient agriculture to commercial production was essential for the welfare of both those in farming and the nation. All the technological breakthroughs to increase production and to fuel the growth of the food chain were the groundwork for the rise of an agricultural giant among nations.

References

Atack, Jeremy, and Fred Bateman. "Self Sufficiency and the Marketable Surplus in the Rural North, 1860," *Agricultural History* 58, No. 3, July 1984, 296–313.

Bailey, L. H., ed. *Cyclopedia of American Agriculture: Farm & Community*. New York: Macmillan Publishing Co., Inc., IV, 1909.

Busch, Lawrence, and William B. Lacy, eds. *The Agricultural Scientific Enterprise: A System in Transition.* Boulder, CO: Westview Press, 1986.

Carrier, Lyman. "The United States Agricultural Society, 1852–1860," *Agricultural History* 11, No. 4, October 1937, 278–288.

Church, Lillian. *Partial History of the Development of Grain Harvesting Equipment.* Beltsville, MD: USDA, Agricultural Research Administration, Information Series 72 (revised October 1947).

Danhof, Clarence H. *Change in Agriculture: The Northern United States, 1820–1870*. Cambridge, MA: Harvard University Press, 1969.

Demaree, Albert Lowther. *The American Agricultural Press 1819–1860*. Studies in the History of American Agriculture 8. New York: Columbia University Press, 1941. Reprinted, Philadelphia: Porcupine Press, 1974.

Drache, Mary Thompson. "Milk's the One." St. Paul: American Dairy Association / Dairy Council of the Upper Midwest, manuscript, 1994.

Edwards, Everett E. "The First 300 Years," *Yearbook of Agriculture, 1940*. Gove Gambidge, ed. Washington, DC: USDA, 1940, 171–276.

Ellsworth, Lucuis F. "The Philadelphia Society for the Promotion of Agriculture and Agricultural Reform, 1785–1793," *Agricultural History* 42, No. 3, July 1968, 189–199.

Ferleger, Lou, ed. *Agriculture and National Development: Views on the Nineteenth Century*. Ames: Iowa State University Press, 1990.

Fletcher, Stevenson Whitcomb. *Pennsylvania Agriculture and Country Life 1840–1940*. Harrisburg: Pennsylvania Historical and Museum Commission, 1, 1950; 2, 1955.

Gates, Paul Wallace. *The Illinois Central Railroad and Its Colonization Work*. Cambridge, MA: Harvard University Press, 1934.

Gras, Norman S. B. *A History of Agriculture in Europe and America*. New York: F. S. Crofts & Co., 1925.

Hampe, Edward C., Jr., and Merle Wittenberg. *The Lifeline of America: Development of the Food Industry*. New York: McGraw-Hill Book Co., 1964.

Helms, Douglas. *Readings in the History of the Soil Conservation Service*. Washington, DC: USDA Soil Conservation Service, Economics & Soil Sciences Division, NHQ, 1992.

Kinney, J. P. *A Continent Lost—A Civilization Won: Indian Land Tenure in America*. Baltimore: The Johns Hopkins University Press, 1937.

Leavitt, Charles T. "Attempts to Improve Cattle Breeds in the United States, 1790–1860," *Agricultural History* 7, No. 2, April 1933, 51–67.

Lemmer, George F. "Early Agricultural Editors and Their Farm Philosophies," *Agricultural History* 31, No. 4, October 1957, 3–22.

MacCleery, Douglas W. *American Forests: A History of Resiliency and Recovery*. Durham, NC: USDA Forest Service (FS-540) and Forest History Society, 1992.

Miller, August C. "Jefferson as an Agriculturist," *Agricultural History* 16, No. 2, April 1942, 65–78.

Neely, Wayne Caldwell. *The Agricultural Fair.* New York: Columbia University Press, 1935.

Osterud, Nancy Grey. "Gender and the Transition to Capitalism in Rural America," *Agricultural History* 86, No. 2, Spring 1993, 14–29.

Phillips, Ulrich B. *A History of Transportation in the Eastern Cotton Belt to 1860.* New York: Columbia University Press, 1908.

Plumb, Charles. *Marketing Farm Animals.* Boston: Ginn & Co., 1927.

Rasmussen, Wayne D. "The Impact of Technological Change on American Agriculture, 1862–1962," *The Journal of Economic History* 22, December 1962, 578–599.

Ricks, Joel Edward. *Forms and Methods of Early Mormon Settlement in Utah and the Surrounding Region, 1847–1877.* Logan: Utah State University Press, 1964.

Riley, Glenda. *Frontierswomen: The Iowa Experience.* Ames: Iowa State University Press, 1981.

Rogin, Leo. *The Introduction of Farm Machinery.* Berkeley: University of California Press, 1931. Reprinted, New York: Johnson Reprint Co., 1966.

Rossiter, Margaret W. *The Emergence of Agricultural Science: Justus Liebig and the Americans, 1840–1880.* New Haven, CT: Yale University Press, 1975.

Ryerson, Knowles A. "History and Significance of the Foreign Plant Introduction Work of the United States Department of Agriculture," *Agricultural History* 7, No. 3, July 1933, 110–128.

Schlebecker, John T., and Andrew W. Hopkins. *A History of Dairy Journalism in the United States 1810–1950.* Madison: University of Wisconsin Press, 1957.

Smithcors, J. F. *The American Veterinary Profession: Its Background and Development.* Ames: Iowa State University Press, 1963.

Street, James H. *The New Revolution in the Cotton Economy: Mechanization and Its Consequences.* Chapel Hill: University of North Carolina Press, 1957.

Taylor, Carl C. *The Farmer's Movement 1620–1920.* New York & Chicago: American Book Co., 1953.

True, A. C. "Popular Education for the Farmer," *Yearbook of Agriculture, 1897.* Geo. Wm. Hill, ed. Washington, DC: USDA, 1898, 279–290.

Wilson, Harold F. "The Rise and Decline of the Sheep Industry in Northern New England," *Agricultural History* 9, No. 1, January 1935, 12–40.

Winther, Oscar Osburn. *The Transportation Frontier: Trans-Mississippi West 1865–1890.* New York: Holt, Rinehart, and Winston, 1964.

Unit III

1860 to 1914

Some Events and Technological Innovations That Affected Agriculture

1861–1865 The Civil War triggered the first agricultural revolution

1862 Homestead Act

1862 U.S. Department of Agriculture (USDA) established

1862 Morrill Land Grant College Act

1862 Railroad land grants

1871 First Granger laws passed

1874 Barbed wire patented

1892 Gasoline tractor produced

1896 Rural free delivery (RFD)

1907 Hog cholera serum developed

1914 Smith-Lever County Extension System

Introduction

From 1860 to 1914 American farmers completed the extension of the agricultural frontier to the Pacific Coast. Prior to 1860 the government had laid the groundwork for many legislative acts that were passed early in this period. The long-delayed response to act on behalf of agriculture, as called for by George Washington, was fulfilled in 1860 and shortly after. The Homestead Act served as a magnet to attract some urbanites and thousands of immigrants to the frontier in search of land. The Civil War created a demand that improved prices and at the same time depleted the labor supply. This prompted farmers to purchase labor-saving machinery at an unprecedented rate.

The trend toward commercialization of agriculture aided in expanding production to meet the increased demand. Society in general profited far more from that improved production than did the farmers. Once the demands of the Civil War ceased, farmers were caught in a decades-long decline in prices and saw little personal benefits from their productivity other than a reduction in drudgery as provided by improved technology. Agricultural education programs were expanded rapidly under the Morrill Acts, which aided rural students in elementary and secondary schools and provided college programs for those continuing their education.

The booming industrial sector and the swelling urban population gave farmers hope that constantly growing demand would improve prices. Continued improvements in machinery, transportation, rural free delivery, telephones, rural schools, and rural roads all served to break down isolation, ease the toil, and enhance the life style of farmers. But virgin land, the constantly increasing number of farmers, and ever-improving methods of production kept production equal to or greater than demand. Thus, prices remained on the decline.

Protest movements were caused in part by the chronic economic problem, even though farmers were living better than earlier American farmers or most other farmers of the world. After the Civil War, agriculture started a relative decline as the industrial and business sectors grew. Farmers sensed their waning influence in a nation rapidly shifting from an agrarian base to an industrial urban base. This made them uncomfortable.

Governmental Activities, 1860 to 1914

During the nation's first 300 years, agricultural conditions were very fluid as farmers constantly left the established areas and moved to a new frontier. This progression continued from early settlement until the 1890s, when the land frontier closed. During that period most farmers lived in relative isolation. They tended to be narrow in their thinking, because the farm was their world. At the same time, frontier life made them resourceful, self-reliant, and hard working.

In the past, social and economic structures had centered chiefly around agriculture. What the small-scale farmer stood for and believed in was thought to be the thinking of the nation. Farmers competed with nature, which created intense individualism and extreme provincialism. They resented outside influence, especially any aimed at changing their way-of-life farming.

The Civil War hastened the forces that were working to change agriculture. Those forces were transfer of the public domain to private hands; completion of the westward movement; invention and adoption of farm implements; improved and expanded transportation facilities; the movement of industry from the farm to the factory; expansion of domestic and foreign markets; and the establishment of institutions to promote scientific agriculture.

The Creation and Early Work of the USDA

President Washington appealed for a department of agriculture, but it was decades before one was established. The work in the Patent Office in 1836, the 1849 transfer of that office to the Department of the Interior, and dramatically increased appropriations after 1855 were part of the slow process that created pressure within the government for a department of agriculture.

The lack of a department of agriculture was not the cause of local and sporadic protests by farmers. Farmers did not enjoy the opportunities open to other groups in society, but they were not unified enough to do anything about it. At the same time, they believed that they were superior to other sectors of society. The farm press was influential in giving them ideas to form societies and clubs, which resulted in an 1841 premature attempt to create a national organization. That goal succeeded in 1852 with the creation of the United States Agricultural Society, which had significant backing to eventually influence legislation for a department of agriculture.

In December 1861, in his first address to Congress, President Lincoln suggested an agricultural and statistical bureau. On May 15, 1862, the Department of Agriculture was created. The timing was bad because the traditional mood was against further centralizing control in the federal government. Lincoln pointed out, however, that agriculture was the largest business of the nation but only rated a clerkship in the federal government.

The objectives of the new department were to collect and to publish useful information; to introduce improved plants and animals; to answer questions of farmers; to test new implements; to conduct tests on soils, seeds, fertilizers, and other materials used in farming; to establish a professorship of botany and entomology; and to establish an agricultural library and a museum. The Department became a major scientific and economic agency of the nation. It not only became a model for state departments of agriculture but succeeded well in unifying national and state programs. It rose in stature, and on February 9, 1889, became an executive department headed by a Secretary of Agriculture.

Weather Bureau

In 1850 the Smithsonian Institution commenced issuing daily telegraphic weather maps, making the United States the first nation to do so. In the 1860s, balloons were used for upper air observations. In 1870, after the Signal Corps introduced meteorological service, it established posts on mountain tops to do the same work. By 1895 upper air observations were made at various spots throughout the nation at heights of one mile using kites 70 square feet in size. The Secretary of Agriculture wrote in 1899 of that work: "For the first time in the history of meteorology we have facts . . . [and] have probably added more to our knowledge of vertical gradients of temperature, humidity, and wind . . . in the lower portion of the atmosphere, than the sum total of all that was previously known on the subject."

The USDA, through the Weather Bureau, was one of the first departments to experiment with wireless telegraphy to send weather messages along the coastline. By 1906 wireless telegraphy was perfected so that it could be used on ships at sea. An underwater cable was laid from Alaska to transmit weather information to the states. The Weather Bureau continued its expansion, and by 1912, 193 weather observation stations were in service. The only problem was that the average farmer did not have the means to receive the information.

Animal Health

Little was done in the early days to protect people from impure foods. Massachusetts took the lead in 1856 when it passed the first law prohibiting the adulteration of milk. In 1859 Boston followed by appointing the first milk inspector, and in 1864 it prohibited the sale of milk from diseased cattle. The movement was slow, however, for in 1909 only 36 cities required dairy inspection, and by 1914 that number had dropped to 9.

By 1850 the animal disease problem was quite well understood, but because of the indifference of public officials, it was not until the 1870s that much was done. Veterinarians were not appointed to boards of health until the 1880s. It was not until 1884, when the Bureau of Animal Industry was established, that a major effort to prevent and control animal disease was effective.

The first major scientific accomplishment of the USDA came in 1889 when it was discovered that Texas fever was transmitted by the cattle tick. The significance of that discovery was the knowledge that a disease-producing organism could be transmitted between animals by a carrier. The result was that in 1906 a tick eradication program commenced. In the meantime, Robert Koch developed the tuberculosis test, which worked on both animals and humans. This meant that locating diseased animals and culling them from the herd could greatly reduce the spread of tuberculosis to humans.

In 1899 the second major event in animal health took place when a successful inoculant for anthrax was developed. This enabled southern farmers to turn to grass and more cattle as soon as more fertilizer, seed, and breeding stock became available. An increase in the number of beef cattle occurred rapidly, followed by increases in the number of dairy cattle, poultry, and hogs.

In 1833 hog cholera was reported for the first time in the United States. In 1861 a major outbreak struck the nation's hog farmers. Farmers were advised that nature could do more for them than medicine. It was not until 1878 that the USDA began research on the disease. By 1893 there were 90 known infected areas. Three outbreaks in the 1890s each killed more than 13 percent of the nation's hog supply. An experiment station eradication project in 1897 saved about 80 percent of the animals. The program continued, and by 1903 the virus was discovered and it was learned that the survivors of cholera were immune for life. In 1907 a serum was developed that greatly lessened the threat of hog cholera, but a 1913 epidemic still killed 10 percent.

Crusading work by members of the veterinary profession led to the passage of the first meat inspection law in 1891, which included mandatory inspection of meat for export. This improved our chances for exporting to Europe. The public was not aware of the connection between human and animal diseases. The problem was compounded because the cities still contained so many animals. Only the major meat packing facilities were federally inspected. By 1900, 148 such facilities in 45 communities slaughtered nearly 35 million head of livestock, of which about 160,000 were rejected. In the short span of nine years the number of inspected meat packing facilities rose to 936 in 255 cities and covered 53 million animals.

Most of the research that led to the passage of the Food and Drug Act of 1906 was done by the Bureau of Chemistry of the USDA. It was then charged with enforcement of the Act. The Adams Act of 1906 greatly stepped up the amount of scientific work carried on by the experiment stations.

Farm Production Studies

The first work in farm economics was done by the government in 1839. In 1859 the Commissioner of Patents conducted surveys on cost and marketing data. The Maine Board of Agriculture, in 1867, pioneered a study of farm management. The study dealt with farm size, fertilizer needs of growing oats, how to grow more hay, how to use oxen efficiently in the winter, and similar topics.

In the late 1880s the USDA made a study of the profitability of agriculture in relation to that of other industries. The study was aimed partly at determining the purchasing power of a crop acre in relation to the cost of the products that farmers needed. The USDA learned that most farmers were not bookkeepers, and it was disturbed that they were not concerned about profit and loss. Farmers' bulletins dealing with the cost of production and profits were produced.

During the 1890s efforts were made to convince farmers to become more efficient as producers because city people no longer grew much of their own food. The USDA wanted farmers to produce "more good food . . . [that could be sold] at reasonable prices to city dwellers." In 1901, surveys and studies commenced on farm management. The more surveys that were made, the more the USDA became concerned that farmers needed help in management. The USDA commenced management work in earnest in 1902, when it created the Office of Farm Management (OFM). Secretary James Wilson wrote in the 1904 *Yearbook* that the USDA had little concrete on-the-farm data on what represented good management. Efforts were made to encourage farmers to visit experimental farms to learn management practices.

A nationwide program of gathering records from farmers was begun. Finding farmers to cooperate was difficult because very few had any records and others did not want the government to know what they were doing. About 4,000 farmers out of the 5.7 million total agreed to cooperate in a farm costs-and-returns study. One hundred farms were selected for a prolonged detailed study to learn how many hours were required to produce crops and animals and to determine the per unit cost of production. The Secretary wrote that little was known about input and operating costs in agriculture. The data gathered from those 100 farms was used in Farmers' Bulletin 511, called *Farm Bookkeeping*, in an attempt to get farmers started in the cost phase of farm management. In 1922 the work in farm management was expanded again, and the Bureau of Agricultural Economics (BAE) was created.

One of the major concerns of USDA leaders was the terrible lack of efficiency of the farmers and their failure to use adequate business methods and "existing agricultural knowledge." In 1913 Secretary Houston wrote that he felt that only about 12 percent of the land was sufficiently cultivated to yield "fairly full returns" and only 40 percent was reasonably cultivated.

One of the best ways of disbursing knowledge to the farmers was through bulletins. In 1912 a total of 1,462 publications were printed into 24,900,557 copies, of which 10,409,000 were Farmers' Bulletins. In the first 16 years of publication the USDA provided 225,000,000 copies to farmers and others interested in agriculture.

As early as 1845, government employees assigned to agriculture started monitoring market potential domestically and abroad because agricultural exports were so important to the nation. That work was particularly concerned with the quality of pork and butter being exported. In the 1890s, the USDA prepared bulletins on the marketing of farm products that dealt with packing and preparing products for the market. By 1911 it had completed 30 years of study on the impact of cold storage on the quality of products, the annual price levels, and the consumption patterns arising from year-round availability.

Virtually every annual report by the Secretary in agricultural yearbooks during the 1890s commented on the need for expanded foreign markets. It was obvious that the United States would have continued surplus production. At the same time, new seed varieties were imported to determine what might be successfully produced here. In 1897, 7 tons of sugar beet seeds were imported and distributed among 22,000 farmers in 27 states to find where the sweetest beets could be produced.

In 1898 funds were appropriated for investigating irrigation. So little was known about irrigation that the Department had to train people in proper irrigation procedures. Even though agricultural products were in great surplus, it was determined that one-third of the country depended upon irrigation for agriculture, industry, or public use. The USDA was thought to be the best department to handle affairs related to irrigation. In 1902 a reclamation act was passed that for decades became the foundation legislation for dams, reservoirs, irrigation systems, and other water projects.

Soil surveying legislation was passed in 1899 as a logical next step to learn about the land and its potential for future production. The United States was one of the first nations to map its soils. By 1906 a total of 461 soil types had been determined.

In 1913 the Office of Markets was created to formally implement a plan of action for the marketing and distribution of agricultural products by the USDA. The plan included marketing, transportation, storage, grading and standards, and investigation of the need for cooperatives as well as determination of access to capital. This eventually led to placing agricultural attachés in foreign capitals and publishing information on foreign agricultural conditions for the benefit of U.S. producers. From that time on, the United States has been involved in world agricultural conferences, institutes, and programs.

The USDA Branches Out

Because agriculture is so essential to the welfare of the nation, any problem that affected the production of food became a concern of the political leaders. This caused the USDA to become involved in a chain of activities that was not imagined initially. One of the earliest of such activities took place in 1892, when the boll weevil was discovered in Texas and the USDA took action to stop the infestation. That much could be expected, but other activity followed when the plight of those affected by cotton monoculture was realized. The Department took immediate steps to change the practices of those farmers and their families. This led into new ventures: boys' corn clubs, calf clubs, pig clubs, potato clubs, followed later by 4-H Clubs, canning clubs, and home demonstration work—a full program for the entire family.

The Homestead Acts

The foundation of the first agricultural revolution lay in the various governmental acts that allocated land to settlers. As noted previously, after 1820 the federal government became more lax in transferring land to settlers. The Preemption Act of 1841 gave squatters first rights to buy land when it was offered for sale. The Graduation Act of 1854 provided for price reductions on federal land in proportion to the time it was on the market.

The key legislation among our national land laws was the Homestead Act of 1862. This enabled any person who was 21 years old or the head of a family and who was a citizen or had filed papers to become a citizen to claim either 80 or 160 acres of land. Direct charges were a $10 filing fee, a $4 commission at the time of application, and another $4 at the time of final proof. Residence and improvement requirements had to be met.

The major impact of this act was that speculators were largely removed from acquiring land. This was distinctly an American act. No other nation had any like it. The federal government was protected, the states were filled with settlers, communities were built, and the chance of social and civil disorders was minimized by giving settlers small tracts of land. However, land craze became big business, especially where the land grants to educational institutions and railroads were involved. These grants were major sources of land for non-homesteaders.

As homesteaders moved into the drier western areas, it became obvious that more land was needed to eke out a living. The Timber Culture Act of 1873 allowed a settler to acquire an additional 160 acres of land provided 40 acres were planted to trees and were cultivated. The 10 years originally required to obtain final proof was later reduced to 8 years, and only 10 acres needed to be in trees.

In 1877 the Desert Land Act offered land at 25¢ an acre if it was irrigated and cultivated for three years. Up to 640 acres could be held under this act. The following year the Timber and Stone Act enabled individuals to purchase land for $2.50 per acre that was valuable only for timber or for stone. The Kinkaid Act of 1904 allowed the homestead to be increased to 640 acres but required five years of residence and $1.25 improvement for each acre. The 1909 and 1910 Enlarged Homestead Acts provided for a 320-acre homestead in most states. In 1916 the Stock-Raising Homestead Act was passed to encourage further settlement in the West by providing a 640-acre homestead for grazing purposes.

In the years from 1860 to 1900, land in farms increased from 407 million to 839 million acres. Of the 432 million–acre increase, only 80 million was in grants directly to homesteaders. The other 352 million came from purchase by farmers from the government, land speculators, railroads, and others. Most of the land was transferred to well-intentioned people, but there were exceptions.

During the years 1875 to 1890, thousands of acres of land were granted to insiders, cronies, and people of influence by the General Land Office. As Gifford Pinchot, a leading conservationist, wrote:

In practice, thanks to lax, stupid, and wrongheaded administration by the Interior Department, the land laws were easily twisted to the advantage of the big fellows, and Western opinion was satisfied to have it so. At the same time there were no adequate safeguards against homesteaders acquiring timber land and "hooking" the timber from it and then abandoning their claim.

The Morrill Acts

After several attempts by individual states to establish agricultural colleges, Senator Justin Morrill, from Vermont, presented a bill to Congress to create a uniquely American educational system for agriculture. On July 2, 1862, the Morrill Act was signed, providing the cornerstone for the education system that greatly enhanced the position of agriculture and made it a world leader.

On August 30, 1890, the Second Morrill Act was passed, providing for 17 separate land-grant colleges for Negroes. Unfortunately, federal and state governments allotted funds in unequal amounts. The uncertain funding prevented any of the schools from offering college-level courses until 1916. But, like the colleges created in the act of 1862, the 1890 schools opened the doors for many who otherwise would not have had an opportunity for an advanced education.

Experiment Stations

The Morrill Acts and the Hatch Act of 1887 did much to encourage the movement toward professionalization of agriculture. But the system still lacked a testing or experimenting base to do work that was needed for adequate knowledge before reaching the farmers. In 1849 the New York Agricultural Society established a chemical laboratory to analyze soils, manures, and plants. The land where the original USDA building was erected was used as an experimental farm until 1862. In 1858 Maryland's agricultural college became the first to establish experimental fields in conjunction with its program.

There were always people who did not think that such work was necessary, but those in favor persisted. In late 1873 the Connecticut Board of Agriculture met to establish an agricultural experiment station based on the European pattern. In 1875 the Connecticut legislature appropriated $2,800 to Wesleyan University to run the "first regular experiment station." Because of popular demand, the first big task was to analyze commercial fertilizer for farmers and for dealers. That same year the California Agricultural Experiment Station was established and, like the one in Connecticut, was modeled after those in Germany.

Agricultural experiment stations had difficulty finding staffers learned in agriculture because chemistry and botany were the only mature disciplines involving agriculture at the time. However, that did not stop other states from going ahead. North Carolina opened a station in 1877, New York in 1879, New Jersey in 1880, and Wisconsin in 1883. By 1886 a total of 12 states had created experiment stations, and by 1899 there were 56 stations. Their growth was in response to the public mood for improvement and the belief that agricultural progress was both necessary and inevitable.

Because of abuse of standards by commercial fertilizer firms, nearly 20 states had appointed chemists by 1885. This enabled chemistry to make scientific inroads into agriculture, which had two positive impacts. First, it helped some farmers shed their reluctance to use new technology. Second, it served notice that the USDA would become a regulatory agency for protecting farmers and consumers.

The Hatch Act provided funds to establish an agricultural experiment station in each state in conjunction with the land-grant college. Agronomy, horticulture, forestry, human and animal nutrition, and agricultural technology were a few of the major areas of study for the stations. Reinforcement by verification and demonstration experiments and dissemination of the information to the public were part of the tasks perceived by the originators of the system. New ideas and discoveries quickly were put into the hands of the farmers, as can be attested to by the fact that in 1899 over 2,000 farmers' institutes were held. They were attended by over 500,000 farmers in 43 states and territories.

A unique agricultural experiment station was developed at Tuskegee, Alabama. George Washington Carver was convinced that agricultural education was the key to freedom for his people. Booker T. Washington had secured funds for the "first all-black experiment station" six years after a land-grant college act for Negroes was passed, but it was Carver who did the work of spreading the information to the people. Carver experimented with crops and determined the practical needs of blacks, who generally were poverty-stricken small farmers of the South. He realized that many could not read the bulletins, so he concluded that the only way to get information to them was through demonstration work in the black communities.

Farm Management

Little definite scientific work was done in farm economics prior to 1900. In 1874 the first farm accounting was taught at Iowa Agricultural College. In 1902 cost accounting research was started at the University of Minnesota, which became the first scientific work on cost accounting. In 1907 the person who taught the farm accounting course in Iowa developed a more extensive farm accounting program at Cornell University.

Farmers showed considerable interest in learning about production costs by the 1880s because of continued low prices and poor returns. The University of Illinois was one of the first schools to answer their request for help when it published a bulletin in the late 1880s on the cost of producing corn and oats for the years 1885 through 1887. In 1893 the USDA sent a questionnaire to 28,000 farmers asking for estimates on the cost of raising corn and oats. The results were compared with those of 4,000 farmers who were agricultural college graduates and published in 1898.

When Minnesota started cost accounting research, it knew that farmers did not keep records and thus it was fruitless to attempt to gain information from them via questionnaires. The researchers sought farmers who were willing to keep records for the purpose of getting accurate on-the-farm costs. They ran into considerable opposition from farmers who wanted no advice from white-collar university people, felt they could run their own businesses, or feared the information would be used against them.

When willing farmers were found, the researchers worked out detailed programs that included all financial aspects of the farm operation and the household. Dairy, diversified, and small grain farms were studied for three years, after which the information was published. Those farmers profited so much from the program that they continued working with the researchers for from 9 to 16 years.

When the chief Minnesota researcher became Assistant Secretary of Agriculture in 1905, he brought the results of the Minnesota cost accounting investigations with him. This led to organization of the Office of Farm Management by the USDA. Other states joined the program, and between 1919 and 1923 the drive for cost accounting peaked. By 1942 there were about 1,200 detailed farm cost accounting programs covering different farm areas of the nation. One of the early discoveries from the records was that about 15 to 20 farmers out of every 100 had incomes of about four times the average for the group. This revealed that one of the real weaknesses of agriculture was the lack of good management.

Henry C. Taylor, who developed the first department of agricultural economics between 1901 and 1909 while teaching at the University of Wisconsin, is the acknowledged father of agricultural economics. In 1905 he published the first textbook on the subject. In 1908 the first formal meeting of people engaged in farm management studies was held. This led to the founding of the American Farm Management Association in 1910, which was later called the American Farm Economics Association.

The Extension Service

The Farmers' Cooperative Demonstration Group, using private financing, was organized in Texas in 1904 to aid farmers suffering from boll weevil infestation. Those involved from the USDA, led by Seaman A. Knapp, helped fight the weevil but also showed the farmers how to diversify. Their major task was to prove to the farmers the benefits of rotation in stabilizing their income and to break the grip of the boll weevil. Rice, corn, and livestock were the best alternatives. The system caught on in the South. By 1911 about 600 agents were employed who had helped approximately 100,000 farmers in 13 states. By then the boys' clubs had enrolled over 46,000 farm boys. It was only through the boys' projects that some of the farmers could be approached.

Corn club members, Tyler, Texas, 1909. (NAL)

About the same time, boys' and girls' clubs were started in northern states. Ohio, Minnesota, and New York were some of the pioneering states. By 1909, 12 counties in Minnesota were involved in promoting agricultural clubs through the county superintendents of schools. Farmers' clubs were started as early as 1903 as a way of improving community living. All age groups participated in the meetings, which involved talks, debates, readings, dialogues, and musical entertainment as a way of

Farmers' club members, Tuleta, Texas. (NAL)

getting the entire family involved in improving agriculture. Later, these clubs became a factor in the cooperative movement.

Chambers of commerce, banks, railroads, the Council of North American Grain Exchanges, and the Better Farming Associations all became involved in promoting agriculture. Money was provided to employ an agricultural agent in each county to help the farmers. Within a few years most of the work of those groups was turned over to the USDA to make up the extension service. Again there was widespread opposition to the extension system from individuals who did not see the need for such activity. It was not until World War I started and the demand for food skyrocketed that most of the opposition ended.

In 1906 Thomas M. Campbell, a graduate of Tuskegee, became the first Negro extension agent. This was the first step in the plans of Booker T. Washington and Seaman Knapp to uplift the black farmers of the South by bringing a traveling school to them. The Movable School of Agriculture was equipped with seeds, fertilizers, dairy utensils, and machinery. It appeared before 2,000 people each month during the 1906 growing season. Some white landowners requested that the wagon visit their tenants. By December a second black demonstration agent was appointed. The USDA soon recognized the importance of the work of the first two agents and made them district agents to supervise others being employed to expand the service. In 1910 the first black women were named home demonstration agents in Virginia and in Oklahoma.

Farmers' Institutes

In 1839 members of the Massachusetts House of Representatives held weekly meetings, open to the public, for the purpose of discussing agriculture. Those meetings continued for several decades, and other states adopted the practice. Because farmers had contacted the land-grant colleges for information, in 1872 the Kansas Agricultural College inaugurated a program of bringing information to the farmers by traveling institutes. This was done in cooperation with the railroads, which saw the need and the potential of such a program. The institutes were a means to popularize agricultural education. As with previous movements, there was farmer opposition to off-campus educational courses, but the institutes grew in popularity. By the mid-1890s, efforts were made to create uniformity in the institutes and to achieve closer cooperation with the experiment stations. In 1901 the federal government became involved with the institutes via the experiment stations.

By 1902 institutes were held in all but three states. They lasted from one-half day to four days, depending on the purpose and the area. In 1902 the number of farmers enrolled in institutes was 20 times the number of students in all courses in all land-grant colleges. By 1911 a total of 71 trains were used in 28 states. These trains were visited by 939,120 persons out of the nearly 2.4 million who attended all institutes.

Special courses for women became some of the most popular that the institutes offered. Commercialized agriculture changed the position of women on farms, because more of their activities depended on the marketplace instead of the home. In addition, farm women became more interested in nutrition, sanitation, hygiene, and better living quarters. The USDA had started work in those areas in the 1890s.

The institutes flourished until extension programs became available during the second decade of the twentieth century. They had served their

Interior of sod house with magazine pages as wallpaper, flour container, coffee grinder, lamp, kerosene stove, and homemade chair, North Dakota, ca. 1880. (State Historical Society of North Dakota; hereafter SHSND)

purpose well in getting the message of better farming to the farmers, and they helped to break the barrier against "book farming."

The natural extension of the institute program was the development of short courses held on agricultural college campuses or at experiment stations. These short courses lasted from a few weeks to two years. In the early decades of the 1900s they filled a void in the land-grant program, because many people felt that the schools of agriculture did not always provide adequate practical agricultural education. Some universities, such as Wisconsin, had short courses before they had a college of agriculture.

Elementary, Secondary, and Post-Secondary Courses

To advance the teaching of agriculture and nature for boys and girls, high school agriculture courses were developed in 1896 by the University of Minnesota. In the 1899 *Yearbook*, Secretary Wilson expressed that there was a growing interest in the teaching of agriculture-related sciences in the elementary schools. Training in those courses was made available for the teachers. Before long some states decreed that agriculture and horticulture should be taught in all schools.

In 1897 not one high school taught agriculture. A big effort was made nationwide to upgrade the image of agriculture—to make agriculture more profitable and productive so people would remain in farming. By 1910, 875 institutions taught agriculture, mostly on the secondary level, while 214 post-secondary institutions provided teacher training for agriculture. By 1913 the number of institutions teaching agriculture climbed to 2,600, of which 600 were on the college level.

In addition to upgrading the image of agriculture, more professional training was taking place because national leaders felt that unless productivity increased, we would soon become reliant upon imports to maintain an adequate food supply. Had this happened, it would have slowed our overall growth. The full impact of the closing of the frontier was upon us, and leaders now recognized that increased production would have to come from improved farming methods. After a decade or so of lagging increase in productivity, farmers again proved that they could meet the challenge when the market sent the signal.

The Rise of the Industrial Society

In addition to the dramatic growth of agriculture, the nation experienced a virtual explosion in its industrial, commercial, and transportation sectors in the decades following the Civil War. Although all four sectors complement each other, the agrarians felt overawed by the fact that they no longer were the dominant single sector that they had been in earlier times. The following topics will relate both how agriculture helped the growth of the other sectors and how those other sectors benefited agriculture.

Exports

Prior to the Civil War, leaders from the southern states speculated that if a war were to take place between the states, they would have the advantage because the northern industrial states would not be able to supply the necessary food and still carry on a war. One of the big surprises of the war was the way northern agriculture performed—increasing production annually throughout the war years. During the war decade, agricultural exports increased nearly a half billion dollars over the previous decade, despite the great drain of farm workers to factories and the military.

Exports virtually exploded during the period of the first agricultural revolution. In 1851 exports totaled $147 million annually, and by 1901 they had grown to $952 million. From the end of the Civil War to 1900, the actual volume of international trade increased fourfold, while agricultural exports increased fivefold. The perceived threat to those in agriculture at that time was that manufacturing had grown so much more rapidly than agriculture.

In 1906 Secretary Wilson wrote, "Agriculture in this country, bad as it may be in its unscientific aspect, has had large economic justification." The undercapitalized farmers had built their operations at the expense of the soil. Fresh land created production faster than domestic consumption and the foreign markets could absorb. From after the Civil War until 1900, exports made up 16 to 20 percent of farm income. Then the percentage declined slightly until World War I. Wheat, corn, pork or pork products, cattle or beef products, cotton, tobacco, and fruits made up about 90 percent of the total agricultural exports during most of this period. This was at a time when farm labor rates were 50 to 500 percent more than farm labor rates in countries with which we were competing. However, American farmers had adopted labor-saving technology, which improved productivity and made the prices of our exports competitive.

Without the foreign markets to help meet the constantly increasing surplus production, the agricultural sector would have suffered more than it did in this period of prolonged price decline. From 1870 to the mid-1890s, the growth in inputs was over 2 percent per year, causing a growth in productivity and a major annual increase in total farm output. Without the growth in nonfarm population and exports, prices would have dropped more sharply.

Ironically, during those years, agricultural exports declined from 77 percent to 65 percent of all exports. The drop in productivity from 1900 to 1925 due to a reduced rate of inputs and no increase in productivity caused national leaders to become concerned about the ability of agriculture to meet future needs. During this period farmers learned to adjust land-saving technologies to the labor-saving technologies they had adopted in the previous half-century.

The slowdown in total productivity caused the prices of farm products to turn around in 1896, and they increased at a faster rate than the prices of all other commodities. By 1909 livestock prices were 193.1 percent of the average of 1896 to 1900. During the next two decades farmers enjoyed unprecedented prosperity.

From 1890 to 1903, agriculture produced a $4.806 billion trade balance. This was sufficient to offset the deficit trade of other sectors and still leave a nearly $4 billion favorable balance of trade for those 14 years. By 1907 the favorable balance had increased to $6.5 billion over the 18-year span. This was a critical turning point for a nation that historically had a deficit balance of trade, and the nonagricultural sectors still had a deficit balance. Without agriculture, our debtor nation status would have continued much longer. Gross farm income was the highest of any nation in history and has remained in that leadership position.

Industrial Markets

Farmers were disappointed by the results that industrialization failed to deliver to them. They had hoped that it would bring prosperity, but instead they found themselves more dependent than ever upon the rest of the economy. Farmers loudly proclaimed that their prosperity was fundamental to the nation. Other sectors in the economy appealed to farmers that building an industrial base would be good for agriculture. Manufacturers appealed on the basis that industry would create a big domestic market and then farmers would not have to be so concerned about exports.

Industrialization caused a rapid growth in commercial agriculture. Subsistence farming had to come to an end some day, but that fact was painful to many who did not want to face the new problems in commercial farming. The idea of an industrial society haunted them. Managing land, labor, and capital, as well as being exposed to worldwide forces, was not a pleasant thought, but each year that they held on to the old methods, the further behind they fell. The industrial sector won out because it was better disciplined, better organized, and better financed.

Business and industry had an advantage in that they could locate near the best markets and concentrate on production. Farmers, however, by the nature of their business, were tied to the land. Nearly all farmers blamed the railroads for their plight, for even though freight rates dropped, the farmers' profits did not increase. Farmers attributed this to the fact that the railroads widened the markets and brought unwanted competition.

Farming was critical to manufacturing. For example, in 1910 meat packing ranked first, with 6 percent of all manufacturing. It had a total business volume of $1.37 billion. Lumber products had a total volume of business of $1.156 billion. Flour milling ranked fifth, with 4 percent of all manufacturing and a total volume of $883 million. Cotton and worsted goods had a combined volume of over $1 billion. Tobacco, butter, cheese, condensed milk, sugar, cottonseed oil, cotton seed, cotton seed cake, and food preparations were all leaders in manufacturing.

Regulating the Marketplace

As cities grew and farmers became more commercialized, it was no longer feasible to sell directly to the consumer. A marketing system had to be developed. Milk was brought to larger centers by rail starting about 1840, and by 1872 New York City received some of its milk from as far away as 250 miles. Most of the major cities, even those in the Midwest, received milk from up to 100 miles away.

Adding middlemen between the producer and the consumer provided opportunity for manipulation of price or quality or both. For example, in 1871 farmers were paid $4.17 million for milk that sold in New York City for $15 million. This implied that adulteration by adding water was taking place. Boards of health took action. News reporters discovered that water, salt, saleratus, chalk, and other products were added to milk that came from the farms. Some milk was skimmed and then water was added. Some companies sold condensed milk that had been diluted with water. It was estimated that New York milk customers received only one quart of actual milk in every three quarts they purchased.

But not all the trouble was caused by middlemen, for many farmers produced milk under unsanitary conditions and did not cool it properly to prevent it from spoiling before it was sold. Some farmers were not willing to take the proper steps to correct their problems. The lack of an association of milk producers made it possible for questionable middlemen to enter the trade to the disadvantage of both the producer and the consumer. Further investigations indicated that price increases for milk were greater in small communities than in large ones. The conclusion was that the milk delivery system was inefficient because of lack of competition or overlapping routes.

On a broader scale than milk marketing, but not as directly consumer related, was the development of the futures trading industry. It is not known when futures trading activity commenced in the United States, but by the late 1840s the daily volume of corn or wheat sold on futures was greater than the amount sold on the spot market. Although speculators generally sustain the futures market, it was the interest of the produce merchants involved with commodities that encouraged its growth.

In 1865 the Chicago Board of Trade established rules regarding futures trading in grains. Some cotton futures trading had been done as early as 1857 in Liverpool, and after the transatlantic cable was laid, it became easier to conduct business with the European markets. In 1870 southern cotton traders joined the New York Cotton Exchange so they could trade in futures on that market.

With the adoption of the steamship in the 1840s and then the transatlantic cable, American farmers were suddenly in the world market. Now the world market and not the local market determined the farmer's destiny. Farmers were forced to become more specialized and more dependent upon the prices of their products sold and of the goods they needed. To cope with these changes they had to shift from extensive to intensive farming. Many were unable to face the challenge of higher production costs, rising land prices, higher wages, and other capital inputs. The industrial sector had learned to use economies of scale, but farmers resisted that move. Farmers did not progress like the rest of the economy, causing them to be discontented. But at the same time, food prices rose because productivity lagged behind demand. This increased the concern of our leaders over resource development in agriculture and conservation. They saw a need for allocating more funds to research and education to increase productivity.

Coping with Increased Domestic Consumption

One of the growing challenges of agriculture was to meet the demands of an ever-growing urban population that no longer was able to produce any of its food needs. Even in colonial days, not all individuals had permanent homes, and others who had to travel needed to be accommodated. To meet that demand, the first U.S. restaurant—a tavern—opened in Boston in 1633. Other commercial eating places were developed haphazardly along routes of travel as needed. Most of them were combination taverns, restaurants, and inns. One of the earliest innovative attempts came in 1876, when Fred Harvey established restaurants at 100-mile intervals to accommodate travelers along the Santa Fe Railroad. This proved to be a major change in food service for travelers.

It is estimated that in 1850 about 58 percent of disposable income was spent for food. As the food industry continued to innovate, this cost declined. At that time nearly all food moved through wholesalers, and their salespeople called on all retailers regardless of their sales volume. Each year the urban population grew, and each year it provided less of its own food, but it expected more of what it purchased to be processed. As the railroads expanded, it became possible to move more farm products to canners, to packers, and to millers. After processing, the commodities went to wholesalers or to chain warehouses, finally getting to the retailers.

Another method of coping with the shift away from urban home production of foodstuffs came through the development of greenhouses. It appears that the first greenhouse in the United States was built in 1764 in New York. Production in greenhouses was slow until 1825, when considerable progress was made in growing plants under glass. Philadelphia was the leader in growing flowers and vegetables in greenhouses, followed by New York and Boston. The demand for those products grew rapidly, and by 1830 heat was used to improve production. About 1835 hot water was used for heating greenhouses, and the industry made rapid progress during the next decade. Lettuce, radishes, cucumbers, and flowers were the chief crops grown for the big city off-season market.

By 1860 commercial floriculture and the forcing of vegetables in hotbeds had become important business, only to be slowed by the Civil War. After 1870 the business revived sharply with a demand for cut flowers and winter vegetables. By 1875 southern vegetables came to northern markets via the railroads. This forced the greenhouse industry to become more efficient. In the late 1880s, iron framework was adopted for greenhouses, which made them more economical than previous structures. The business continued to expand at a steady rate.

After the turn of the twentieth century there was concern that the meat supply might not meet future needs. The ratio of livestock to population declined. Meat exports increased. Cold storage and refrigerator cars made fruits and vegetables available from as far away as 3,000 miles. The consumption of canned goods, wheat, and sugar increased. The development of dry farming techniques caused grassland to be converted to farm land. High corn prices made it uneconomical to fatten beef. Fish was considered an alternative meat, but there was little knowledge about commercial fishing. Fish was not popular among the foreign born who came from fish-eating homelands. They now preferred red meat.

The first patent on freezing fish was issued in 1861. By the end of the nineteenth century, ice-and-salt freezers were used extensively at fishing ports on the Great Lakes, in New England, and in New York. The same technique later was used to freeze fruit that was processed into pies, ice cream, or jellies. By 1908, when the USDA made its first thorough study of fish and fish products, the industry caught 2.1 billion pounds of fish, with a value of $66 million. Further progress was delayed until means of preserving the fish were improved and uncertainties involving the fish harvest were reduced.

Growth of the Food Chain

The opening of the Chicago Union Stock Yards in 1865 marked the beginning of meat inspection by local officials. This was several years before the first federal meat inspection law was passed in 1890. The Chicago packers wanted to have meat inspection to maintain the dressed pork trade they had developed with England. Pork packed in salt and shipped in bulk arrived in England in good condition and at a low cost. In 1873 the first shipments of beef for slaughter arrived in England in fair condition, but storms caused heavy losses, and most export shipments stopped by 1885. It was not until the 1890s, when refrigerator ships were introduced, that any great volume of beef products was exported.

In some respects what happened at the Chicago yards was a key to the development of the food chain because meat was a major food product. In the 1861–62 season Chicago surpassed Cincinnati as a packing center and retained that leadership until the 1930s. By 1890 the Chicago Union Stock Yards was the world's largest stockyard, receiving 311,557 carloads of livestock that year.

Refrigeration spurted growth of the packing industry and enabled highly centralized large-scale plants to take over. In 1860, 992,310 hogs were slaughtered at central markets. By 1880 the figure was 11,001,689, and by 1900 it was 28,742,551. The volume of the slaughtering and meat packing industry grew from $29 million in 1860 to $790 million in 1900. It was not an easy task, because the buying public initially opposed refrigerated meat, as did the railroads, local butchers, and cattle shippers. In 1877, when G. F. Swift shipped his first dressed beef east, he had to build his own refrigerator cars. He also had to get eastern slaughterers to cooperate with him so he could enter that market.

During the 1880s and 1890s, Swift and Armour established their own branch houses throughout the country to control marketing directly to the retailer. This eliminated the wholesaler. The large-scale packers had such an advantage with economy of scale that they could sell meat at lower prices and still make a profit. When consumers experienced the lower prices, their resistance to refrigerated meats declined, and local butchers gave way to change.

At the same time, the large packers created centralizers to buy cream from dairy farmers in areas of sparse milk cow concentration. Such areas had a difficult time justifying building creameries, and the centralizers captured most of the cream output. By the early 1900s, the packers entered areas where creameries existed. They were able to pay higher prices for cream than the local creameries because they shipped to plants that operated on a larger scale and produced butter more economically.

South St. Paul, Minnesota, Union Stockyards covering nearly one square mile, including massive Armour (right) and Swift (background) packing plants. Built in the 1880s. (Minnesota Historical Society; hereafter MHS)

To combat the centralizers, local creameries developed their own pickup routes, but that was not popular with the farmers.

On the retail end of the food chain the A & P, founded in 1859, started a revolution by creating chain stores. By 1910 chain stores had captured 20 percent of the retail grocery market. In 1912 the A & P created another change in the industry by eliminating credit and delivery. The stores appeared in most neighborhoods. They remained one-person operations with a minimum of fixtures. The stock was selected for maximum consumer appeal and was sold at a narrow margin. The average daily volume was $175. Later, the stores became combination stores, which sold meats, groceries, bakery products, fruits, and vegeta-

The Rise of the Industrial Society / 159

Food vendors in New York, 1906. (NAL)

bles and were much larger than the traditional stores. Their numbers grew rapidly.

Consumers complained about the rising cost of food during the years from about 1895 to 1915 that resulted from a reduced rate of productivity in agriculture. Secretary Wilson blamed part of the increased retail cost on consumers who wanted someone to pick up the orders and then deliver the groceries. He also pointed out that there were far too many small retail establishments that were of little value to the consumer and the proprietor "when one store could do the business of 20." He understood the problem, and soon the market forces made the needed corrections.

Progress in Farming

Muscle to Mechanical Power

In 1860 the agricultural area of the country consisted of 163 million improved acres, 244 million unimproved acres, and 1.45 billion acres in undeveloped territory. Land was still abundant and cheap, while labor was scarce and dear. But some individuals expressed the idea that the United States might soon be a grain-importing nation because we were not taking care to renew soil fertility. Others pointed out that we were no different from any other country and that farmers would farm better as soon as they determined that it was necessary and profitable. The term "exhausted soil" was the most commonly used phrase in agricultural literature, and yet no phrase was more vague and indefinite.

By 1860 agriculture was in transition in some of the older sections where the soil had lost its original fertility. But many farmers in the older areas refused to change and continued the old ways, while farmers on the fertile western land adopted horse-powered machinery as rapidly as they could. The scarcity and cost of labor was a major reason they mechanized so rapidly. Most of them also realized that the more land they worked, the greater their profit potential. Those who did not change were beaten in the marketplace.

One agricultural magazine editorial writer suggested that as long as land was still available for homesteading, farmers tended to be complacent and self-sufficient and had a feeling of moral superiority—always reluctant to adopt new ideas and technology. Those attitudes changed under the influence of the land-grant colleges, extension service, tractors, and all the other innovations that impacted agriculture after 1900.

Compared to European farmers, Americans were very wasteful, because land was available for the asking. Agriculture far outstripped the growth of the rest of society but produced only a living wage for those involved. The rapid expansion in all sectors of the nation forced everyone to economize on labor. The strong demand for workers in industry caused a relative shortage on the farms, which hastened mechanization. From 1870 to 1900, over 3.7 million workers left agriculture for employment elsewhere, but in no way was the efficiency of the farm worker impaired. During those years the per capita production of livestock and crops increased and exports expanded. As one authority from the USDA wrote in 1908, "Instead of being a great drawback to industrial expansion, the scarcity of labor has been its greatest stimulus and a blessing not only to America but to the whole world, because it has been the incentive for the development of labor saving tools." From 1900 to 1910, both manufactured products and agricultural products doubled in dollar value. By 1910 manufactured products totaled about $20.7 billion in contrast to $8.5 billion for agricultural products.

Wind Power

Today, except for remote places too far from electric lines to be economically served, the windmill is nowhere to be found. In its day, however, it performed an invaluable service to its many users. In 1854 Daniel Halladay invented a windmill with paddle-shaped blades that turned out of the wind when the velocity was too great. Originally he built a manufacturing plant in the East, but because of the local lack of demand, he moved it to Illinois, closer to the major market.

By the 1870s, windmills became very popular in the Midwest, and within the decade on the southern plains. Large-scale production commenced in 1873. Over 200,000 were in use by 1880, 400,000 by 1890, and over 1 million by the 1920s. Innovations were added regularly until the small one-cylinder gasoline engine with a pump jack or the electric motor proved much more reliable, especially on hot, calm days. But to those farmers with large herds of cattle or hogs, to those ranchers without watering holes, and to the railroads, which needed to refill steam engines at intervals along the tracks and to supply livestock at stockyards, the windmill was a boon.

Windmill pumping water for a stock tank, Colorado. (NAL)

Steam Power

Individuals dreamed about the use of steam power for farming in the days before horses replaced oxen. The horse, which was more costly than the ox, worked much faster and was able to outproduce the ox. It is generally recognized that horses operated at about 20 percent efficiency, but because they were needed only about one-fifth of a year, they were expensive. The steam engine, however, was only 2 to 3 percent efficient, but there was no maintenance cost when it was not in use.

Between 1807 and 1849, skid-mounted stationary steam engines were used to thresh grain, gin cotton, and saw wood. During those years these engines replaced treadmills and rotary sweeps. In 1849 the first portable steam engines were produced in the United States. These "Forty-Niners" came in 4-, 10-, and 30-horsepower sizes and were moved around by horses. During the 1850s, several attempts were made to plow by pulling the plow across the field with cables mounted to the stationary engines. The results were not practical.

Throughout the 1840s and 1850s, substantial prizes were offered for individuals who could develop successful steam engines. At least 90 separate steam plow inventions were made up to 1877. Joseph Fawkes designed a steam plow that was accepted, and in 1859 he won $3,000 and a gold medal at the United States Agricultural Society fair. The Civil War stopped his efforts to manufacture it. In 1859 Abraham Lincoln noted that farmers were waiting for the day that the steam engine would pull a plow, doing the work better, cheaper, and more rapidly than horses. Some even forecasted that plowing, hauling to market, and ditching with steam power would be done by individuals who would do only that for their living.

By 1875 the steam engine became self-propelled. Steel gears, better-designed driving wheels, and larger water and fuel capacities had to be developed before the steam engine was successful. It was a day farmers had been waiting for. The new engines were heavy, required two and preferably three people to keep a plow rig in operation, and moved about one mile an hour, but they could work around the clock. They were used most extensively for breaking sod on the open prairies and for threshing. Prairie and straw pile fires were one of the great dangers involved. Occasionally explosions occurred caused by dynamite left in the coal by the miners. By 1880 about 24,000 steam engines were produced. Their numbers peaked at about 72,000 around 1910. When the internal combustion engine made the tractor possible, the number of steam engines declined rapidly.

Tractor Power and Trucks

Europeans had developed gasoline engines as early as 1860, so it was inevitable that the day would come when such engines would be applied to a running gear for whatever purpose the inventor had in mind. In 1889 a U.S. firm built six gasoline tractors by mounting engines on Rumely steam tractor chassis. In 1891 William Deering and Company mounted an engine on a New Ideal mower to make it self-propelled. In 1892 John Froelich mounted a gasoline engine on a Robinson running gear, which received recognition as the first successfully operating gasoline tractor. Froelich's tractor completed a 50-day threshing run powering a Case thresher in temperatures of –3° to 100°F. The following year Froelich incorporated the Waterloo Gasoline Traction Engine Company, which eventually became John Deere.

In 1899 S. S. Morton placed an engine on a truck, and the International Harvester Company became involved. This was the beginning of the International tractor line. In 1902 C. W. Hart and C. H. Parr built a tractor for drawbar work. In 1905 they established the first business in the United States exclusively for the manufacture of tractors. In 1906 the Hart-Parr Company gave popular use to the word "tractor" in its advertisement for the gasoline traction engine. The Hart-Parr eventually became the Oliver.

The first tractor school was held in 1906 in St. Paul, Minnesota, and others soon followed. In contrast to the steam engine, the gasoline traction engine was popular with farmers. The first tractor demonstration in the United States was held in 1911 in Omaha, Nebraska. That event led to the formation of the Nebraska Tractor Test, which was established to test tractors to determine whether they were capable of what their companies promoted.

Tractors were adapted so that they could be used for road work as well as in the field. By 1912 four-wheel drive and four-wheel steering were produced. Other attempts were made at four-wheel power in the 1920s, but it was not until the 1950s that four-wheel drive became accepted as a popular concept.

In most cases, the early tractors were too large for the average farmer. However, the Wallis Cub, produced in 1913, caught the attention of smaller farmers. It was the first frameless-type one-piece tractor. All the moving parts, except the final drive gears, were enclosed. The newer, smaller models became more popular as the price of horses continued to rise.

The farmers' ideal was a tractor that could replace four horses, even though some people in the industry advocated as early as 1913 that the small tractor was a passing fad because the big tractor would eventually be what farmers needed. This proved to be correct, but the promoters were 40 years ahead of the farmers in their thinking.

To answer the desire for a small tractor, Henry Ford produced the Fordson in 1917. This tractor dominated the tractor market for most of the next decade, losing its lead when other tractor makers came out with innovations that met the needs of farmers while Ford was not willing to adapt. In 1915 about 25,000 tractors were on American farms, and in 1925 over 500,000. That marked the end of the era of the steam engine, but it was not until 1954 that tractors outnumbered horses on farms, even though tractor horsepower far exceeded the horsepower of horses by World War II.

By 1910 about 25 percent of the horses in our 30 largest cities had been replaced by trucks, but horse numbers continued to grow. By 1912 about 200,000 horses had been replaced by 50,000 trucks on the streets of our cities, marking the end of an era. However, because of the prosperity in agriculture during the World War I era, horse and mule numbers grew to over 25 million in 1920. A decline set in, dropping numbers to 12 million by 1930. In 1962 horses no longer were counted in agricultural surveys.

Mechanized Farming

The development of steel production made a material available that could withstand the stress required of farm machines and yet was inexpensive enough to be affordable. Steel was quickly adopted by the U.S. market after 1870, and particularly after the Centennial Exposition of 1876 at Philadelphia, which made many people aware of what was available and also what was needed. Ohio led the North in the production of agricultural implements during the 1860s and early 1870s, but that does not imply that Ohio farmers were the most mechanized. States farther west were much quicker to adopt labor-saving machinery, because of their acute labor shortage.

The value of agricultural machinery rose from $6.8 million in 1850 to $43 million in 1870 to $101 million in 1900. This was all part of the process of division of labor and the transferring of jobs to those who could do them more economically. By 1890, 42,544 people were employed in 910 manufacturing plants producing implements and machines for agriculture. Their average wage was above the average net income of farm families for that day. G. F. Warren, a pioneer farm economist, wrote in 1910 that the age of machinery impacted the farm as well as it did the city, and the city machinery makers made it possible for a small number of farmers to supply food for a large urban population.

Warren pointed out a truism that has remained to this day when he noted that machinery investment on 51- to 100-acre farms was $6.37 per acre, while on farms of 200 acres and over it was only $3.13 per acre. A 1912 study by the USDA also pointed out how much labor time was reduced by the use of machinery. A bushel of barley required 127.2 minutes of labor in 1830 and 5.4 in 1895. In 1855 it took 274 minutes to produce a bushel of corn, and in 1894 only 41 minutes. A bushel of wheat required 192.8 minutes in 1830 and only 8.9 minutes in 1896. The reduction in time to produce a ton of hay was from 35.5 hours in 1860 to 11.34 hours in 1894. In every case the farmer's net margin per unit of production increased.

An interesting sidelight to the time-and-cost studies came from a six-year trial in Minnesota. The cost per hour of horse labor ranged from 7.32¢ to 9.25¢, while that of hired labor on the same farms ranged from 7.44¢ to 9.96¢. The cost of keeping horses was higher because horses had to be cared for year round, while labor was generally employed seasonally. Interest and depreciation were the next highest costs.

The Binder

The Civil War skyrocketed the demand for reapers to about 90,000 a year by 1864. The peak production of 120,000 was reached in 1869 and continued at that level to take care of the new acres opened every year. But the reaper was too labor intensive, and something had to be done about binding the grain.

In 1872 a successful wire-tie binder was marketed. Now the potential for harvesting was limited only by the endurance of the horses and the mechanical stability of the machine. Horses were exchanged at noon to maintain a steady speed for a full day. With a 4-horse, 8-foot binder, 16 acres was normal production.

But the wire-tie binder presented a new problem. Farmers complained because they lost livestock that ate straw containing wire.

Millers complained because wire in the wheat was harmful to the machinery. Consumers complained when they ruined a tooth from wire found in the bread or other flour products they used. In 1878, when the Illinois Millers Association formally objected to wire-tie binders, the price of wheat dropped sharply for those farmers who used them. Other millers took the same stand, but they were reluctant to make a public announcement for fear that the market for northwestern flour would be hurt.

Binder manufacturers worked to solve the problem and in 1875 produced the first successful twine-tie binder. One person operating a twine-tie binder could do the work that once required eight people. In 1879 twine-tie binders proved their worth when 155 of them were used on the Dalrymple bonanza in the Red River Valley of the North. In the meantime the price of twine fell below that of wire, and the twine-tie binder took the country by storm. Farmers junked their previous equipment because it had no resale value.

When manufacturers saw how eager farmers were for the new twine-tie machines they overproduced. From 1880 to 1885, more than 250,000 were sold, and the price had fallen to $100. The new twine-tie binders were so much better and more labor efficient that western farmers were encouraged to plant more acres. This made it virtually impossible for small-scale eastern farmers using cradles to compete. Those who did not discontinue growing wheat hired neighbors with the new binders to harvest for them. The only changes from this date on were upgrading the mechanisms and enlarging the platform until 1919, when the binder became the first machine to utilize a power takeoff.

The Combine

The first combined harvester-thresher was patented in 1828, but there are no records to indicate that the machine was ever built. In 1835 Hiram Moore, of Michigan, developed a harvester-thresher that harvested grain at a cost of $0.82 an acre versus $3.12 for cradling, shocking, and threshing. Michigan's moist climate was not suitable for testing such a machine, nor were adequate power units available there. The machine was further tested in California, and in 1854 it harvested 600 acres.

That machine, purchased by John Horner, served as a model until 1867, when the first California combine was developed. The combine, as Horner rebuilt it between 1867 and 1869, took three operators—a driver to guide it, another person to raise and lower the platform on uneven ground, and a third to sew the sacks as they filled and to drop them onto the ground. It harvested 1,400 acres in its first year. The combine caught on quickly because of the dry climate, which prolonged the effective harvest season, and the psychology of the area, where new ideas were easy to sell.

By the 1880s, about a dozen combine manufacturers existed on the Pacific Coast. A major breakthrough occurred in 1885 when the tight-gear system of transmitting power was replaced with link chains and V-belts. In 1905 a small 7-foot horse-drawn combine was developed for use in the West, and in 1918 an engine-mounted, horse- or tractor-drawn combine was introduced to the prairie states.

Again, it was the tight labor supply that forced farmers to accept the combine. The sharp drop in wheat prices in the mid-1880s caused farmers to turn to the combine to cut costs. By 1886 combines did most of the harvesting in California. A combine with an 18-foot cutter bar required 20 horses or mules and 4 workers—one to drive the horses, one to tend the header, one to watch the threshing machine, and one to sew the sacks. This machine could harvest 36 acres a day. Later, cutting bars were increased to over 20 feet, requiring 40 horses and a crew of 8. These combines were capable of up to 45 acres a day.

In the late 1880s, steam engines were adapted to operate the combine. Auxiliary engines, powered by steam from the pulling engine, were used to run the sickle and the thresher independent from the ground speed. With this advance, sickle lengths were increased to 42 feet, requiring a crew of 8, and could cut up to 125 acres a day. Steam-powered rigs were no apparent labor savers compared to the horse-drawn machines. The added danger of fire with the steam rigs kept them from being as popular as the horse-drawn combines.

In the 1890s, after the bonanza farms of California were broken up, combines remained in service on the smaller farms because using them was so much more efficient than any other method of harvesting. In 1912 gasoline engines powered the combines, and later tractors were used to pull them. The next major advancement in combines came in the 1930s, when power takeoff models with 4-foot sickles operated by one person were produced. The use of 4- and 6-foot models became very popular because it was so much more economical than threshing.

The Drill   In the earlier discussion of seeders and drills, reference was made to two-wheeled box-broadcast seeders that dropped the grain onto the ground. In the 1860s, hoe attachments were added to the broadcast seeder. To avoid the investment in a seeder, many smaller farmers continued for several decades to use the endgate seeder attached to the end of a wagon box.

The first drills were developed in the late 1850s, but it was not until the 1880s, when the shoe drill was introduced, that they became popular. Shoe drills worked much better than hoe drills where there was considerable trash on the land. The old-style hoe drill plugged too frequently. The single-disk drill was popular where grain was seeded in stubble that had not been worked from the previous year.

In the late 1850s, a seeder that required less power than the force-feed drill was marketed. Instead of disks, shovel openers were used to open the soil for dropping the seed, which was then covered by chains that dragged over the dirt. A 2-horse, 8-foot seeder could plant 12 to 16 acres a day. Seeders as large as 14 feet were built and remained in use into the 1930s.

A report of the Commissioner of Labor noted that in 1893 on the Red River Valley bonanza farms using the latest equipment, 3 hours were required to raise a 15-bushel wheat crop and 5 hours and 46 minutes were required to secure it. This compared to totals of 10 to 12 hours, and even as high as 25 hours on small farms, and up to 42 hours per acre in New England, where primitive equipment was still in use.

Corn Equipment

The two-row, horse-drawn, checkrow planter, introduced in 1853, was standard equipment in the Corn Belt. In the 1920s it was adapted to the tractor. In 1877 the knotted wire was perfected to enable checking corn for cross-cultivation, which was necessary until chemicals were introduced.

In 1882, the lister was introduced in Kansas for use in drier areas. It consisted of a small double moldboard for opening the furrow, followed by a subsoiler, a seed dropper, and a covering wheel. By 1902 about 60 percent of Kansas corn was listed. Elsewhere, by 1900 the steel two-wheel, two-row, two-horse machine, with seed boxes, checkrowers, drill, and force-feed attachments capable of row adjustments, was standard equipment. Other than structural improvements and perfecting, little was done to alter the corn planter until 1915, when fertilizer boxes were attached.

In 1856 a patent was issued for the first practical straddle-row horse-drawn cultivator. This was 123 years after such a machine had been demonstrated in England. In 1893 the rotary hoe was introduced, which greatly reduced the amount of time needed for cultivation, but it was not widely used until several decades later.

As in the case of small grain, harvesting was the big bottleneck in producing a corn crop. If the farmer wanted to harvest both the grain and the stalk, a corn knife was used. This was extremely hard work, and the most that could be cut in a day was one to two acres. In 1886 sled cutters were patented, but three to four hours were still required to harvest an acre. Harvest time was cut to about 1.5 hours per acre in 1892, when the self-binding corn binder was introduced. Then it became possible to use the silo filler or shredder for larger-scale silage or fodder production. With each invention the cost of producing a unit of crop was reduced.

Working the Land

The tremendous advances in mechanization, transportation, education, and communication brought about major changes in the life style of those living on farms. The transition was ongoing, but not all farmers accepted change rapidly or willingly. Isolation of farm families was reduced for those in the more settled areas, but farmers in general still lived independently and separated from the mainstream of society.

Female Farm Work

In 1871 the Commissioner of Agriculture conducted a study on women in agriculture. This proved to be a challenge for those conducting the survey because there was little data on what women did on the farm. A general conclusion was that the "outdoor work of white women on farms of medium or better sort has greatly declined from early days, and the decline was especially marked after the Civil War."

Farm wives and daughters generally no longer milked cows, worked in the fields, or cared for livestock. They spent less time gardening, working in the orchards, and making butter. Cheese making had become a lost art. They still cared for poultry and bees and gathered vegetables for the table. Primarily they did domestic work.

Man and woman using oxen to plow. Note zigzag fence. Ca. 1870. (NAL)

In New England, women helped to spread and to rake hay because labor was scarce and that work was still performed without machinery. Girls sometimes aided in dropping seeds during planting. In general, women's work outside the house had declined greatly from previous generations, and female outdoor labor was no longer considered compatible with New England institutions.

Girls were hired almost exclusively in hop picking because they were considered superior to boys and men, but they were paid only about one-fourth of what men earned in the hop yard. Girls were also the chief workers in cranberry planting, weeding, and picking. Here they were paid rates equal to men and were considered more efficient. In New Jersey, girls and women picked berries by the quart and often made more than the men.

As late as 1871, women in Pennsylvania still did field and livestock work, especially in the areas of large recent immigrant population. In Delaware, women worked extensively in harvesting corn fodder, which substituted for hay. They produced as much as the men but generally were paid only three-fourths the wages.

In Maryland, among the poorer classes, women planted, hoed corn, weeded tobacco, and raked grain. Sometimes they were paid as much as men, but most of the time they received about three-fourths as much. In Virginia, women on the small farms aided in worming, suckering, and stripping tobacco. Even though they were often more efficient than men, they received only half to two-thirds the pay.

Black women declined most requests to work on farms because they regarded such work as a relic of slavery and not "suited to ladies." Less than one-fourth of former slave women were still doing farm work in 1871. In North Carolina, women helped out only on small farms. However, in South Carolina as much as 20 percent of the farm work was done by women—both black and white—in some districts. They did the same work as men for half the pay.

In Georgia, about the only farm work done by women was picking cotton. The big exception there and in the Gulf States was in the case of widows, who managed and did all the farm work without any help from males. Most of the gardening was done by women. Missouri was a contrast, because more white women and fewer black women worked on farms than prior to the Civil War. A favorite saying there was, "One woman in a garden or at the sorghum kettle is considered equal to two men."

In the states of the Ohio Valley and the Great Lakes, most American-born women did farm work only in an emergency. They engaged in gardening or fruit and hop picking for their own convenience or because of necessity. A higher portion of immigrant women worked on the farm for the first few years after arriving. As they became more Americanized they reduced their work on the farm. In Minnesota, female immigrants worked extensively in all branches of farming. They were equal to males in binding and shocking grain but received only $2.50 to $3.00 daily, while men were paid $3.00 to $3.50.

A 1912 USDA study pointed out that in Kansas, the farm wife did only gardening, while a Nebraska correspondent wrote, "The day is passed in progressive Nebraska for the `weaker vessel' to get less pay than men for the same work." Women did less work on farms in the Pacific States than in other areas "on account of their comparative paucity of numbers."

The above-mentioned study revealed that the national monthly wage rates for domestic labor of women on farms hired by the year were $8.69 in 1902, $10.80 in 1906, and $10.39 in 1909. For the same work, if the women were hired by the season, the rates were $9.71 in 1902, $11.95 in 1906, and $12.02 in 1909. Room and board were included as part of the pay. Day labor rates for those years were 62¢, 76¢, and 77¢, respectively. In 1909 the monthly pay rates varied by states from $6.15 in Virginia and $6.40 in South Carolina to $23.33 in Montana and $25.00 in Nevada.

In the first decades of the twentieth century, there was great resistance among both rural and urban girls to do domestic hired labor. They considered it unrespectable. Farm women also were more reluctant to do tasks performed by earlier generations. Making soap, candles, and lye and canning were not done on farms to the degree that they had been earlier.

In July 1881, the *Farm Journal*, a leading farm magazine, took a stand on behalf of women's rights and the inequality of the laws. By 1910 the *Journal* became quite emphatic that women ought to have suffrage. That position did not harm its circulation, for in 1882 it had 100,000 subscribers and in 1910 about 500,000.

One of the noticeable changes in the early part of the twentieth century was that each year more women were enrolled in agriculture courses at the land-grant schools. In many cases they were the leaders in adopting new farming methods. The census figures verified that the number of women farmers was increasing steadily.

Isolated cases relate accounts of exceptional input by females. For example, a seven-year-old 4-foot 2-inch girl in Kansas bridled and saddled an Indian pony and went to locate a lost calf. Then she marked off a field and hand planted corn, becoming so tired by the end of the day that she could not drop the seeds steadily into the openings. The men were all gone to the Civil War, and her mother and two siblings had died, so she had to take their place. When she was nine, she was in charge of herding 25 cattle and of tending a flock of 135 chickens from hatching to maturity. She had to harness a team of horses and hitch them to a wagon. She harnessed them by standing on the pole between them in the stall.

An Ohio woman of the 1880s and 1890s noted that her butter and egg sales made an important contribution to household expenses. She milked three cows and took care of the chickens twice daily. Purchases at the store were determined by the value of the eggs and butter she had to trade in. Total family expenses in 1888 for six people were $615.51, of which $130.41 were paid by butter and egg sales. That year a local rural school teacher earned $182.

The eggs and butter sold were what was left after the family had satisfied its needs, entertained company, and donated to the church bazaar. The woman concentrated on dairy and poultry for 30 years and hired a neighbor woman to do the family laundry for $1 a week, had

Typical farm flock found on most farms in the East, South, and Midwest from the 1700s until the 1940s. (June Tweten)

another do sewing, and had still another do spring house cleaning. She was fortunate that she lived in a settled area of Ohio where neighbor women were available to help.

Women improved their position considerably from 1860 to 1914, but they still had much to gain. Historian Robert Dunbar wrote:

> No one has mourned the passing of farm self-sufficiency less than the farm woman. The new division of labor has relieved her of such tedious tasks as spinning, weaving, and sewing. The churns and the candle molds have been relegated to the attic or antique shop. There is still plenty of work on the farm, but the endless drudgery of earlier generations has passed into discard.

The Labor Shortage

Farm labor emerged early as a social and an economic problem. The good laborers used the agricultural ladder to get ahead, while the poor ones wanted to get by with as little effort as possible. The farmer had to find ways to reduce labor cost.

Farm labor was in short supply since the days of original settlement—a pattern common to the industrial nations of the world. Originally, it was a problem chiefly of quantity, but once industrialization took hold, it also became a problem of quality. A USDA official wrote in 1912 that labor quality failed to keep up with the need for more skill and intelligence. Farmers complained that workers always migrated to the city, but they refused to recognize that rural wages and living conditions did not match those of the city.

The family farm with a live-in hired hand was the pattern throughout much of our history. It was thought to be an ideal arrangement. This changed by the 1870s, when the number of larger-scale farms increased. Progressive farmers had learned that by applying mass-production techniques to farming they could reduce their costs. The farmer with the largest machinery could pay better wages than his small farmer neighbor, and he could get help more easily.

From the late 1860s on, prices of farm products were in a steady decline. Many farmers knew that the best way to defend themselves was through mechanization, which lowered their production cost. Between 1880 and 1900, the average value of machinery on farms nearly doubled. Farmers often did not know what they could afford to pay for labor. That may not have made much difference, for the drift of young people to the city was as much a problem of psychology as of economics. The glamour of city life was too attractive.

Historically, local neighbor boys who wanted to farm some day made up the bulk of the agricultural labor force. The next most readily available source was recent immigrants. In 1890 foreign-born workers still made up a large percentage of white male agricultural laborers. In many states they accounted for over 35 percent of the hired farm workers. After the turn of the twentieth century, immigration was greatly reduced and most newcomers went to the city. States tried to induce immigration to agricultural land but with little success. Many immigrants did not have good memories of farm life from their homeland and did not want to repeat the unpleasant experience.

G. F. Warren, pioneer agricultural economist, determined that one of the greatest drawbacks of farmers in determining what they could afford to pay labor was their poor management of labor. Most farmers did not have good labor plans for either the short term or the entire year. Warren argued that in the early 1900s, the wage rate was not the real problem. The problem was that most farmers had not yet adjusted to modern machinery and size of business. Warren emphatically stated that the best way to make more profit from labor was to have each worker drive more horses on larger machines.

Labor in a Changing Agricultural Structure

The emancipation of slaves freed millions of people who had to be absorbed into a bankrupt plantation economy. Most of them became tenants and sharecroppers and had a standard of living far below that of the traditional hired laborer and industrial worker. This problem continued for seven or eight decades. In 1900 the 11 states of the Old South had over half of all the farm laborers in the nation. The South continued its practice of labor-intense farming.

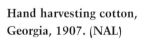
Hand harvesting cotton, Georgia, 1907. (NAL)

Other areas of the nation reacted to the new industrialized society in different ways. The older eastern states shifted to specialized production for local urban markets. Highly mechanized and capitalized farms were created in California and in the Northwest. The factory system of operation was adopted by the bonanza farms of the Red River Valley of the North and by eastern Washington and Oregon farmers. Cattle ranching operations in the Southwest became highly centralized large-scale businesses.

Cowboys eating at a chuck wagon, Kansas, 1890s. (NAL)

Large-scale agricultural ventures faced labor problems far different from those encountered by traditional family farms. The close relationship between employer and employee disappeared. Farmers did not work with their laborers on the repetitive jobs, and the workers no longer lived with the farm families. Seasonal employment became the standard for most workers on large farms. This led to an unrest not previously experienced among agricultural workers.

One of the first conflicts caused by labor discontent occurred in the early 1880s when several hundred cowboys near Lascosa, Texas, struck seven cattle-ranching firms. Cattle rustling, gunfighting, blacklisting, and high labor turnover caused the problem. In the 1890s, the Sheep Shearers Union of North America formed the first stable union among agricultural workers. Because sheep shearing was a very specialized labor and because clipped wool, a perishable commodity, was involved, it was less difficult to form such a union than where diverse groups of workers were involved.

In 1865 the Central Pacific Railroad started employing Chinese immigrants, and within a short time 11,000 were at work. Most of the Chinese had no intentions of becoming farmers, for their aim was to become miners or to have businesses of their own. However, they soon realized that they could make a steady living by farming. By the 1850s some Chinese had gone into fruit and vegetable farming with rice as a grain crop in the rotation. Next, they were involved in hog and poultry production.

Because they were hard workers and good managers, the Chinese paid higher rent than other farmers on truck gardens in the suburban areas. They also profited by direct selling of their produce. Most of their original farms were between 5 and 60 acres, but by 1870 they averaged over 100 acres, with some as large as 550. The Chinese included more field crops to balance labor demands and to improve crop rotation.

Many Chinese worked for white farmers, who liked them because they were industrious and not militant. By the 1880s, however, many Chinese had entered other industries. This depleted the supply, which enabled those remaining to unite for higher wages. They formed brotherhoods or protective associations called "tongs." The tongs were an effective part of the labor-contractor system, unless the labor contractors took advantage of the workers they controlled.

By the 1890s, the Chinese shifted to other work and were replaced by Japanese. By 1898 the Japanese were the most numerous workers in new crops, such as sugar beets, rice, strawberries, and other labor-intensive crops. By 1909 they were the most numerous labor group in California and had organized into work gangs under one spokesperson.

In 1903 the Fruit Workers Union local was organized in California within the American Federation of Labor (AFL). It was not successful but deserves recognition as the first such attempt to form a union. The AFL was reluctant to get involved with farm workers because it was not comfortable with non-whites.

By 1906 the Industrial Workers of the World (IWW) became active among California farm workers and within a few years reputedly had 10,000 to 12,000 members. In 1913 a hop ranch near Wheatland, California, had twice as many workers as it needed, and conditions became deplorable. Sanitation, food, and pay were bad, which was compounded by the fact that 27 different nationalities were involved. The IWW members, plus two organizers, united the hop pickers and protested the conditions. This became known as the Wheatland Riot, which was a highly publicized and significant event in California agricultural labor history.

The Agricultural Ladder

Tenantry was not a major concern as long as public domain was available. Free land tended to act as a lever causing landowners to give rather generous rental contracts. Figures on farm rentals were not kept until 1880, when 25.6 percent of the farms were tenant operated. By 1910 that figure had increased to 37 percent. The other noticeable change was that the length of time for being a tenant increased from 4.9 years in the 1880s to 11.1 years by 1910.

These figures were disturbing to some who thought that in a few decades there might be more tenants than landowning farmers. Others saw tenantry as a process of climbing the agricultural ladder toward eventual farm ownership. In the late 1800s and early 1900s, it was popular to think that the agricultural ladder led from apprentice, to hired laborer, to share renter, to cash renter, and finally to farm owner.

Some individuals argued that tenantry affected the stability of rural communities and that a system of tenantry would have serious social and economic consequences. They based their arguments on the fact that they believed that tenants would not have as much interest in the community as landowners. Others were afraid that the shift of background of immigrants after 1890 to southern and eastern Europe might lower the status of tenantry to that of peasantry, because they believed farmers from there to be less skilled.

Some suggested that having more tenants would probably lower the intellectual and moral standards because tenants were not as interested in education and were indifferent to religion. From an economic point of view, many people felt that tenants would not be as good farmers as landowners because their chief interest was to make a rapid return for their efforts.

Others argued that the agricultural ladder was never really effective. Many tenants were people who once had owned farms and then failed. Others maintained that the agricultural ladder was never accepted in the Old South, where the labor system and opportunities for homesteading prevented the ladder from working. But after the Civil War, tenantry was most prevalent there. Of the 893,370 black farmers, over 73.6 percent of them were tenants at the time when 30 percent was the norm for the nation. The tenantry system from 1870 to 1920 was much different in the South than elsewhere in the nation. That was not corrected for several decades.

When the above debate was taking place over the relative merits of the agricultural ladder, farm economists were discovering that generally the farmers who showed the highest returns were those who both owned and rented land. Those farmers knew the secret of using another's capital. The surveys also indicated that they had a higher level of education than the average of all farmers.

Many attempts were made to combat the changing structure, to keep people on the land, or to get new people into farming. One of the first big efforts was the Farmers' Cooperative Demonstration Group, which was started in 1904. This work eventually was absorbed by the extension service. Farm management programs were started. Then came movements to instruct country children in agriculture from the elementary grades through college. It was hoped that such efforts would lead to successful farming and that more children would want to return to farming.

It was popular to write about a back-to-the-land movement by those who feared that the nation was facing a decline in agricultural production. Others hoped that such a movement would relieve the congestion, want, and misery of the cities. Similar movements had been attempted in Europe and by the U.S. Salvation Army and the Jewish Agricultural and Industrial Aid Society of New York City. Several Italian agricultural colonies were formed specializing in horticulture and viticulture. Most of the above failed.

An agricultural economist of the early 1900s wrote that the only back-to-the-land movement that had sustained itself was of wealthy people who wanted a country home. That movement was not primarily agricultural and extended itself as much as 100 miles from the city. The same writer suggested that the extension system and educational programs "may be depended upon to save for our agriculture all the labor it will need for the maintenance of our national self-sufficiency."

Commercial Agriculture

The Civil War was a major watershed for agriculture. Prior to the war, the South had a system of commercial agriculture based on slaves producing cotton. The North had a few specialized dairy areas but otherwise consisted of small diversified farms. When the war came, southern planters were reluctant to stop growing cotton to plant crops for food and did not change their methods of production. In the North, many farmers started using commercial fertilizers and replaced their former hired workers with machines. The shift to animal power and commercial farming took place rapidly.

After the war, the South shifted to share renting with the former slaves and diversified to livestock. In the North, the farm economy changed rapidly even though many farmers were not aware of what was taking place. Mechanization and large-scale farming were promoted by agricultural exhibitions. Illinois had farms up to 40,000 acres using 225 plows, 45 corn planters, and a ditching plow that required 68 oxen and 8 men to operate. These were owned by investors and/or speculators. They were forerunners of the large farms that later came to the Red River Valley of the North and to Kansas.

Some of these large-scale farms were operated by the owners, but many were leased on share or for cash rent. Resentment built against the large landholdings, and legislation against absentee landholding was enacted, but it had little impact on the movement. The large farms of the 1850s and 1860s played a major role in Illinois agriculture. They hastened the development of the state; they popularized the use of new machinery; they spearheaded the development of corn monoculture; they led in importing purebred livestock from Europe; and they did extensive drainage of lands, which political units refused to develop and small-scale farmers could not afford to do. Generally, large farmers used more scientific methods and had higher yields per acre. More importantly, they set the tone for much of what was to follow in Corn Belt and Great Plains agriculture.

The farm depression of the late 1880s and 1890s stimulated interest in learning more about the cost of production so there could be better understanding of the farm distress. Farmers had to know more about operating more efficiently because of the importance of low-cost food to the nation. Secretary Houston wrote in 1914 that probably not more than 10 or 12 economists in the country could qualify as experts in rural economics. He noted that in the previous two years no agricultural businesses or farmers had requested aid from an economist.

W. J. Spillman, a leader in agricultural economics, pointed out the advantages of larger operations, especially their success from improved management ability. He continued that the "utter lack of system in the management of farm enterprises on many farms" proved to be the real surprise of those making the 1912 farm survey. Farms with the best profits had proven the importance of adapting enterprises to soil and climate and using the proper improvements and equipment. When land prices rose between 1900 and 1910, most farmers were unconcerned about the increase because they felt that the price of land did not affect them. Land prices rose even faster after 1914, and many farmers purchased land without any thought of what it would do to their cost of production.

After 30 years of declining prices, farmers prospered after 1897, but still they did not keep up with the advances in the rest of society. This was of little concern to the public in general because of the feeling that agriculture could take care of itself. Our farmers had proven their leadership in productivity per person. By the early 1900s the USDA wanted to increase production per acre, then educate farmers in distribution, organization, cooperation, credit, and improvement of their standard of living.

Changing Farm Practices

It was mentioned earlier that farmers started using commercial fertilizers in the 1840s. Justus von Liebig brought the world's attention to guano and its relation to plants. By 1896 steam shipping and railroads had made fertilizer economical to use, and U.S. farmers purchased about 1.9 million tons. By then they also had learned more about the value of barnyard manure and made better use of it.

In 1906 Secretary Wilson wrote:

> The economic revolution in the art and science of agriculture, which became noticeable in this country half dozen years ago, has continued during 1906, with tremendous effect upon the nation's prosperity.
>
> Crops so large as to be beyond any rational comprehension have strained the freight carrying ability of railroads.

Wilson continued that if farmers had been any more scientific they could have raised so much wheat, corn, cotton, and other crops that they would not have been worth harvesting. He revealed a look into the future when he predicted that doors were being opened so that farmers could increase production so much that the market would be glutted. Farm prosperity had provided farmers with money to drain land and make other capital improvements. Farmers also employed labor-saving devices to reduce the need for labor and the cost of production.

Each year during the first decade of the 1900s, farmers increased their income, nearly doubling it within the decade. Production responded to the demand, and in 1909 there was a record corn crop, the second largest cotton crop, and the third largest wheat crop. This had a profound effect on the lives of farm families as they improved their level of living and broke out of their isolation.

Ideal conditions in 1910 gave the nation's farmers an all-time high in production, but prices continued to climb. In 1911 adverse weather reduced production, but demand remained strong and prices increased. That year the U.S. corn crop was three-fourths of the world's production. Production increased to a new high in 1912, and the price index reached 202.1 based on 1899 value. Corn was the number one crop, followed by hay, cotton, wheat, oats, and potatoes.

Corn Development

The first account of white settlers working to improve corn dates to 1716, when Cotton Mather demonstrated that crossing occurred between two varieties of corn. He observed the effect of pollen on cross-fertilized seeds.

In 1813 John Lorain mentioned that fields of Indian corn without manure yielded 15 bushels per acre, while manured fields yielded 90 to 100 bushels, and one field yielded 138.5 bushels. One field that was planted at 26,880 seeds per acre yielded 118 bushels. Lorain suggested that if it had been seeded at 20,000 plants, it probably would have produced 200 bushels.

By then, corn varieties were purposely crossed to produce a more vigorous stock, indicating that at least some farmers were aware of the value of crossing to the corn plant. In 1846 Robert Reid crossed a late Ohio variety with an early flint variety, which resulted in Reid's Yellow Dent. This became a contributor to later hybrids.

In 1876 Charles Darwin published a book that discussed the improved vigor of crossed varieties. At the same time, William J. Beal, working at Michigan Agricultural College, learned that crosses exceeded parent varieties by nearly 25 percent. Beal probably was the first experimenter to cross varieties to improve yields via hybrid vigor. He also experimented with detasseling to control breeding, a practice still in use. By 1877 his hybrid corn yielded 50 percent over open-pollinated varieties.

Other breeders continued the work on corn. They proved that the selection of desired qualities could be made, and they established the principles leading to later hybrid corn research. By 1905 the drive for a hybrid corn was started at the Carnegie Institute in New York, at Illinois Agricultural College, and at the Connecticut Agricultural Experiment Station.

The Sugar Industry

The first known sugar produced in the United States was in 1751 from cane grown in Louisiana by Jesuits.

Attempts to raise and to process sugar beets were numerous after 1830, but it was not until 1879 that the first processing plant, which had been in operation for nine years, was proved successful.

The beet sugar industry began experimenting after 1879, and by 1888 production started in earnest. The USDA sent beet seed throughout the nation to farmers who were willing to experiment with raising sugar beets to determine where the best beets could be grown. By the late 1890s it was determined that the corn land of the northern states seemed to be best suited. In 1896 sugar processing from beets and cane was started because of the rapid increase in demand for sweeteners and the growing world deficit of sugar. In 1907 production reached 500,000 tons and continued to climb. By 1912, 67 sugar plants in 11 states were

Loading sugar beets from a field wagon to a rail car, California, 1900. (NAL)

in use and produced one-fifth of our national consumption. In addition to studying the use of sugar, the USDA learned what the Europeans were doing with beet pulp and molasses—by-products of sugar beets—for livestock feed.

While the above activity was taking place on the mainland, the United States acquired the Hawaiian Islands. The first commercial sugar was produced there in the 1830s, and by 1880 about 30,000 tons were produced. Sugar production continued to expand, and in 1898 each of the 60 plantations had its own sugar mill to process the raw sugar.

Agriculture was the leading industry of the Hawaiian Islands, and sugar made up 96 percent of the total exports. Coffee was the other export. Rice, the second-ranking crop, produced by Chinese farmers, was consumed by the local population. A few large ranches produced an adequate supply of beef for local needs. A few large dairy farms satisfied local needs for dairy products.

Dairying   The period from 1866 to 1886 is sometimes considered the golden age of animal agriculture. Cattle numbers increased from 25 to 45 million, horses from 8 to 12 million, sheep from 40 to 48 million, and hogs from 27 to 46 million. Livestock numbers easily kept pace with the booming population. Cheese factories opened in the 1850s, the refrigerator car was invented in 1871, and silos were developed in 1875.

After the recessions of 1866 and 1873, interest in specialized commercial dairying increased because dairy farmers, more than any other group, had the best survival rate during those two downturns. The regions along railroad networks around larger metropolitan centers became recognized as major dairy production areas.

The solid and stable demand for dairy products encouraged dairy farmers to upgrade their milking herds. In 1869 the first purebred herd of Holsteins came to New York from Holland. A heifer from that herd was the first cow to be registered in the *Holstein Herd Book of America*. Out of that herd, the top three cows had a six-year average of 8,738 pounds of milk production in contrast to an average of less than 3,500 pounds for the nation. The top year for one cow was 14,027 pounds, and by 1885 one animal had achieved 23,775 pounds.

In 1877 the American Jersey Cattle Club began publication of a monthly bulletin. Other specialized merchant/producer journals were started. *Hoard's Dairyman,* founded in 1885 in Wisconsin, proved to be so significant in getting information to the producers that by 1889 it was recognized as the leading dairy journal nationwide.

The development of the silo in 1875 at the University of Illinois provided dairy farmers with a practical way to store feed for improved milk production in the winter, when prices were the best. Because of the added investment in machinery required for making silage, silos did not become popular immediately.

The invention of the centrifugal cream separator in 1879 solved another problem for dairy farmers. Now the cream could be removed for butter production and the skim milk retained on the farm for calf and hog feed. By the early 1880s separators were popular with leading dairy farmers.

The 1890 invention of the butterfat test by S. M. Babcock proved to be another step in encouraging farmers to do a better job of breeding. Farm magazines played a leading role in fostering the sale of cream based on the butterfat test.

The next significant step in improving dairy herds was the formation of cow testing associations in 1905, but farmers were slow to join. By 1909 only 741 farmers with 11,686 cows in 27 associations in 9 states were involved. The results from cow testing encouraged the progressive farmers to do more upgrading. However, others remained satisfied to continue their old ways. The only attention they gave to breeding was to make sure that they purchased a bull from outside of their herd.

New Challenges

During colonial times, when farmers were quite isolated and livestock numbers were not great, there was a relatively lesser problem with disease in both livestock and plants. But as livestock numbers grew on some farms and monoculture crop production took place on others, diseases and pests multiplied.

In 1880 the first known shipment of fruit affected by the San Jose scale was exported from California. This insect, which attacks the bark of deciduous trees, originally came from China, but by 1893 it was established in the eastern states.

The boll weevil was first discovered in 1894, and by 1903 it was in Louisiana and by 1907 in Mississippi. By the 1920s it was in all the cotton-producing areas.

The Colorado potato beetle, believed to be a native of the Rocky Mountain area, multiplied rapidly in the 1860s and soon covered the eastern potato-growing states. Paris green, which was imported to combat the beetle, was mixed with water and applied to plants with brooms. Later, it was also used to control cankerworms and the gypsy moth.

The above are examples of threats to crops and livestock that added to the risk of farming. Battling weeds, insects, pests, and diseases was done without the aid of any scientific input. In most cases, farmers were totally helpless and had to let nature take its course.

The problem of draining land to eliminate potholes and sloughs has challenged farmers from the first day of tillage. The Romans knew the value of digging drainage ditches and planting crops on the higher ridges along the ditches. The English did trenching with a spade to get improved tilth of the soil. This writer has been on fields in England that were tiled every 75 feet across, with coarse gravel laid above the tile nearly to the surface.

In 1763 George Washington had the Dismal Swamp of North Carolina and Virginia surveyed to determine the feasibility of that land for crops and transportation. By 1790 the first drainage laws were enacted. The Swamp Land Acts of 1849 and 1850, which came after 20 years of Congressional debate over draining public domain, were the first important federal drainage legislation. Federal land was given to the states with the stipulation that the income from the sale of the land be used to build the necessary drainage. About 64 million acres were eventually transferred to the states, and the federal government freed itself of the problem of draining public domain.

With hand tillage or even with the early ox or horse rigs, small water bodies were not the problem that they are today. However, some American farmers, using the crude tools available to them, drained land from the earliest times.

In 1835 a New York farmer, who had immigrated from Scotland, brought patterns of clay tile with him. He molded tile, which he laid for drainage, marking the beginnings of tile drainage in the United States. By 1880 Illinois, Indiana, and Ohio had 1,140 tile factories. In 1882 Indiana had more than 30,000 miles of tile, and by 1884 Ohio had more than 20,000 miles of public ditches. Illinois, Indiana, and Ohio needed the drainage or much of their land would have been unfit for tillage.

In the days of horse farming, soil compaction was a problem because the pressure per square inch on the ground by horses was equal to or greater than that by tractors; but this was not understood by farmers. By the 1890s some farmers and leaders in the USDA realized that continuous plowing also caused compaction at plow depth. Soil scientists suggested that there was need for a method to stir the soil without compacting the subsoil. They were well aware of the benefits of trenching by a spade but knew that such a method was not feasible on a large scale.

Little was known of the value of subsoiling, but as early as 1875 farm magazines advocated that it be done. By 1895 subsoiling had proven to be highly valuable in semiarid regions. In areas of heavy, wet soil, subsoiling was not beneficial unless the soil was first tiled. But enough was known by then to aim at getting 16 to 18 inches below the surface without bringing new soil to the surface. Subsoiling had to be done well before the planting season to enable the spring rains to soak in. The big challenge was the lack of adequate power.

Sharecropping

Both 1865 and 1866 were poor crop seasons, which compounded the already destitute condition of southern agriculture. Much of the South experienced food shortages. It was not until 1868 that conditions had improved enough so that the federal government did not have to distribute food. A report from the Commissioner of Agriculture indicated that the plantings of 1866 and 1867 did not pay share workers for their labor, food, or clothing. The report continued that it had not been expected that the freedmen would produce the same results that the compulsory labor of 1860 had.

The freedmen apparently had expected to receive land of their own and were disappointed when that did not happen. They feared going back to the plantations, either as wage laborers or as sharecroppers. A sharecropper furnished all the labor and half the seed, and the planter furnished everything else. The proceeds were split 50:50. At the same time, the former planters, who had lost much of their wealth, were unhappy with those prospects and considered hiring white or Asian labor. By 1867 nearly 40 percent of the cotton-field workers were white, and in some states whites were in the majority.

The former planters lacked the capital to hire workers and operate the farms on their own, so they were forced to go to either share tenant or sharecrop arrangements. The share tenant supplied everything but the land, had title to the crop, and received 75 percent of the proceeds. Most freedmen became sharecroppers and rented land from their former masters. Many freedmen were lured to the West, and others no longer wanted to farm.

A shortage of available labor further reduced the possibilities of operating the former plantations with hired labor. In addition, former overseers realized that they could use their talents to better advantage by managing their own farms rather than managing for plantation owners. Many owners had neither the ability nor the desire to manage freedmen, so they discontinued farming.

The lien laws enabled freedmen to secure financing to put in crops and run their own farms. Unfortunately, this put the tenant croppers under the control of the merchants, who commonly charged high rates for financing. The freedmen were forced to manage their new freedom. Some planters also sold land to their former slaves on contract-payment basis. Many freedmen became so hopelessly in debt that they lost their farms to the merchants who had financed them or to the former landowners.

By 1880 there was a 65 percent increase in the number of farms in the 13 former slave-holding states. This created other problems. First, most farms were too small to provide an adequate living for the family even under the best conditions. Every state had thousands of farms that were less than 10 acres. Second, there were about a half million poor white farmers who did not cooperate with other groups. They lived on wastelands, which they halfheartedly worked, and survived by hunting and stock raising. They intensely hated the freedmen. Third, many of the blacks not only were destitute but were totally unprepared to cope with their new-found freedom. The government had done little to help bring them into the new society. By 1900 the South Atlantic states had 44 percent tenancy and the South Central states 49 percent tenancy. By 1910 Georgia had 65.4 percent tenancy. In the entire South, only 17 percent of the former slaves had acquired land.

Revival of the South

As soon as the Civil War was over there was a renewed demand for cotton. The demand was so great that it created a strong desire to return to the single cash-crop system. This caused an increase in cotton acreage, both in the older areas and in the newer cotton areas of Oklahoma and Texas, where a new class of small farmers emerged. The new cotton producers used better management, applied more fertilizer, had fresher soils, and were more open to new ideas. This gave them a decided advantage over the older regions still using sharecroppers and tenants. It also gave them an edge in competition when the price of cotton dropped.

The South lacked an efficient, low-cost transportation system, which also slowed recovery. Fortunately, the superior quality of southern cotton was able to overcome all the handicaps, and the United States recaptured the lead as a cotton exporter. Unlike other agricultural regions, the South did not readily adopt new technology that was used elsewhere.

Changes off the farm helped the South revive. In 1880 only 13 percent of the cotton grown was manufactured into cloth in the South, but by 1900 that figure increased to 50 percent. The demand for cotton continued strong in Europe, and Asia was developing a textile manufacturing base. All this helped to stabilize the southern economy. Improved marketing because of new cotton exchanges, the use of futures, and better connections to the world market all helped to level out the cycles in cotton marketing.

Workers carrying two-bushel sacks of grain aboard a river boat at a landing, Mississippi, 1906. (NAL)

The growing demand for cottonseed, which provided oil, meal, hulls, and linters, boosted the cash income from cotton. After 1910 the rapid expansion of irrigated areas in New Mexico, Arizona, and California increased production sharply. The irrigated farms produced greater yields. The farmers of these lands adopted new technology rapidly and made significant use of migratory labor during peak seasons.

One of the most promising signs of a change occurred on March 25, 1911, when the largest number of black farmers ever to assemble met at North Carolina Agricultural College. Here, over a three-day period, agricultural specialists spoke on all topics of interest to practical farmers. About 53 percent of all blacks were still in agriculture, of which 64 percent were farm laborers and 36 percent were overseers or independent farmers. Most of those who left the farm generally went to larger cities in the North and the East.

Transportation and Communication

Railroads

Importance to the Livestock Industry

After 1860 the railroads helped the livestock sector become one of our leading industries. Livestock no longer had to be driven to market on foot, and distant cities could be assured of a supply of animal products. After the Civil War, individuals realized that cattle could be driven to northern markets and sold for several times their cost in Texas and the expense of driving them. The first cattle drive took place in 1866. By the end of the year, over 250,000 head were driven to Sedalia, Missouri, the nearest railhead. The initial shipments of livestock by rail were costly because animals were not handled properly

Construction crew of Great Northern Railroad, North Dakota, 1887. The wagon team on right is hauling ties to the front of the rail line. Rails were skidded from flat cars and spiked down as the train moved ahead. (SHSND)

and because traffic management did not take their needs into consideration.

In 1867 the Kansas Pacific Railroad arrived in Abilene, Kansas, which became the first rail cow town. The drive from Texas to Abilene covered 700 miles in two months and required eight men. Direct costs were $2.00 per head, plus an additional $0.40 for loss from theft, stampedes, or death along the route. From 1867 through 1871, about 1.4 million cattle arrived in Abilene. They came in 164 drives with an average of 2,343 head and 12 men. Average loss on a drive was 1.6 percent. In addition to Abilene, the Kansas towns of Newton, Ellsworth, and Dodge City were the leading rail cow towns in succession between 1866 and 1879, when the Long Drive from central Texas to the Kansas railheads terminated. During those years about 4 million head were driven to the rail cow towns. The old cattle drives were dramatic, but the savings in time, reduction in conflict with people along the route (free pasturage was no longer available), and less stress on the animals were all pluses.

By 1860 the first double-deck livestock car had been built. A roof was installed not only to protect livestock from the sun but also to keep sheep from jumping over the sides. By the 1870s solid trains of cattle cars were formed. Provisions for watering and feeding in the cars were made. Some routes had rest-stop feeding stations, and occasionally pastures were available. Laws were enacted requiring trains to water, feed, and rest livestock every 36 hours. By 1908, 98 percent of livestock was shipped by rail. The remaining animals were transported on river boats.

Railroads made large terminal markets and packing plants possible. From 1905 to 1907, the four largest markets received 8 million cattle a year. Chicago received 8 million hogs, while the other three received 2

Typical loading scene at a railroad siding stockyard. This was part of a shipment of 6,300 cattle from a western North Dakota station, 1910. (North Dakota Institute of Regional Studies; hereafter NDIRS)

to 4 million each. The yards at the terminal markets were so organized that a 40- to 50-car train could be unloaded in 15 minutes. Cities continued to grow as the industrial revolution progressed, and the demand for animal products grew with them.

Promoters of Agriculture

Railroads quickly realized that to make a profit they had to stimulate agricultural production. Farmers also were buyers of the railroad land grant. In 1851 the Erie Railroad was the first railroad to become involved in immigration when it ran an immigrant train from New York to the West. In 1870 the Union Pacific Railroad advertised 12 million acres of land for sale and promoted land exploration tickets. If land was purchased, the price of the fare was credited toward the cost of the land. After 1871 the Northern Pacific built reception houses along its lines to accommodate new arrivals and potential land buyers.

Railroads cooperated with federal and state agricultural agents. Sometimes railroads supplied agricultural experts, who were available to respond to farmers' calls. Agricultural instruction trains were widely used in conjunction with the farmers' institutes. In some areas railroads established model farms. In others they set up demonstration plots on private farms for the purpose of showing one specific experiment.

Railroads motivated farmers by giving cash prizes at fairs and at crop shows, or they gave prize animals to encourage better strains. Scholarships for short courses in agriculture were given extensively to winners in many contests sponsored by the railroads. Railroads promoted agricultural associations to encourage better overall management, to increase diversification, or to adopt more intense livestock production. Marketing groups were also promoted for specialty crops or livestock.

Railroad station agents often acted as contact persons between farmers and state officials to obtain labor. When the railroads brought people to new areas, they actively aided in opening those lands. They especially promoted irrigation, drainage, or the sale of cutover timberland. They published large amounts of literature on areas of settlement, specialized production problems, and community development programs. A railroad stood to profit whenever the programs it offered were adopted by farmers, but so did the towns along the road, and eventually the nation. In 1910 grain, hay, cotton, and livestock made up 20 percent of all the railroads' revenue, so their efforts were not in vain.

Transportation of Perishables

As early as 1607, Europeans had experimented with ice and salt, saltpeter, or other chemical agents for cooling drinks and foods. In 1762 Fahrenheit used ice and salt for cooling, but no practical application was made of this knowledge. Starting in the 1850s, work was done using insulated buildings to store fruits and vegetables. Within a decade ice and salt properly stored in insulated buildings produced a workable system of cold storage able to freeze large quantities of poultry and game. Until then the fruit industry, except for wine production and sun-dried products, was not of much importance outside of areas where the produce was grown. After transportation became more rapid and regular, with the application of steam to both ships and railroads for power and of ice for cooling, the fruit industry started to expand. In 1868 mechanical refrigeration was successfully used in railroad shipments of fruit and vegetables. The first successful shipment of fresh fruits from California to Chicago was made the following year. In 1869 refrigerated cars were successfully applied to meat and dairy products.

During the 1870s, mechanical refrigeration for storage was adopted in the United States and grew rapidly because of the demand from breweries and for iced water. By 1874 public cold storage warehouses were being constructed. In 1889 Union Cold Storage and Warehouse Company of Chicago brought refrigeration technology to the Midwest. Chicago became a leader in the cold storage industry for fruits and vegetables.

The apple trade gained most quickly from refrigeration. Other fruit growers saw the advantage, and in 1888 a carload of fruit was sent from California to New York without re-icing. The market for fruits and vegetables was expanded in 1892, when fresh vegetables were shipped from Florida to England. The shipment was a success, but the cost was too high and the technology had to be improved, which delayed more shipments for a few years.

With these advances the ice business grew rapidly because technology made it feasible to use ice for prolonged storage. The availability of many perishable commodities was extended throughout the year because they did not all have to be sold at harvest. Meat packing became a year-round industry and expanded greatly. Icehouses were built along rail lines to resupply the cars as they traveled across the nation delivering fruits, meat products, and vegetables. The American diet improved considerably with these advancements.

Highways

Better transportation was needed to haul the increased production of agricultural goods to market. Until 1800 only the more affluent could afford the farm carts pulled by oxen or the chaises drawn by horses. In 1830 light one-horse wagons appeared, and by 1840 they were quite common. Buggies and trotting horses were beginning to be used, but generally they were too costly for farmers.

Road development progressed slowly. Heavy freight wagons needed better roads, but it was not until the bicycle era of the 1870s and 1880s that the lack of good roads received widespread public attention. The cyclists worked to convince farmers that better roads would save them time and energy in marketing their products. Farmers seemed slow to realize that good roads would improve their income and living.

As the nation became more urbanized, it was critical that better farm-to-market roads be built. This led to state and eventually federal participation in road building. In 1893 the USDA became involved in the good-roads movement. The Office of Public Road Inquiries of the USDA began experimenting on roads in all parts of the country in response to nationwide pressure. In one experiment steel tracks were laid where there was no road. It was learned that 1 horse could pull 11 tons on the rails, while it took 20 horses to pull the same load on the adjacent dirt road. But steel rails were not what the public wanted.

By the early 1900s, there were 2.1 million miles of road in the nation, of which over 1 million miles were in the trans-Mississippi West. The need for better roads for marketing and social purposes was obvious, but when school consolidation started, good roads became mandatory.

A conflict arose when local units of government wanted to retain control of road building. This was impractical, for as much as 30 to 40 percent of the funds allocated for roads were wasted or misdirected due to a lack of experienced road engineers. Less than half of the states had expert highway commissioners. Secretary Wilson confronted the issue in 1911 when the country was in the midst of a national readjustment with regard to road improvement. New methods of road construction and maintenance were essential. That meant that road planning had to become centralized.

Ironically, some farm magazines, which usually were progressive in their outlook toward change, were slow to recognize the advent of the automobile and, hence, the increased need for roads. A December 1903 article in one magazine stated, "The auto is a fad . . . this is certain." The editor opposed the automobile on the basis of energy, commenting that oats, corn, wheat, and hay would be grown "for ages to come" but that coal or oil might not be in permanent supply. He suggested that humans would disappear when the horse did. The auto was only for those who had interest in the racing craze.

When automobile ads appeared in the 1908 farm magazines, the editorial policy became more enlightened. A 1911 farm magazine article commented about 10,000 farmers who all went to work at one time to build a 380-mile road across Iowa. The same issue carried an ad for an International Auto Wagon with a pickup box to "deliver produce to town." In 1919 the magazine promoted a 3,633-mile, 5-week auto trip to visit farms and observe conditions from Buffalo, New York, to Lincoln, Nebraska, and back east to Philadelphia. By the 1920s the federal government used grants to aid the states in a coordinated road construction program.

Rural Free Delivery

When rural free delivery (RFD) was first promoted, predictions were that it would cause a rural transformation far more significant than the grain harvester had produced. It would cause a social revolution. Where the idea of RFD originated is uncertain. It has been suggested that a female member of a western Grange first conceived of it and brought it to the National Grange. Others say that in 1879 a newsperson first proposed the idea.

Congress was slow to act on RFD because it feared the cost. In 1890 the Postmaster General first proposed the idea and realized that he had to approach farmers through their organizations to obtain support for it. That year, village delivery, an extension of city delivery, was first recommended. It caused a movement for RFD that Congress could not ignore. Strong protests and petitions from many of the 60,000 rural postmasters, 16,410 star route contractors, and local merchants, all of whom felt endangered by RFD, delayed RFD for a brief period. But a counter campaign was successfully waged.

Congress finally approved funds for 44 routes in 29 states, and on October 1, 1896, RFD was first tested in West Virginia. The experiment showed that RFD would require Congressional underwriting because it could not pay for itself. That did not stop the program, however, for in the next year 44 routes were functioning; in 1899, 634; and by March 1901, 3,391 routes were in operation, and applications were on file for 4,517 more. About 3.5 million rural people were being served by 1901. By March 1902 the number of routes had increased to 8,458, with 9,904 pending, and by 1903 the numbers had increased to 14,741 and 9,930, respectively. By 1902 RFD was a permanent and very costly part of the postal system.

RFD was a great timesaver for farmers. It put them in daily contact with the world and enabled them to get more up-to-date market information. At the same time, it fostered the cause for better roads. An automobile was used on RFD for the first time in 1912. Next came agitation for delivery of parcel post, with the promoters advocating that

Mail carrier's rig, Jamestown, Kansas, 1899. (USDA)

the increased business would help eliminate the postal deficit. Parcel post went into effect in 1913 and had a significant impact on rural life. By 1920 over 35,000 small post offices had been eliminated because of RFD, and with them many rural communities disappeared.

Frontiers

Indians and Eskimos

Once the United States became formally organized, efforts were made to determine the direction that should be taken regarding the future of the Native American. Because farming was the prevailing occupation of the early settlers, it was natural for them to assume that farming was also the proper occupation for the Indians. The early treaties indicated as much. A treaty with the Cherokees in 1791 specified that implements be given to them in the hope that they might be directed to become herders or cultivators. In 1793 an Indian trade act authorized Congress to allocate $20,000 annually to purchase domestic animals and farming implements for the Indian nations.

Treaties during the 1800s specified that funds be provided to purchase cattle, implements, and seeds and to employ instructors in agriculture. The Annual Report of the Commissioner of Indian Affairs in 1878 commented on problems that were experienced while getting agricultural pursuits underway on three reservations in Texas. At that time local agents suggested that Indian agriculture would be more successful if the tribes were broken up and each family had a farm. It was hoped that the families would take pride in having a personal possession and would realize that their livelihood was dependent upon their efforts.

In the 1880s the Bureau of Indian Affairs opened Indian schools, where agriculture was learned through actual practice with farmers living adjacent to the schools. Within a few years the land under cultivation and livestock numbers increased sharply. Probably because of those results the General Allotment Act of 1887 provided heads of Indian families 160 acres for livestock or 80 acres for crop farming.

While the above activities were taking place in the states, the Eskimo population of Alaska showed signs of starvation. Commercial whalers had found their way into the Bering Sea and the Arctic Ocean and competed heavily with the Eskimos. Next came the American canneries, which fished the waters to provide for their factories. About 1890, firearms were introduced, which seriously reduced the reindeer population and cut further into food and clothing supplies. Most of the Eskimos' food came from whales, walruses, and seals, supplemented by fish, birds, caribou, and reindeer.

In 1891 private funds were collected to purchase reindeer in Siberia, move them to Alaska, and give the Eskimos a chance to become herders to preserve their food supply. In 1892 the Teller Reindeer Station was established to teach management and propagation of domestic reindeer. Siberians, and later Lapps, were employed to teach the Eskimos. The

Lapps proved to be excellent teachers, and reindeer farming became a successful venture. Some early leaders suggested that the reindeer would provide an excellent export product.

Partly because of the above efforts, which were carried on by church organizations, the USDA became interested in Alaska. Reports as early as 1897 indicated that agriculture could be developed as far north as the Yukon Valley. One USDA article stated that about 3 million acres of land were potentially tillable, but the cost of clearing and ditching was projected at $200 an acre, making use of the land impractical. In 1899 the first information on crops, soil, soil moisture, and livestock was collected and the first experiment station was built at Sitka. By 1901 four stations were in operation in Alaska.

Cattle Frontiers

The Great Plains   Few frontiers have had as much written about them as the short-lived but dramatic cattle frontier of the Great Plains. The cattle kingdom got its start in the 1700s from animals that were brought into southern Texas by the Spaniards. American farmers began moving into the area in the 1830s, and their cattle mixed with the Longhorns. As early as the 1840s some of those cattle were driven to mining communities and other distant markets. Then the Civil War interfered, and the cattle were left to multiply.

In 1864 it was discovered that dried bunch grass of the western plains was very nutritious. This encouraged cattle operators to move their herds north. A strip of land about 200 miles wide stretching from Texas to Montana was the drive path. By 1880 about 7 million cattle were on the northern ranges and basically left to fend for themselves.

Hereford and Angus bulls were crossed with the Texas Longhorns to produce a winter-hardy but heavier and more tender beef animal. After railroads penetrated the West and brought cattle to Chicago, there was a growing market. Initial profits led to overstocking the range, and in

1882 the market started to decline. At the same time, input cost increased. The cold winter of 1885–86, followed by a hot, dry summer and the "killer" winter of 1886–87, put an end to the open-range cattle industry.

By 1890 the full impact of homesteaders and railroads was felt, which led to settlement of the Great Plains. Sheep herds provided additional competition, as did forest reserves established by the federal government. By that time farmers in the Corn Belt were fattening cattle, which provided a more desirable animal for the packers. The range-cattle business became more restricted and sedentary. Using forage for winter feeding became part of the routine. The annual calf crop was sold to Midwest feeders instead of being held on the range until maturity.

California Ranching and Wheat Farming

At the same time the cattle kingdom rose and fell on the Great Plains, another lesser cattle frontier based on Spanish stock rose and fell in California. When California was acquired in 1848, several hundred individuals presented claims to the United States for land on which they held title. Of that number, 604 claims averaging 13,245 acres were honored. Most of the claimants were in the cattle business.

Because of the lack of a local market and the lack of transportation, the Spanish cattle were killed for their hides and tallow, which were exported. Only the choice cuts were saved for local consumption. When gold was discovered in California, American-bred cattle were driven from Oregon and New Mexico and readily sold because they were superior to local cattle. But the demand for cattle was strong enough that even the Spanish breeds brought good prices, and the California ranchers became accustomed to extravagant living.

When the demand caused by the gold rush subsided, the ranchers borrowed funds to maintain their life style. Then, as in the Great Plains, the weather interfered. A drought from 1863 to 1865 starved out most of the remaining cattle. The ranchers were unable to finance the purchase of new breeding stock, so they had no choice but to lease their land to wheat producers.

Wheat production exceeded local demand by 1854, but during the Civil War, California was forced to buy wheat from Europe. Railroads came after the war, which led to the expansion of the wheat-growing area. The surplus now could be shipped to eastern markets.

Like the cattle ranchers before them, the wheat producers soon were replaced by development colonies and small-scale farmers. The land was divided into 20- and 40-acre plots for irrigated crops. The first irrigation water was used to produce alfalfa for dairying. The early Spanish missions had used irrigation, but the ditches fell into disrepair when the missions declined. By 1877 the first claims for water were filed, but it was not until 1896 that canals became successful. In 1897 a sugar beet factory was built, and sugar beets grown on the irrigated acres became a second major crop.

The Farmers' Frontier

New Englanders started leaving the land by the 1840s, and by the 1860s much of the abandoned acreage had reverted to pasture or to forest. Those farmers who survived had done so by converting to dairying, orchards, or truck farming. Sheep farmers also used the abandoned land to good advantage. During the 1850s to the 1870s, New York was the nation's leading agricultural state. It then slipped into a severe economic decline, and thousands of farmers anxiously moved west.

Everywhere squatters preceded those who preferred to stay within the bounds of the law. Professional squatters were the greatest abusers of preemption and preferred to move on a regular basis. They contributed to settlement, but, unfortunately, they sometimes added to the problem because they often caused trouble with the Indians. Squatters were every bit as important as speculators. In some respects the legitimate settlers were also speculators, because many of them frequently applied for more land than they knew they could farm. They hoped to resell the additional land at a higher price after more people arrived.

From the end of the Civil War to 1900, about 500 million acres of public domain were disposed of. Of that amount, 80 million acres were disbursed under the Homestead Act, 108 million by auction, and 300 million in grants to railroads and states, which then resold the land to farmers and speculators.

When settlers first reached the Great Plains, they were faced with a new challenge—a lack and unpredictability of rainfall as well as a scarcity of trees. This was a new experience for immigrants who came from the wooded areas of Europe. This writer has interviewed children of homesteaders who moved onto the Plains but returned to the wooded areas. They felt that areas without trees must have deficient soil. Settlers were also faced with the threat of cattle ranchers, who felt that they had prior rights.

A 160-acre homestead could be secured for less than $20 filing and locator fees, but the cost of fencing ran as high as $1,000. In addition, there was the expense of annual maintenance. The lack of economical

Breaking sod at the rate of 1 to 1½ acres per day, Sun River, Montana, 1908. (NAL)

fencing material continued until J. F. Glidden developed twisted wire with barbs attached. In 1874 he patented the idea, which was an instant success. States passed laws requiring that farmers had to fence in their crops if they expected to receive damages from roving livestock. Railroads also were required to fence in their right-of-way.

The great demand for barbed wire encouraged mass production, which caused the price to fall from 20¢ a pound in 1874 to 2¢ by 1894. With each decline in price, the demand rose. Eventually, cattle ranchers resorted to fencing the water holes and public domain before the homesteaders moved in. Fences brought an end to the cattle drives and forced ranchers to use the railroad. Once they were protected by fences, farmers were able to practice better crop rotation, graze their cattle in stubble after harvest, and upgrade their breeding. This resulted in a marked increase in the number of cattle as well as improved weights and productivity.

By the 1880s a change had taken place on the farm frontier that reflected a change in thinking of society regarding the legal rights of women. Women were always scarce on the frontier, but the problem was particularly acute on the Great Plains. That fact played a key role in liberalizing the law to make life there more appealing to women. The two watershed dates in that respect are 1862 and 1869. The Homestead Act of 1862 was a major piece of democratic legislation because it gave women an opportunity to homestead. The word "person," and not "male" or "female," was used. A "single woman" meant a woman who was unmarried, widowed, or divorced or who was a runaway wife. In 1869 Wyoming Territory granted unrestricted equality to women. That move was further enhanced in 1888 when six western states came into the Union with full recognition of women's suffrage.

A woman became a homesteader for several reasons: she wanted to add to her father's, husband's, or brother's land; she wanted to farm herself; she was looking for adventure and fun; she was seeking a husband or escaping from an unhappy marriage; she wanted financial security, independence, or equal rights; or she was seeking other opportunities. Some women filed a claim and then sold to a neighbor.

Here is an example of one woman homesteader, who had been a launderer. This widow with a small daughter became a housekeeper for a rancher in Wyoming. She homesteaded an adjacent quarter section because it seemed the obvious thing to do. Later, she wrote enthusiastically about women homesteading, because she felt it was a more satisfying way of raising a family than supporting children with the normal domestic work that was available to most women of that day. She realized that homesteading was not for all women, but she enjoyed the fun and experience of the effort.

Agricultural Communes

The United States was a beacon to many Europeans, some of whom chose to come in groups rather than individually. Some desired to remain in groups once they came here, partly for security reasons and partly for financial support that sometimes was available to groups and not to individuals. Others, like the Mormons, preferred to distance themselves from society at large and sought settlement where they would be left alone. The following accounts describe the trials, defeats, and successes of three diverse groups.

Jews In the early 1800s, about 1,500 Jews left Austria for the United States. They arrived here without funds, work, or contacts. Whether it was because of that experience or not, the Hebrew Immigrant Aid Society of the United States was formed to help people of the faith who came here after they had suffered persecution in other countries.

In 1820 the first effort to form a colony of Jewish farmers took place when a group settled on Grand Island in the Niagara River in northern New York. Little is known about the colony, for it lasted only a few years. A second attempt took place in 1837 when a Jewish farm settlement was started at Shalom, Ulster County, New York. By 1847 this settlement also ended in failure.

In 1881 a wealthy Jewish merchant with aid from the Alliance Israelite Universelle and others established a colony of 124 Russian Jews on Sicily Island, Louisiana, in the Mississippi River about 400 miles north of New Orleans. This was formerly the site of a plantation of about 3,000 acres. None of the people had farmed previously, and few were accustomed to manual labor. The colony was wiped out by flood in 1882. Some of the group went to South Dakota to homestead. The Hessian fly and drought destroyed their crops, and many left after the second year. Those who remained were either forced to quit after hail in 1884 or drought in 1885.

Another Jewish colony was founded in 1882 at the foothills of the Rocky Mountains in Colorado. It was pronounced a success but soon disappeared.

A colony started in 1882 by the Hebrew Emigrant Aid Society was located at Vineland, New Jersey. The families lived in separate homes and branched into nonfarm enterprises. Those enterprises provided work and enabled them to process their farm produce, which subsidized the farming operation. This colony was still in existence in 1926.

A successful farming colony, founded in 1882, was settled by several Jewish-Russian groups at New Odessa, Oregon. This 760-acre farm on choice land benefited from good organization and financial backing. By the end of its first year, the colony had paid on its mortgage and had erected the necessary buildings. Even though its farming was successful financially, its members could not agree on a socio-political organization, and by 1885 the experiment ended.

A second and final attempt of the Alliance Israelite Universelle came in 1885 in South Dakota. A fully collectivist community was founded by single men who called themselves Bethlehem-Jehudah. They lacked farming experience and had no trained leadership. Strife and discontent, plus natural disasters, caused the colony to fail in 18 months.

Several colonies were established in North Dakota after 1882. As long as they received outside help they survived, but their numbers dwindled through the years, and by 1907 the last remaining farmer of any of the groups left. The same was true of the Palestine Colony founded in 1891 in Michigan. It survived because of a flow of aid from Jewish groups in Detroit. Unlike the other Jewish groups, several members of the Michigan group remained to become successful farmers.

In 1908 the Jewish Agricultural Society founded an extension department, which provided itinerant field instruction, held group meetings, conducted demonstrations, and organized bull associations, credit unions, farmers' clubs, correspondence courses, and night schools. The department also provided scholarships for students wishing to pursue agricultural education and published *The Jewish Farmer*.

In 1918 the Farm Settlement Department was founded to help Jewish farmers. The Department helped the Toms River Colony of New Jersey get started in 1920. The colony eventually had a school for religious instruction and a community building. It survived because of its auxiliary businesses and proved to be an excellent example of an agro-industrial community.

The Jewish Agricultural Society made hundreds of loans annually that were secured by signatures to first mortgages. Over 95 percent of the loans were to farmers. After 26 years of financing, it averaged an annual loss of only one-sixth of 1 percent. In 1925 there were about 75,000 Jewish people living on farms, chiefly as a result of the Society's work. Many individuals from all walks of life had joined the groups with the desire for a more simple life, but most of the efforts ended in failure.

Hutterites

The Hutterite Brethren, a small cooperative religious group, moved from Russia between 1874 and 1879 and established its first settlement in South Dakota. Three colonies and a few private farms were founded by 1877. By 1913 South Dakota had 17 Hutterite colonies. By 1934 all but one had moved to Canada. In 1935 the South Dakota legislature allowed communal colonies to incorporate. This encouraged the Hutterites. As a result the original Bon Homme Colony started a second colony, and another colony returned from Canada.

Besides their communal living, the Hutterites are known for their distinctive dark clothes. They practice agricultural self-sufficiency as much as possible, partly for reasons of economy and partly to minimize contact with the outside world. Simplicity of living and efficiency in farming are two main goals of the Hutterites, but they make extensive use of labor- and cost-saving technology.

Colonies are divided into departments, each headed by a specialist. However, labor can be shifted between departments as the workload demands. Colonies have extensive livestock and poultry enterprises as part of their self-sufficiency philosophy and to maximize profits. The average colony contains about 5,000 acres, with nearly 3,000 in crops.

After the population of a colony increases to between 100 and 150 residents, it splits and a daughter colony is established. The colonies no longer make brooms, shoes, soap, and clothes or do canning, gardening, or beekeeping. Their very advanced, intensive livestock and poultry enterprises make better use of the available labor. Today, the Hutterites have about 45 colonies in South Dakota and a total of 131 in Minnesota, the two Dakotas, and Manitoba. Their colonies are high-technology enterprises, but the communal system is maintained, and they prefer to isolate themselves from the mainstream of society.

Mormons Late in the 1850s some Mormon families left the northern Utah settlement to determine if they could raise subtropical crops in the Virgin River Valley farther south in the state. A formal settlement started in 1861, and by 1866 about 20 villages were located there. First, the Mormons produced food for living, and then they experimented with subtropical crops.

Initially, the villages were tightly knit units, but as farming increased, land farther out was developed. Subtropical crops, such as cotton, sorghum, tobacco, wine grapes, and other fruits, were initially a success, as were sheep. That changed when transportation improved and cheaper goods were shipped in. Then the irrigated lands were used for alfalfa, chiefly for dairy cattle but also for beef animals. The economy, however, remained sluggish until 1920, when an adjacent national park brought in tourism. Although this meant prosperity, it upset the Mormon leaders, who were disturbed about a cultural change.

Irrigation and Dry Farming

After the closing of the frontier in the 1890s, one of the major issues confronting the nation was how best to use the semiarid lands of the West that were still under governmental control. Better farming methods had to be learned to combat the negative influence of the dry climate. A long battle took place between the nation's agricultural leaders as to what was the best way to farm on the Great Plains—by irrigation or by dry farming.

To combat the lack of moisture in some areas of the Great Plains, settlers realized that they had to construct ditches and dams to collect and apply water before any attempt was made at crop production. In a few areas ground water was available at 10 to 20 feet, but in most cases it was down 100 to 300 feet. Some farmers hoped to survive by extensive dry farming, speculating that a periodic large crop would overcome the losses of dry years.

From the Mormons' experience with irrigation after 1849, it was clear that their system worked only because they were a tightly knit group disciplined by their religious ideas and excellent leadership. Cooperation was the key to their success. That spirit continued in the Union Colony established at Greeley, Colorado, in the 1860s. Cooperation was a premise upon which the colony was built. The success of the colony led to the creation of other projects at Boulder, Longmount, Loveland, and Fort Collins, Colorado.

During the same period, irrigation projects with different goals were started at Anaheim and Riverside, California. They originated as cooperatives but soon became incorporated and speculative. Stock was sold to individuals who had visions of becoming water millionaires. This type of venture soon became the dominant factor in promoting the growth of western agriculture. By 1870 irrigation districts and companies were formed to govern the construction of, payment for, and use of water projects.

The corporate structure of organization was necessary because the amount of irrigation needed was too large for individual and cooperative enterprises. The larger projects focused on water-rights and land-rights issues. Most of the corporate ventures were money losers because of the length of time necessary for settlers to arrive, to obtain financing, and to improve the land. Thousands of settlers wasted much of their lives and resources trying to make the arid land pay. The problem was compounded by the fact that public land laws were not applicable to irrigation. In addition, many of the landholders were absentee or speculative, and they did little to cooperate with those who wanted to irrigate.

The earliest intensive use of irrigated land in California was by missionaries who planted vineyards. But it was not until the 1850s that French and German immigrants gave proper attention to using irrigation with grape and wine culture. The warm, dry climate caused the Mission grape to become too sugary to produce good wine. Other imported varieties grew well, and soon there was a surplus of grapes. Dealers labeled the poor wine as California wine, and the good wine as French or German wine.

By 1880 the grape industry had stabilized. There were over 200 wine cellars and wineries, ranging in capacity from 10,000 to 3.5 million gallons. By 1886 overproduction became so severe that the price of grapes dropped from $25 and $35 a ton to $8 and $10. At the same time that the market was depressed, thousands of acres of grapes were destroyed by disease. By 1892 the industry was in such bad financial condition that the growers banded together in the Wine Makers' Corporation. Soon it controlled 80 percent of the output and was able to maintain prices. The same applied to the production of table grapes and raisins.

In 1894 the Carey Act was passed for the purpose of opening the arid regions for settlement. It was followed in 1902 by the Reclamation Act, which provided that the proceeds from the sale of land in the western states be put into a fund to build irrigation systems. A 1910 study indicated that the 160-acre grant in irrigation districts was very adequate. Because of the intense labor demands, most family operations that were irrigated consisted of only 80 acres.

With the technology then available, it was not profitable to pump water from depths of more than 50 feet. Windmills were the most important source of power for lifting water for irrigation. Windmills could not be too large for pumping, even though they lost efficiency with size. Once a windmill was erected, the cost of operation was virtually only the upkeep of the mechanism. Ponds and storage tanks had to be built to hold the water until it was time for irrigating.

By 1915 over 7,300 miles of canals had been dug providing water for over 1 million acres on 14,200 farms. The results were mixed, because many inexperienced farmers did not produce the income that was anticipated, costs were higher than projected, the government was not receiving returns on its investment, and some land was not suited to irrigation. At first, wheat was the dominant irrigated crop, but it was gradually replaced by fruits, vegetables, and mixed farming, which helped to raise the income level from irrigated land.

In 1879 Hardy W. Campbell homesteaded in what is now South Dakota and practiced what was later called "dry farming." Campbell recognized the need for packing the topsoil to keep it from blowing, and in the early 1890s developed a soil packer. In 1897 he managed 43 farms for 4 railroads in 5 states to prove the merits of dry farming. During that time he introduced the technique of planting grain in rows wide enough to permit cultivation. In 1898 he proved the value of summer fallow. He advocated not using the moldboard plow more than every four to six years.

University specialists, working with the railroads, agreed with much of what Campbell had learned about the worth and flexibility of dry-land farming. However, the USDA maintained a doubtful attitude toward farming on the Great Plains. The ideas promoted by Campbell persisted, and in 1907 a Dry Farming Congress was held, which opened the way for a farming method that could partially cope with the dry area. The congresses continued for 10 years, and by that time most of the area that would ever come under the plow had been settled.

The Timber Realm

Initially, forests provided everyone with firewood and fencing and with building materials, and most people thought that the supply of timber would last forever. By the late 1700s that attitude changed. In 1791 the Philadelphia Society for the Promotion of Agriculture offered awards for tree production. In 1817 Noah Webster warned of the need to curtail consumption of timber and to increase its production. The following year Massachusetts offered premiums to plant oaks and other trees necessary to maintain a supply of ship building material. Several

Removing stumps using a horse-power sweep, Minnesota, 1870. (Drache)

states imposed stiff penalties for cutting timber or setting fires on public lands. In 1828 President John Adams set aside a live-oak reservation to insure the navy of a supply of oak. Soon other reserves were set aside.

In 1865 a USDA report predicted that there would be a timber famine within 30 years. The author of that forecast advocated planned forest-management research. Part of the result was a management program for federal timberlands, the creation of forest experiment stations, tree planting, and public education on the need for forest conservation.

The above work was a good start, for in 1872 Nebraska designated Arbor Day as a day for tree planting. This was recognized as an effective means of making people aware of the value of trees. New York State proposed a forest reserve, and Congress set aside 1.3 million acres for Yellowstone Timberland Reserve (Park). A national movement in forestry resulted in the Timber-Culture Act of 1873, followed by funding for the Forestry Service in the USDA.

In 1875 the American Forestry Association was formed to promote forest management for better social and economic use. In 1876 a survey on conditions of U.S. forests commenced, and the Minnesota State Forestry Association became the first state association. With aid from the state legislature, the Minnesota organization had as its sole purpose the promotion of planting forest trees. After the 1894 Hinckley fire, which took 418 lives and created a national stir, Minnesota again led the nation by appointing fire wardens to help prevent and suppress fires. In 1899 it established forest reserves.

In the 1870s Cornell and Yale universities became the first schools to offer courses in forestry, followed by Michigan in 1881. Iowa may have been the first state to offer courses in practical forestry. It was not until the late 1880s that more schools offered courses in forestry. In 1898 the first training center for foresters was opened in North Carolina.

In the late 1880s the Daughters of the American Revolution became involved in the conservation movement. Then the National Audubon Society and the Federation of Women's Clubs became involved. They proved to be significant forces on behalf of conservation. In 1891 the federal government took two steps that aided the preservation of forest land: the President was authorized to set aside forestry reservations from the public domain, and the Division of Forestry was created within the USDA.

The discovery of coal and the advent of coal stoves reduced the need for forests as fuel, but the expanding economy created a strong demand for timber products. After the railroads expanded into the West, timber

Lumberjacks cutting a tree in a virgin forest, Oregon, 1900. (NAL)

could be hauled anywhere, so the pressure on forests continued. In the treeless areas of the Great Plains, settlers became particularly aware of the value of forests.

President Harrison set 13 million acres aside for forest reserves. He was followed by President Cleveland, who added another 20 million acres. By 1900 over 50 million acres were in reserves. In 1901 President Theodore Roosevelt upgraded forestry to the Bureau of Forestry, and in 1905 he renamed it the Forest Service. By the end of Roosevelt's term, the forest reserve had grown to 148 million acres.

In 1892 a program of systematic forestry was begun on private land in North Carolina. Timberland owners were shown how they could profitably cultivate trees. In 1904 the Secretary of Agriculture wrote that not enough was known about the economics of long-range forest management, but states began to act on their own. By 1913, 20 states had state foresters and 33 had enacted forestry legislation, indicating that the public was aware of the need for a national forestry policy.

Shortly after 1900, wood preservatives were used for the first time on a large scale. Railroad ties, telephone poles, and telegraph poles were among the first products to be treated. Soon bridge timbers and other timbers that were exposed to the elements were treated. These were all used in large quantities, and treating them significantly extended their useful life, which reduced demands on the forests.

Other Conservation Issues

The need for conservation called attention to grazing on public lands, which caused concern in the cattle and sheep industries. Public grasslands were surveyed, and it was determined that grazing could be continued but under a regulated basis. A fee would be charged according to the number of cattle or sheep that were grazed. In 1906, despite rancher objections, the fee permit system was established. Until the 1920s, fees from grazing exceeded the revenues from timber.

Early tree-planting rig, with a box of seedlings mounted between the two workers, USDA plot, 1910. (NAL)

The White House Conference of Governors of 1908 was a landmark meeting for conservation. The first result was a survey of natural resources. That led to the withdrawal of an additional 1.6 million acres of forest land and 16 western rivers for public domain. Next came the Weeks Act of 1911, which sought to preserve forest watersheds for the purposes of protecting navigation waters, retarding flooding, and providing better drinking and irrigation water. For the first time the government reversed the trend of disposing of public lands to acquiring land to be set aside in the public interest.

Major Movements, 1860 to 1914

The Civil War

The immediate effect of the Civil War on the South was that the blockade stopped the export of cotton, which reduced funds available to carry on the war. The ban on the North's exporting food and other supplies to the South caused suffering in southern cities and eventually in the countryside. On the other hand, the North could not get cotton, so hemp and flax were substituted for clothing raw materials. The lack of sugar from sugar cane forced the North to turn to maple sugar, honey, and sorghum as sweeteners.

A far greater portion of farm men in the South went to the military than in the North, but in both sections women had to do field work to keep production up. Northern farmers were free to adopt labor-saving machinery and continued to increase their production. From passage of the Homestead Act in 1862 to the end of the war, 2.5 million acres of land were opened under that act alone, plus additional land that was sold by the railroads. These acres were nearly all in the West, and because of the new rail lines, western products got to the market.

In the North, the shortage of labor, more land in farms, strong markets, and the doubling of prices of farm commodities were strong motivations for farmers to adopt labor-saving horse-drawn machinery. Most of the machines were available earlier, but profits in agriculture had not justified their use. The use of machinery hastened the inevitable trend to commercial, market-oriented farming. By contrast, in the South the lack of transportation, markets, and machinery forced a shift to diversified agriculture relying on labor. This caused a decline in production throughout the wartime period.

After the Civil War, community life changed sharply in both the North and the South. In the North this was the result of rapid continued adoption of machinery, the change in the style of living in the eastern areas, adoption of scattered farmsteads because of the Homestead Act, and more stress on cash cropping. Also, farmers became more mobile and did not pass down their farms to succeeding generations as frequently as in the past. This was probably because of increased education, which gave farm children opportunities off the farm rather than limiting them to farming.

In the South, plantation owners found that their money was worthless, that their former slaves were gone, and that they had little or no livestock and machinery. They were left with large landholdings that had little value because they had no means to work them. Many farms were abandoned, leaving thousands of acres available to be settled by poor whites, freed slaves, and some outsiders. The makeup of the southern population was similar to that of the western frontier. Conditions were primitive, and now all were in a state of equality. A massive revolution in rural society had taken place as a result of the Civil War. The former plantation owner experienced the same poverty as the rest of society.

The credit system of the South changed, and bankers shirked their responsibility as lenders. Bankers were replaced by the supplying merchants, who were protected by lien laws that had been passed by the legislatures for their benefit. The crop lien system became universal. Unfortunately, this perpetuated cotton monoculture and made the

South dependent upon the North for food products. This was despite the fact that the South was 80 percent rural, in contrast to 35 percent for the rest of the nation.

The production of the so-called one-mule farmer was far less than that of any northern counterpart. As late as the second decade of the twentieth century, in one southern state the average weekly egg production was less than two a week, the average daily butter production was two-thirds of an ounce, and the average daily milk production was two-thirds of a pint. Annually, the one-mule farmer produced 1/3 of a hog, 1/12 of a beef, and 1/100 of a sheep. The output in cotton production could not buy the food that had to be purchased. Blindly sticking to cotton was a basic reason for the lack of progress in the South.

In 1871 the Tennessee Association was organized in an attempt to rebuild. It had delegates from 11 states representing 40 different agricultural societies, but little resulted from its efforts. In 1899 the Commissioners of Agriculture of the Cotton States Association created the Association of Southern Agricultural Workers. It promoted improving the services of the USDA; diversifying farming; teaching more agriculture in rural schools; building cotton factories and mills; setting up more experiment stations; and, in general, improving agriculture. Again, little was accomplished, for the bulk of the farmers preferred to stick to their outmoded ways. Recovery would have to wait for other events.

Migration of Agricultural Workers

The Homestead Act, the creation of the USDA, the establishment of land-grant colleges, and the development of transcontinental railroads all aided in settlement of the West. The unrestricted movement of immigrant labor to the farm frontier was as important to farming as cheap raw materials were to industry. The boom in agriculture caused by the Civil War speeded up the movement. German and Irish immigrants came before and during the war to ease the labor shortage and to establish new farms. By 1863 the northern states raised as much wheat as the entire nation had in 1859.

After the Civil War, the movement into Kansas and Nebraska was rapid. Veterans were allowed to apply their service time against the five years needed to satisfy homestead requirements. Railroad companies offered bargains to prospective settlers to encourage development of the area.

The number of farms increased from 2.044 million in 1860 to 4.564 million in 1890 to 6.366 million in 1910. However, some long-range trends became apparent before the total number of farms peaked. One was a decline in the number of farms in some of the older established areas. Another was a sharp decline, both relatively and absolutely, in the number of women engaged in agriculture. A third was that immigrants no longer swelled the ranks of agricultural workers, for by 1900 they made up only 258,479 of the 3,038,884 white farm workers. A fourth was that blacks were leaving the land for the city. A fifth trend was a strong movement of workers from the farm to the city, with only a small number of city workers moving to the farm. Between 1870 and 1900, about 3.7 million farm workers left the land.

Early farm economists felt that the movement to the city had to continue as long as farmers desired to become more efficient. Some argued that continued migration of foreigners to work on larger farms was a decided disadvantage to the small farmer, who had to compete with the large farmer.

Agrarian Discontent

From May 1862 to July 1897, 529,051 homestead entries encompassing 70 million acres were patented, with another 32 million acres pending. In addition, states and railroads had sold about another 32 million acres. Roughly 350 million acres of new land were put into production in little over four decades. Overproduction continued, and in 1897 the USDA calculated that 14 million surplus acres were in corn, wheat, and oats.

On every frontier there was a boom that caused farmers to overborrow and overexpand. The boom was generally followed by a quick decline. At the same time, farmers saw that industry was rising rapidly while they were falling behind in the rate of progress. Farm output increased 135 percent from 1870 to 1900. During this period the price index dropped from 119 to 53. Farmers who had purchased land and machinery on fixed contracts at higher prices could not meet their obligations. Those who practiced wheat monoculture were faced with declining yields and were forced to seek alternatives.

Farmers had remained very individualistic and preferred to blame the railroads, monopolies, low prices, tariffs, debts, and governmental monetary policies for their problems. Farmers were bothered by the closing of the frontier and felt that governmental policies favored industry. They were hampered by traditionalism and negativism, while other sectors were free of cultural restraints and were better prepared to face the changing social structure.

When prices of farm goods dropped because of overproduction, farmers produced more. They ignored the fact that the export market was expanding and demanded protective tariffs. This reduced the farmers' chance to export and at the same time increased the cost of goods they purchased because it shut out competition to American manufacturers. They blamed high freight costs for their problem, even though rail rates had dropped substantially. Railroads also brought city advantages to the farm, had an impact on the rising value of farm land, and delivered farm products to distant markets. Agriculture was an occupation in which those of mediocre ability could still eke out an existence.

Farmers had difficulty uniting into a single group, and some farm organizations promoted cures that ran counter to what other farm groups fostered. The tax system worked a hardship on the farmer because property taxes were relied on for schools and roads. Urbanites with larger incomes were taxed little in comparison. Ironically, the federal government failed to recognize that there was a farm problem at the end of World War I. Secretary Morton wrote that legislation could neither plow nor plant. "The intelligent, practical, and successful farmer needs no aid from the government. The ignorant, impractical, and indolent farmers deserve none."

The drought of the 1880s in Kansas, Nebraska, and the Dakotas caused many settlers to realize that they had gone too far west. In the 1880s, 20 towns in western Kansas were abandoned, and in the 1890s, 26 counties in South Dakota lost population. The drought and the end of the frontier frustrated the settlers, causing many to leave farming. The really discontented ones were those who remained on the land and suffered. They became targets for the agitators.

Studies by agricultural economists early in the twentieth century indicated that one-third of the farmers received less income for their labor than the average urban worker. The top one-third, however, made considerably more for their labor. Housing and farm produce were extra benefits. Even during depressed times there were always those who did well. Most farmers on small units would have been better off lending out their capital and taking jobs elsewhere. The studies indicated that to provide a good family living, farms should be large enough to require work by two or three family members.

In the late 1890s, farm prices started to improve at a faster rate than the general price level. In addition, leading farmers recognized the benefits of the work of the USDA, the power revolution, and the improved outlook for demand.

Government studies determined that there were several obstacles to farm profitability, which led to discontent. Few farmers kept records, and most had no idea of their cost of production. Many used poor seed or seed not adapted to their area. At least in part because of their failure to rotate crops, they were unable to protect crops from disease or insects. They experienced losses in marketing because of the lack of marketing information. They were unable to get perishables to market at the proper time. They exhibited ignorance of, or lack of attention to, the details of marketing, demand, quality of crops, or handling crops and livestock.

They had a general lack of good management understanding and procedures.

The conclusion of President Theodore Roosevelt's Country Life Commission suggested that many of the above weaknesses had changed after the frontier came to an end, because land was no longer for the taking. Farmers were better off in the early 1900s than ever, but not as well off as they could have been when all possibilities were considered. A century of change had left agriculture behind because other industries adjusted so much faster. The unequal development came about without any direction, but the problems resulting from it had to be remedied.

G. F. Warren wrote that the nation had more than achieved agricultural self-sufficiency and had a huge surplus of commodities as well as a large surplus of rural population. That reserve was far more than what was needed to maintain national agricultural self-sufficiency. One obvious solution to ease the suffering was to continue to move people out of agriculture.

Farm Organizations

The economic and social revolution in agriculture that started with the Civil War was welcomed by the progressive farmers but resented by the majority who feared change. Farmers were no longer tied to the local community where they were a dominant force. Because of that, they lost much of their independence. Again, the more aggressive liked the opportunity, but most did not.

Some historians have maintained that during the years 1870 to 1900, farmers suffered more from lack of status than from economic deprivation. They were disturbed by the prospect of living in an urban industrial commercial society, which caused them to fight to preserve the past. Farmers believed that the nation could not be prosperous without a prosperous agriculture and that it was the greed of others that prevented them from enjoying success.

They no longer were the dominant independent class of the past. Only the agricultural press, farm societies, and politicians remained to promote the concept of the virtuous farmer. At the same time, farmers' sons and daughters led the migration from the farm to the city. The lack and high cost of credit, decades of declining prices, the isolation of farm life, the destitution of many farmers, and the harsh physical conditions of the Great Plains caused farmers to unite in search of a better way of life. They had to organize if they were to catch up to the rest of society.

The Grange

In 1858 farmers met in Illinois to protest freight rates, but that was forgotten with the boom of the Civil War years. Similar protests were renewed after the war, and western farmers joined in because they could not pay the freight on their grain shipments. These protests, which started about 1867, served as part of the seeds for the Grange movement and gained momentum during the Panic of 1873.

Oliver H. Kelly, a Minnesota farmer, had faith that technology and enhanced knowledge through reading farm publications would improve agricultural conditions. He felt that agriculture should be commercialized and profitable but was disturbed by the anti-book traditionalist farmer.

In 1866 Kelly was appointed as agent in the USDA to gather statistical information. He was shocked by the farmers' great lack of interest in progressive agriculture. He learned that for every farmer who read agricultural books and papers, 10 did not. Methods of farming were usually handed down from previous generations, and 90 percent of farmers had no knowledge of scientific agriculture.

Kelly conceived of the idea of a secret society, because he felt that it would bind people closer together. He also hoped that it would break down the distrust that farmers from various sections of the country had for each other. His niece introduced the idea that women had to be included in such an organization. Kelly needed a cause to excite the farmers. The Patrons of Husbandry (Grange) was organized in 1867, and in 1869 Minnesota had the first state Grange.

Educational work was encouraged to stimulate farmers' minds. Cooperatives were encouraged, and meetings helped the farmers and their wives to break the isolation. This was critical to the well-being of those on the farm. Many farmers felt so removed from the daily life of greater society that they were inadequate to cope.

Cooperative buying direct from the factory was started by the Grange. It was an initial success, and membership skyrocketed. It is claimed that the Grange was such a lucrative discount market that it was the direct cause of Aaron Montgomery Ward issuing his first mail-order catalog in 1872. Soon other retailers met Grange competition, and the disorganized Grange, which had experienced many retail branch failures, underwent a sharp decline. In 1875 its membership peaked at 858,050.

The efforts of the Grange had long-term results. It was one of the first organizations to promote a program for good roads. In the 1870s it started a movement for rural free delivery. In 1887 it advocated parcel post. In 1891 it recommended a farm credit system. In 1914 it stressed the need for rural electrification. But its opposition to the railroads led to its greatest early accomplishment—passage of the Granger laws, which culminated in the Interstate Commerce Act of 1887.

The Grange began to decline when it became involved in political activity starting in 1873, even before it reached its peak in membership. Internal fighting and the failure of business ventures and of a third-party movement relegated the Grange to a society of farmers conducting social and educational activities. When farm prosperity reappeared in the late 1890s, most farmers forgot about their previous difficulties.

The Grange's position regarding women proved to be one of its lasting social contributions. The Grange was a family organization, and men and women were never divided into separate orders. In 1873 the Grange constitution was revised and women were admitted to state and national units. By 1900 they held many offices and may have held the majority of local Grange Lecturer positions.

Women within the Grange effectively made it a rural branch of the women's rights movement. The Grange received support from feminist leaders in other organizations. Many Grange women also took part in the temperance crusade, which acknowledged women's rights and suffrage. The women of the Grange were similar to those who joined women's clubs or the Women's Christian Temperance Union (WCTU). Eliza Gifford, who was active in the New York State suffrage movement, the WCTU, and the Grange, stated that it was the "greatest equality club the world has ever known."

The Alliance

As depressed times continued, several farmers' political organizations were formed that eventually evolved into the Farmers' Alliance. After the Alliance became stabilized, it settled on two basic aims: the government should build warehouses to store surplus grain, and the government should lend money to farmers without interest.

The Farmers' Alliance was created almost spontaneously at various places. Probably the first Alliance was organized in Texas between 1873 and 1875, just at the time the Grange was losing momentum. The original purpose of the Texas Alliance was to catch horse thieves. It was followed in 1877 by the Northern Alliance.

In 1880 the Louisiana Farmers Union was organized for the purpose of cleaning up a graveyard. Its constitution also called for an effort to gain favorable agricultural legislation, including equitable taxation, lower interest rates, and lower cost of government. In 1887 the Texas Farmers' Alliance merged with the Louisiana Farmers Union and became the National Farmers' Alliance and Cooperative Union of America.

Soon after the formation of the various Alliances, protests against the railroads were augmented with protests against middlemen and bankers. This became the key for the rapid increase in the number of organizations that wished to join the battle. Ironically, opposing philosophies also developed at this time. One faction held that farm problems were economic and social and that farmers could solve them by their own organizations. The other faction agreed that farm problems were economic and social but held that they could best be solved by political action. An agreement was never reached.

In 1882 the Arkansas Agricultural Wheel was organized for the purpose of improving the theory and practice of agriculture. Within a year there were 462 chapters. By 1885 it absorbed the Brothers of Freedom, which was also founded in Arkansas. In 1887 the combined units merged with the National Farmers' Alliance.

In 1886 the word "white" was removed from the eligibility rules, and Negroes became active in forming associations. In 1886 the Alliance of Colored Farmers of Texas was the first to be formed and was quickly followed by organizations in other states. By 1891, 30 states had associations founded by blacks, with 1.25 million members.

In 1887 the North Carolina Farmers' Association was formed to absorb many of the state farm groups. About the same time the Northern Farmers' Alliance and the Farmers' Mutual Benefit Association were organized by northern and western farmers, and both grew rapidly. In 1889 the Alliance was joined by the Farmers' and Laborers' Union of America. However, the Alliances continued to function as usual without becoming involved with labor problems. By 1890 the Farmers' Alliance produced its first agrarian party platform. This was a key in agrarian protest. At the same time, it solicited the Knights of Labor, the Greenback Party, and the Single Tax Party to join it, but without success.

Next, the Alliance formed the People's Party, and in 1892 the Populist Party. The Populists held their first convention in 1892, and that year they received over 1 million popular votes and 22 electoral votes. The major political parties took heed. In 1896 the Populists were strong enough to control the Democratic Party and brought about the nomination of William Jennings Bryan. After Bryan's defeat, the Populist Party waned, and by 1900 it had ceased to be an effective organization. By 1908 it was virtually dead.

Worldwide prosperity, gold rushes in the Klondike and in South Africa, a rapid expansion of industry, and an increase in immigration all helped farmers as demand for their products grew and prices increased. Prosperity in agriculture commenced in the late 1890s and lasted to 1920. This caused the Populist Party to fade into the background and also was responsible for the loss of zeal for the Alliance.

Other Farm Organizations As noted from the above discussion, the farmers' movement peaked during the late 1880s and early 1890s. By their sheer numbers and all their organizations, farmers were able to attract the attention of other segments of society, including newspeople and politicians. Unfortunately, at the height of their power the Alliances became intensely political and lost their grassroots support.

The Farmers' Educational and Cooperative Union in the South and the American Society of Equity in the North were both formed in 1902. These general organizations quite effectively addressed most issues that disturbed farmers. Their basic aims were to help farmers manage better, regulate elevators, and cooperate to control supplies, especially in years of large crops. They felt that farmers needed to be more aroused against conditions that kept them down.

The American Society of Equity was effective from 1902 to 1934 in states of the Ohio Valley and in the grain-producing Northwest. Its major impact was on some problems of distribution. A basic weakness of the Equity was that it never had a large enough membership to control supplies.

The Decline of Agrarianism

After the decline of Populism, the last great mass movement of farmers, agrarianism lost its democratic character. In the past it spoke for all common people, not just farmers. Historically, the assumption was that everyone was a farmer and that democracy meant agrarian democracy. The Populist movement included all the people. It included organized labor and was a movement against the powerful. In the past, agrarian movements bordered on intolerance because of the attitude of superiority by those of the soil and the belief that farmers were the chosen people.

During the long period of declining prices after the Civil War to the 1890s, farmers lost much of their optimism and were quick to blame almost anyone for their predicament. As soon as prices improved, farmers became more optimistic and forgot about supporting their organizations, which suggested that farm groups were "fair weather" organizations. The frontier is sometimes linked with agrarian democracy, and once the frontier disappeared, agrarian democracy lost some of its zeal. The fact that farmers no longer were the majority sector of society had a negative impact on them.

Limitation on the number of acres a farmer could obtain under various governmental acts was a reflection of agrarian democracy. It was an excellent example of trying to distribute land as broadly as possible to the landless. The expansion of technology ran counter to agrarian democracy because it enabled increased production with fewer farmers. This movement started in the 1840s and has continued to the present. This sealed the fate of the small farm.

Cooperatives

In the early 1900s Theodore Roosevelt wrote that he believed it was essential to the welfare of the American people that farmers adopt the cooperative system. He noted that there were many examples where cooperatives had already proven successful and that the USDA should start such a movement.

In the 1780s, forerunners to cooperatives appeared when farmers formed societies to import cattle and later to conduct community cattle drives to coastal cities. Threshing rings, bull and stallion rings, cheese rings, local mutual insurance companies, etc., were natural to meeting the needs of rural people.

The cooperative movement covers several periods. The first period spans from 1810 to 1870. In 1810 cooperative dairying was established in Goshen, Connecticut, and by 1867 there were more than 400 dairy processing cooperatives. In 1857 the Dane County (Wisconsin) Protective Union was formed for the purpose of marketing grain and livestock. Other cooperatives were formed in Wisconsin and Illinois, which included purchasing and/or marketing organizations. New York enacted the first legislation providing for cooperative mutual insurance companies. In 1865 Michigan passed pioneer legislation recognizing cooperatives for buying and selling.

The second phase of cooperatives appeared between 1870 and 1890, when the Grange sought to solve some farmers' problems through cooperative buying, marketing, or both. More than 20,000 Granges were organized to do business for members. Grange stores bought and sold groceries, clothing, general farm supplies, hardware, and implements. Marketing cotton received major attention in the South. Tobacco, grain, and wool were also marketed. Grange banks were established in two states, and the manufacture of farm machinery was also undertaken.

When farm prices improved, many of the Grange enterprises failed. However, enough knowledge had been gathered to prove that the Rochdale system of cooperatives, was sound. This system, developed in England, involved selling at prevailing prices and then paying a dividend based on savings. One of the New York Granges eventually was merged into Agway, a large eastern cooperative that is still in existence.

Late in the 1880s, the Farmers' Alliance replaced the Grange and founded many cooperatives. The Florida Fruit Exchange was one of its major successes. By 1890 there were about 1,000 cooperatives, of which about 75 percent were in dairy products, 10 percent in grain, and another 10 percent in fruit and vegetables. About 1886 farmers formed cooperatives to counteract railroads and grain merchants. That movement was strong for about 15 years and then declined.

The third phase of cooperatives came between 1890 and 1920. During these decades cooperatives were formed in most states for the marketing of products in carload lots. By 1895 California farmers using irrigation raised citrus fruit, walnuts, lima beans, almonds, grapes, deciduous tree fruits, and many other specialized crops. Because of the perishable nature of their crops, they were in a weak bargaining position with most of them. These California farmers were 3,000 miles from their major markets. With production increasing constantly and poor shipping practices prevailing, goods often arrived on the market in poor condition and could be readily rejected. Since the markets were easily glutted, existing marketing practices were risky. To make matters worse, farmers did not understand market conditions and were haphazard marketers. The commission firms they used operated on the farmers' capital but took none of the risk.

Starting in 1895, farmers realized that they had to organize to control the marketing of their products in the hopes of getting better prices. California Fruit Growers' Exchange was their first major cooperative. This later became Sunkist Growers, Inc. Farmers had to learn that unless they banded together to withhold products from the market, they could not effectively change the price.

Other cooperatives were organized following the example of the fruit growers. California Almond in 1910, Challenge Cream and Butter Association in 1910, Diamond Walnut in 1912, and Sun Maid in 1912 were among the early producer cooperatives. Nearly all of them had decades of successful operation. Throughout the nation farmers were accustomed to cooperatives because of their experience with creameries, mutual insurance companies, elevators, and rural cooperative telephone lines.

In the early 1900s the American Society of Equity, the Right Relationship League, and the Farmers' Educational and Cooperative Union were all created to give farmers a more unified bargaining position.

The American Society of Equity had many local cooperatives in the North Central States. Its major interest was in marketing livestock, grain, potatoes, and general produce. Eventually, it got into production supplies and after 1913 sponsored several livestock packing plants. Unfortunately, like many of the cooperatives that started before it, it failed because of mismanagement or political activity of its leaders. Equity stockholders lost heavily in its failure.

The most lasting outgrowth of the Farmers' Alliance came in 1902 with the founding of the Farmers Union, which emphasized cooperative purchasing and marketing services. It eventually expanded into 13 states in the central section of the nation. In the South, the Farmers Union stressed storing and marketing cotton and improving the credit and mortgage system. It became the agent for purchasing supplies for the local elevators, creameries, livestock shippers, stores, service stations, and other agriculture-related businesses.

   Livestock shipping associations were one of the simplest forms of cooperative organizations. They started spontaneously throughout much of the nation. Isolation and small volume caused farmers to join with their neighbors in the hopes of improving their marketing position. By the 1880s, farmers realized that local buyers who bought cattle by lump (by the animal) were taking "rake-offs." There was virtually no defense against this practice, because the local markets were often too small to attract competition.

Farmers delivering livestock to what is believed to be the second cooperative shipping association formed in the nation, Litchfield, Minnesota, 1908. (University of Minnesota Archives; hereafter UMA)

Probably the first livestock shipping association was established in 1904 at Postville, Iowa. The second appears to have been founded in 1908 at Litchfield, Minnesota. Others were established shortly after in Iowa, Minnesota, and Wisconsin. Farmers' cooperative elevators often aided in forming the farmers' shipping associations, or they sometimes ran livestock shipping as an adjunct to the grain business. Cooperative creameries and breeding associations also influenced the formation of livestock shipping groups.

The period from 1900 to 1920 was the heyday for the formation of local independent cooperative associations for both buying and selling. It was also the period for the creation of large-scale regional cooperatives, federations, and centralized associations. The centralized organizations

Typical of the creameries that were found throughout the nation's dairy areas in the late 1800s and early 1900s. Generally, three or four farmer neighbors took turns hauling cream to the creamery. By rotating their delivery, the creamery received cream on a daily basis. (UMA)

Farmers using a horse-power sweep to load grain prior to the erection of a local cooperative elevator on a new rail line, North Dakota, 1911. (Roy Smith)

usually handled a single product. The peak years for the formation of cooperative shipping associations were between 1904 and 1916.

By the 1920s the number of marketing associations reached an all-time high of 12,000, with 2,100 supply cooperatives. Local shipping associations reached their peak of 4,000 nationwide and then started to decline with the advent of truck transportation. Minnesota with 700 associations, Wisconsin with 600, and Iowa with 300 were the leading states for livestock shipping associations. Cooperative creameries and elevators probably each had nearly 3,000 associations at their zenith. Producers of fruit, produce, eggs, and potatoes combined had about 1,000 associations.

In the early 1900s, the USDA studied the needs for standards for various commodities, warehousing, and futures, in an attempt to obtain a degree of uniformity in agricultural marketing. The extension system was very active in creating many of the cooperatives, from rural telephone lines to shipping associations. Much of the time, lack of capital prevented getting things done faster, but lack of business knowledge by farmers presented a major obstacle to securing financial help.

Changing Rural Life Styles

Farm Life   A major article entitled "Health of Farmers' Families" in the Commissioner of Agriculture's 1862 report gives insights not generally aired about life on the farm. The opening sentence states, "The impression pervades all classes of society that the cultivation of the soil is the most healthful mode of life, and gives the highest promise of a peaceful, quiet, and happy old age." The article continued that the facts did not bear out the popular conception, because farmers and their families had some of the highest incidence of insanity. Also, farmers did not have the greatest longevity, as most people generally assumed.

It was suggested that a farmer's "incessant thinking" on his subject tended to craze the brain; it "unhinged the intellect of multitudes." Farmers were compared to inventors, students of prophecy, and others with "sharp pointed memories" who had high rates of insanity and suicide. The reason given for these high rates was that farmers' thinking was limited to too few ideas, their life was "a ruinous routine," and there

Female bundle haulers, western North Dakota, 1910. Typical of many midwestern farms of that era. (Larry Dahl)

was a sameness and a tameness about their life, with "a paucity of subjects for contemplation," that was dangerous to mental activity.

Farmers were seen as persons who had no breadth of view and who could not sustain a conversation beyond a few comments about the weather, the crops, the markets, and the neighborhood news. "And it is worthy to note that their remarks on these subjects are uniformly of the complaining and unhopeful kind, as if their occupation and thoughts were on the same low depressing level." The proposed remedy for the sad effects of the plodding routine was a higher standard of general intelligence and more attention to book farming, which the writer knew was not respected by most farmers.

Special comments were made about the farm wife, who "is a laboring drudge; not of necessity but by design." It was suggested that on three farms out of four, the wife worked harder than the husband, the

Hand-cranked washing machine and wringers in Red River Valley of the North, Summer 1912. (June Tweten)

This Minnesota farm woman was fortunate to have the pump so near to the house. Most wells were located adjacent to the barn for ease in watering the livestock. 1911. (UMA)

farmhand, or the hired kitchen help. Repairing the house and making things more convenient for the wife were the last things done on the farm. Bringing in water and bringing in wood were two of the most overlooked hardships.

From 1860 to the 1890s, technology reduced the drudgery of farm work, increased the productivity of labor, and improved the economic opportunities of farm workers, as well as their income potential. At the same time, virtually nothing was done to reduce the drudgery of the farm wife.

Frequently the strongest proponents of the agrarian way of life were persons who had left the farm and were making a living in another enterprise. A farm writer of the 1840s commented that those who spoke most highly of agriculture never did anything to prove their sincerity, and at the same time they advised their sons to seek another profession.

Woman and girls sawing wood. Typical of the many jobs women did on the farm prior to the availability of electricity and appliances. Minnesota, 1905. (MHS)

As the nineteenth century advanced, it became clearer that farming could not compete with industry and urban life.

Several reasons were given for the problem. Most farmers had a low social status because it was felt that anyone could farm. Farming was hard work. Only the sentimentalist wrote that farming was elevating and purifying, but farm children knew that was so much chaff. Isolation, lack of a good education, and scarcity of social and cultural advantages added to the dullness of farm life. Farming not only entailed more risks than other jobs, but it also lacked intellectual stimulation.

Farm youth wanted a greater challenge and more opportunities, but despite more labor-saving machines, RFD, better roads, and increased income, they still left the farm. The overall profitability was better in industry. Between the mid-1890s and 1914, farmers experienced the so-called Golden Age, but that did not stop the outmigration. Ironically, supporters of the Country Life Movement, to be discussed later, who professed that they wanted to make farming more efficient and wanted to improve rural living standards, were more interested in cheap food and in keeping down the cost of urban living.

The toil of the farm wife did not decrease as farms became more mechanized. One agricultural historian noted that for some reason "farm women endured the drudgery without rebelling." They did not seriously consider a less demanding life. To the farm wife, "work was a way of life . . . and she recognized that without hard work from every family member the family and the farm might not survive."

There are numerous accounts of farm women who had not been away from the farm for a year or more at a time, nor had they been to the nearby town for two or three years. This writer interviewed one individual whose mother had not gone to town in 10 years and another whose mother had not gone to the town only 20 miles away in 25 years!

The Country Life Commission, of 1908 to 1910, pictured the life of the farm woman as one of poverty, isolation, and lack of labor-saving devices. The government thought that by improving the attitude of the farm woman it would do much to revive agriculture. Therefore, massive educational and extension programs were started to convince the farm woman of her importance to the success of the farm.

Home economics programs were successful in reducing the isolation, but they concentrated on housework and created a gender gap when it came to farm duties. Most tasks of the farm woman were nonmarketable but vital to sustaining the farm. This tended to undervalue the economic contribution of the woman.

Health and living standards on the farm, which were often assumed to be superior to those in the city, had fallen behind city standards by 1900 because too many farmers refused to adapt. It seemed that the best educated and most ambitious people left the rural areas and generally became successful. This gave rural people a sense of inferiority, because they soon realized that the cities formulated social values and expectations. The more isolated the rural areas were, the more defensive they were about their status.

Farmers opposed school consolidation because they feared the changes that would take place if their children became better educated. They also feared losing local control. Ironically, some rural teachers opposed consolidation because they feared professional standards would be raised.

In a 1910 speech Theodore Roosevelt hit directly at the great weakness of most farmers. He advised them to practice more conservation and to make their life more attractive. He felt that the two were essential for a prosperous future. "Little good is done by the farmer who refuses to profit by the knowledge of the day; who treats any effort as absurd, refuses to appreciate what he regards as newfangled ideas and contrivances, and jeers at all book farming."

Roosevelt advised that the farmer's wife should be given more consideration. It was important to think of new machines to minimize her labor. He continued: "The welfare of the woman is more important than the welfare of the man; for the mother is the real Atlas, who bears aloft in her strong and tender arms the destiny of the world."

In a USDA survey of the wives of 55,000 volunteer crop reporters, the women complained that they could not get exposure to lectures, travel, educational films, demonstrations of labor-saving devices, etc., that would help them improve their living conditions. They complained that whenever there was extra cash, it was invested in the farm business rather than in the home. The women felt that the home should be considered part of the business.

Farm wives objected to buying more land, because all it did was make more work for them. This was particularly true if more land meant extra hired help at the table and in the home. The women generally resented having live-in hired help, because often the workers were not the kinds of persons that they cared to have around the family.

After prices started to rise in the mid-1890s, more funds were spent on improving the farm home than in several previous decades. This greatly improved farm living, but the gap between rural and urban conditions was still immense. Farm leaders noted that the average farm did not keep enough livestock to add income and also to contribute to a better family diet. Farm women frequently objected to having more animals, because the additional work fell on their shoulders.

Overwork added to farm women's loneliness and isolation, for they were not free to get out and meet people the way the men could. This same complaint came from all over the nation in letters to the USDA. Lack of good educational facilities was the next most noted complaint. The women relied on extension people to give them direction for women's clubs and on cooperative societies, because they did not feel capable of acting on their own.

The Country Life Movement

When the people/food ratio turned around at the beginning of the twentieth century and the price of food increased with no corresponding rise in production, people questioned the cause. The USDA was unable to provide an explanation, because it lacked crop reporting and market information service. Some thought that the problem was the costly distribution system. This was partly correct. At the same time, most people realized that farming was still very inefficient, hence, costly.

In 1908 Theodore Roosevelt appointed the Country Life Commission to look into why food prices were rising while farm life remained less than prosperous. He was the first president to be sympathetic to modern farm problems. This was at a time when most members of Congress were unaware that there was a farm problem. Because he was such a strong advocate of conservation, Roosevelt felt that the soil was the most important resource. He was concerned that the "right type of men and women be on the soil." The attitude of farmers was extremely important, and he wanted to know what they thought.

Judged by European standards, American farmers were prosperous, but that was not so when compared to other U.S. citizens. Roosevelt felt that their material resources were limited, their social life was barren, and their political influence was relatively small. He added, "American farmers have been used by politicians, but have still to learn how to use them."

Roosevelt was concerned by the exodus of youth from the farm, particularly since the brightest and most enterprising were leaving. The Commission was the first national attempt to study farming and farm life. It was particularly concerned with the fact that most of agriculture was too disorganized to help itself.

Starting in November 1908, under the leadership of Liberty Hyde Bailey, 300,000 questionnaires were sent to farmers throughout the nation to learn about all aspects of rural life. Hearings were held in 30 states. The survey and hearings revealed the farmers' concerns. Good roads were a top priority everywhere. Then came the need for a more effective education system. Farmers complained that education took rural children away rather than preparing them for farming. Rural free delivery, parcel post, postal savings banks, health and sanitation needs, a more adequate supply and better quality of farm labor, and, in some areas of the country, more easily accessible land were other concerns of the farmers. Many felt that the exodus of the best farm youth was caused by the lack of intellectual challenge in rural areas.

The conclusions of the Commission as to why agriculture failed to be profitable were as follows: lack of knowledge, lack of training, lack of transportation, poor farming methods, lack of good rural leadership, a weak agricultural credit system, shortage of labor complicated by intemperance, burdens on the women, a poor public health system, and the inability to deal against the established systems.

The Commission determined that transportation costs and middlemen were two of the most misunderstood aspects of the business life of the farmer. It felt that the labor shortage was basic to democracy and never would be solved, although it realized that much labor was wasted on the farm because of poor management and a lack of machinery. That was also true of the waste of other resources.

The liquor problem was considered to be one of the most serious of rural ills. It was attributed to the barrenness of farm life, particularly that of the hired hand. It was extremely serious in the South, and it complicated the racial problem. This eventually became a major factor in the prohibition movement.

Many leaders of the Country Life Movement believed in the yeoman myth—that farmers were the backbone of society. They wanted to blend what they thought were the political and social virtues of the early agrarian society with modern society. This could not be done, for the strength of industrialization was too powerful to turn back the clock. Many of these leaders hoped that they could regenerate rural society with a better educational system oriented toward agriculture.

On the other hand, many farmers refused to become part of the urban-industrial age. They opposed most attempts to convert them to scientific agriculture. Although they were not totally satisfied with rural living conditions, they were more disturbed with the economic and technological changes, which many did not understand.

A contemporary writer stated that the biggest rural problem was that farmers were too content although they realized that the rest of society was so much better off. There was a large spread from the top to the bottom in productivity and profitability of farms. A great diversity existed within regions and an even greater diversity between regions. The South lagged behind the other areas. Within the South the 925,000 black farmers, who, on the average, farmed only half as much as the neighboring whites, were the most destitute.

The immediate results of the Commission were that the USDA and professional leaders in agriculture and business made an active and financial effort to help agriculture. Even though Congress did not adopt the ideas of the Commission, most of them eventually were carried out. Many needs disappeared with the passage of time, because farm numbers dropped sharply, which tended to reduce some of the root problems.

By 1920 two movements had spun off the Country Life Movement. Those who wanted to support farming as a way of life founded the National Country Life Association. Little of lasting value resulted from that group. Others who wanted to direct farmers into taking a more businesslike attitude toward farming organized the American Farm Bureau Federation.

Denting the Isolation

When U.S. agriculture shifted away from farmers living in villages, as in early New England, to living on individual farmsteads, farmers were relegated to a life of isolation. Once the Homestead Act was passed, it was virtually assured that people would live on individual farmsteads. Annual fairs, irregular church services, school programs, an occasional shopping trip, and visits with neighbors "across the section" made up the social life of most farm people.

After 1870 fairs changed some of their goals. Their numbers continued to increase, but their mission was changed to conform to rural needs. In part, fairs were replaced by agricultural schools, cooperatives, departments of agriculture, and farm organizations. Between 1870 and 1900, fairs evolved as a place for the various farm organizations to display what they were doing. This was also an era when fairs attempted to get the government involved to solve temporary problems and long-term development needs. By 1900 most fairs had acquired permanent grounds and facilities and were tied with the local communities.

By 1900 the number of fairs had stabilized, and frequently professional management became part of the scene. From 1900 to 1930, the agricultural fair experienced four main trends: (1) state fairs grew stronger while county fairs stabilized; (2) specialized fairs and trade shows appeared; (3) community fairs and noncommercial fairs relied on input from the neighborhood; and (4) fairs that specialized in boys' and girls' clubs emerged as a result of all the youth club activity during those decades.

One of the greatest innovations to change rural life was the telephone. Mutual farmer-built and -maintained lines brought farm people in touch with each other. The telephone was a boon as a device for both learning the neighborhood gossip and conducting business. The telephone was first displayed in Philadelphia in 1876, and in 1878 it was used on a bonanza farm in Dakota and possibly on a farm in Iowa. In the late 1890s, rural telephone lines were being built. By 1902 there were about 2.2 million subscribers, of which 55,747 were on 295 independent rural lines.

Farmers built these lines and bought shares of stock, which funded the purchase of phones, wire, and other equipment. Assessments were made when the association needed the money. It was not unusual for a rural party line to have 20 or more subscribers, all of whom heard all the rings as they came through and could "rubber" in on the call.

Often local merchants gave one long "general" ring to advertise specials. In case of an emergency, a general ring was used to alert everyone on the line. This writer grew up in a home that had two local lines. The local banker acted as the repairer and sometimes left the bank to fix telephones or lines. About the only time long-distance calls were made was when there was a death or a serious medical problem. Costs to operate lines were virtually zero, except for the pay of the central operator, in whose home the switchboard was located. For example, one line with 20 subscribers had a total income for a year of $21, all of which was for long-distance calls.

As mentioned earlier, boys' and girls' clubs were very popular in the opening decades of the twentieth century. In 1902 a group of elementary students in Ohio organized to grow corn, test soils, have small gardens, and grow flowers and small shrubbery. Pictures were taken throughout the season, and the students made presentations at a farmers' institute. In 1903 the three-H idea—head, heart, and hand—was developed. A local experiment station provided seeds for a second year for the group.

In 1905 a program for elevating the rural standard of living was developed. Clubs were formed in seven midwestern and southern states. In 1910 the fourth H—health—was added. The idea rapidly spread to other states. The chief aim of 4-H was to improve the attitude of young people toward country living. It gave them a chance to meet with their peers on a regular basis. Each club had a leader, and generally parents came to the monthly meetings for an evening of socializing mixed in with the 4-H lesson.

Such programs eventually became a vital part of the extension system. They led to the Smith-Lever Act, which formalized cooperative education work. The task of extension people became one of reorganizing rural society, educating the youth about modern agriculture, and making them more scientific farmers and homemakers. The biggest problem was that most farm adults opposed many of the suggested programs of extension. This often caused apathy on the part of children and handicapped the chance of getting good leaders, which hurt the planning and the organization. If the extension people were not good at pushing the programs, little happened.

Rural Schools

Farmers often had a negative outlook about education. Children were essential to the operation of the farm, and many parents preferred that they work on the farm rather than attend school. Fortunately, some individuals (generally the women) were willing to push for better education. During the 1800s and the early 1900s, the major subjects of rural schools in order of importance were reading, arithmetic, writing, and spelling. Other subjects were grammar, history, geography, literature, and physiology.

Schools usually were established soon after communities were founded. Many small independent school districts were created so the children did not have too far to walk. The length of the school term depended upon how much the people wanted to tax themselves to pay

Sod school in western North Dakota built by a family at no cost to the district. The teacher (in dark suit) received $15 per month, plus room and board. All the children came from two families. 1899. (Christine Olson)

for education and how willing the parents were to free the children from farm work.

In 1896 a committee that was formed to improve the country school expressed the view that funding rural schools was a poor way of spending money on education. Professional educators agreed that most rural school districts were too poor and had too few pupils to provide a good education. Local jealousies often caused bickering among the members of the district, who then split away and created even smaller districts.

Professional educators had discussed school consolidation since the 1850s. Their initial goal was that rural schools be consolidated so that there was one school per township. The number of pupils attending rural schools dropped sharply after 1900 because of migration from the farm. Rural education reform became one of the main themes of the Country Life Movement. Nature study and vocational courses, especially agriculture, were to be established wherever possible.

Interior of rural school in northern Minnesota. The teacher (right rear) was a local farmer and logger. 1904. (James Plummer)

After 1900, when rural school consolidation began to take place, farmers proved quite opposed to the movement. It was not long before states helped the cause of consolidation with monetary inducements. In 1911 Minnesota passed a bill that encouraged consolidation, and 60 consolidated schools were formed in the first year. Other states had similar experiences.

National leaders were determined to improve agriculture in order to keep food production up. They knew that farmers' institutes and extension work would be important to that effort but realized that the real key was in the primary school system.

An adjunct to the education program was the establishment of traveling libraries. New York pioneered the movement in 1882. Generally, about 100 books were placed at one location for six months and then rotated to another location for six months. By 1899 there were 2,500 traveling libraries, with 115,000 books. By 1908 the number had grown to 5,000 libraries in 22 states, with 600,000 books. The traveling library wagon cost about $175 and was an inexpensive way of providing books for rural people. Most of the books were placed at rural post offices, general stores, or designated homes. They were not placed in schools, because they were closed too much of the time.

Migration from the Farm

Between 1820 and 1910, farm population dropped from 83.0 to 32.4 percent of the total U.S. population, despite the fact that it had grown in absolute terms. The movement from the farm was most noticeable in New England, the Middle States, and the North Central States east of the Mississippi River. The greatest movement was near the urban areas. This was part of a worldwide trend that was first recognized in the Middle Ages. From 1790 to 1900, the percentage of the population in places of over 8,000 people had increased from 3.4 to 33.1. By 1900 Rhode Island had an urban population of 91.6 percent.

Abandonment of cultivated acres, begun in the East by the 1830s, spurted between 1840 and 1860. Most of the 34,000 men from Vermont farms who went to the Civil War decided not to return to them. The movement accelerated again after that war, as more people became aware of another way of life. Ironically, the isolation caused by the 160-acre homestead may have been an accelerating factor, but the trend would have continued regardless.

Worn-out soil, poor returns from farming, and the decline of small-town manufacturing as most of it moved to larger centers were all causes of the migration. The small factories, mills, wagon shops, blacksmiths, etc., had all provided job opportunities for farmers seeking additional income. Many times farm building sites were left empty when people retired to town and none of their children wanted to farm. The land was annexed to a neighboring farm or left to return to forest.

In 1889 the New Hampshire Agricultural Commissioner surveyed 154 townships and found 1,342 abandoned farms with "tenantable buildings." Vermont had more than 1,000 abandoned farms on which the soil was good. Massachusetts had 1,461 abandoned farms, of which 772 had buildings. Similar situations existed in other states. A large number of farmers who remained were not profit oriented and continued to farm as they had in early days. But in most cases the children continued to move to the cities. No matter what was tried, nothing could stop the movement of youth from the farm. By the election of 1896 the returns showed that the economic power had shifted away from agriculture.

The USDA was particularly concerned about the flow from New England farms, because those farmers had the best access to markets, educational institutions, and other "desirable environments." By 1900 some of the most prosperous agricultural states, like Iowa, saw a noticeable drift from the farms. In 1906 New York had 20,000 farms for sale. The great expansion of all lines of business was given as the major cause.

In some respects the movement was good, because where labor-intensive agriculture was still practiced, production would have been higher and prices may have slipped even more. The overall causes, in addition to those mentioned above, were mechanization, the switch to commercial agriculture, urban industrial opportunity, rural underemployment, railroads that favored the larger centers and thus caused industry to move, centralization of most businesses, and social concerns, such as entertainment, better schools, and better religious advantages.

The great fear was that the cream of the country was always the first to leave. Some sociologists suggested that this would have a long-term negative effect on the nation. They believed that the city would not be the great molder of Americans that the farm supposedly had been.

Summary

The period from 1860 to 1914 in U.S. agriculture, when compared to the history of agriculture worldwide, could be called explosive. The stage was set by governmental action through the Homestead Act and all the associated activities of the government, particularly the creation of a department of agriculture. This department expanded into education, experiment stations, farm management, and all that those factors imply.

The Civil War was the catalyst of change. From the boom of those years there was no turning back of commercial agriculture. Industrial society expanded and, along with rising exports, kept increasing the demand for farm products. Unfortunately for the farmers, they always managed to produce more than the expanding market could absorb.

When the frontier was closed and no new land was available, many leaders expressed a fear that we would have to become an importer of food. Action was taken to make sure that an abundant supply of low-cost food would be available, a philosophy that has continued to dominate agriculture. At the same time, the great wave of people leaving the land, which commenced in the 1830s, continued to accelerate and continues to this day.

References

Allen, R. H. "The Spanish Land Grant System as an Influence in the Agricultural Development of California," *Agricultural History* 9, No. 3, July 1935, 127–142.

Andrews, Frank. *Railroads and Farming: Some Influences Affecting the Progress of Agriculture.* Washington, DC: Bureau of Statistics Bulletin 100, USDA, October 19, 1912.

Arbour, Marjorie B., ed. *The South on the March.* Association of Southern Agricultural Workers, S.I., 1953.

Baker, Gladys L., Jane M. Porter, Wayne D. Rasmussen, and Vivian Wiser. *Century of Service: The First 100 Years of the United States Department of Agriculture.* Washington, DC: Centennial Committee, USDA, 1962.

Bauman, Paula M. "Single Women Homesteaders: In Wyoming, 1880–1930," *Annals of Wyoming* 58, Spring 1986, 39–53.

Billington, Ray Allen. *Westward Expansion: A History of the American Frontier.* New York: The Macmillan Co., 1949.

Bizzell, William Bennett. "Farm Tenantry in the United States: A Study of the Historical Development of Farm Tenantry and Its Economic and Social Consequences of Rural Welfare with Special Reference to Conditions in the South and Southwest." New York: Columbia University, Ph.D. dissertation, 1921.

Boss, Andrew. "Forty Years of Farm Cost Accounting Records," *Journal of Farm Economics* (also *American Journal of Agricultural Economics*) 27, No. 1, February 1945, 1–17.

Bowers, William L. *The Country Life Movement in America 1900–1920.* Port Washington, NY: National University Publishers, 1974.

Brooks, Robert Preston. *The Agrarian Revolution in Georgia 1865–1912.* Madison: University of Wisconsin Press, 1914. Reprinted, Westport, CT: Negro Universities Press, 1970.

Chan, Sucheng. *This Bittersweet Soil: The Chinese in California Agriculture, 1860–1910.* Berkeley: University of California Press, 1986.

Cox, LaWanda F. "The American Agricultural Wage Earner, 1865–1900," *Agricultural History* 22, No. 2, April 1948, 95–114.

Danbom, David B. "The Agricultural Experiment Station and Professionalization: Scientists' Goals for Agriculture," *Agricultural History* 40, No. 2, Spring 1986, 246–255.

Danbom, David B. *The Resisted Revolution: Urban America and the Industrialization of Agriculture, 1900–1930.* Ames: Iowa State University Press, 1979.

Danbom, David B. "Rural Education Reform and the Country Life Movement, 1900–1920," *Agricultural History* 53, No. 2, April 1979, 462–474.

Dunbar, Robert G. *The Farmer and the American Way.* New York: Oxford Book Co., 1952.

Fairbanks, Carol, and Bergine Haakenson, eds. *Writings of Farm Women 1840–1940: An Anthology.* New York: Garland Publishers, Inc., 1990.

Fite, Gilbert C. *American Farmers: The New Minority.* Bloomington: Indiana University Press, 1981.

Fuller, Wayne E. *The American Mail: Enlarger of the Common Life.* Chicago: University of Chicago Press, 1972.

Fuller, Wayne E. *The Old Country School: The Story of Rural Education in the Middle West.* Chicago: University of Chicago Press, 1982.

Fuller, Wayne E. *RFD: The Changing Face of Rural America.* Bloomington: Indiana University Press, 1964.

Gray, R. B., compiler. *The Agricultural Tractor 1855–1950 in the United States.* Beltsville, MD: USDA ARS, Agricultural Engineering Research Branch, Farm Machinery Section, 1 and 2, June 1954.

Hall, W. W. "Health of Farmers' Families," *Report to the Commissioner of Agriculture for the Year 1862.* Washington, DC: GPO, 1863, 453–470.

Harpstead, D. D. "Man-Molded Cereal—Hybrid Corn's Story," *Yearbook of Agriculture, 1975.* Jack Hayes, ed. Washington, DC: USDA, 1975, 213–224.

Herscher, Uri D. *Jewish Agricultural Utopias in America, 1880–1910.* Detroit: Wayne State University Press, 1981.

Hicks, John D. *The Populist Revolt: A History of the Farmers' Alliance and the People's Party.* Minneapolis: University of Minnesota Press, 1931.

Hines, Linda O. "George W. Carver and the Tuskegee Agricultural Experiment Station," *Agricultural History* 53, No. 1, January 1979, 71–83.

Holmes, George K. *Supply of Farm Labor.* Washington, DC: USDA Bureau of Agricultural Statistics Bulletin 94, 1912.

Horner, J. T. "The United States Governmental Activities in the Field of Agricultural Economics Prior to 1913," *Journal of Farm Economics* 10, No. 4, October 1928, 429–460.

Jamieson, Stuart. *Labor Unionism in American Agriculture.* Washington, DC: U.S. Department of Labor Bulletin 836, 1945. Reprinted, New York: Arno Press, 1976.

Jones, Allen W. "The South's First Black Farm Agents," *Agricultural History* 50, No. 4, October 1976, 636–644.

Kirk, Marvin Steward. "A Study of the Jews' Contribution to Land Settlement and Land Credit with Special Reference to North Dakota." Fargo: North Dakota State College of Agriculture & Mechanic Arts, M.A. thesis, June 1926.

Kranzberg, Melvin, and Carroll W. Pursell, Jr. *Technology in Western Civilization: Technology in the Twentieth Century.* New York: Oxford University Press, 1 & 2, 1967.

London, Joan, and Henry Anderson. *So Shall Ye Reap.* New York: Thomas Y. Crowell Co., 1970.

Malin, James C. "The Background of the First Bills to Establish a Bureau of Markets, 1911–1912," *Agricultural History* 6, No. 3, July 1932, 107–129.

Marti, Donald B. *Women of the Grange: Mutality and Sisterhood in Rural America, 1866–1920.* Westport, CT: Greenwood Press, 1991.

McConnell, Grant. *The Decline of Agrarian Democracy.* Berkeley: University of California Press, 1953.

Nourse, Edwin G., and Joseph G. Knapp. *The Cooperative Marketing of Livestock.* Washington, DC: The Brookings Institution, 1931.

Periam, Jonathan. *The Groundswell: A History of the Origins, Aims, and Progress of the Farmers' Movement.* Chicago: Hannaford & Thompson, 1874.

Rasmussen, Wayne D. "The Civil War: A Catalyst of Agricultural Revolution," *Agricultural History* 39, No. 4, October 1965, 187–195.

Riley, Marvin P., and James R. Stewart. "The Hutterites: South Dakota's Communal Farmers," Rural Sociology Department, South Dakota Agricultural Experiment Station, Brookings, 1966 and 1980, 1–31.

Ruttan, Vernon A. "Agriculture in the National Economy," *Yearbook of Agriculture, 1963.* Alfred Stefferud, ed. Washington, DC: USDA, 1963, 135–138.

Saloutous, Theodore. "The Agricultural Problem and Nineteenth Century Industrialism," *Agricultural History* 22, No. 3, July 1948, 157–174.

Schafer, Joseph. "Some Enduring Factors in Rural Polity," *Agricultural History* 6, No. 4, October 1932, 161–180.

Stewart, Elinor Pruitt. *Letters of a Woman Homesteader.* Lincoln: University of Nebraska Press, 1961.

Warren, G. F. *Farm Management.* New York: Macmillan Publishing Co., Inc., 1913.

Warren, G. F. "The Origin and Development of Farm Economics in the United States," *Journal of Farm Economics* 14, No. 1, January 1932, 2–9.

Washington, Jefferson, Lincoln and Agriculture. Washington, DC: Bureau of Agricultural Economics, USDA, November 1937.

West, Terry L. *Centennial Mini-Histories of the Forest Service.* Washington, DC: USDA Forest Service FS-518, July 1992.

Woods, Thomas . *Knights of the Plow: Oliver H. Kelley and the Origins of the Grange in Republican Ideology.* Ames: Iowa State University Press, 1991.

Unit IV

1914 to 1954

Some Events and Technological Innovations That Affected Agriculture

1914–1918 World War I

1916 Federal Farm Loan Act

1921 Farm Bloc organized

1924 Successful all-purpose tractor marketed

1926 Commercial hybrid seed corn sold

1933 Soil Conservation Service (SCS)

1935 Rural Electrification Administration (REA)

1939 DDT formulated; arrived in U.S. in 1943

1939–1945 World War II

1947 General Agreement on Tariffs and Trade (GATT)

Introduction

During the period from 1914 to 1954, American agriculture went through a continued and accelerated rate of change. The two world wars greatly impacted the industry. World War I caused a dramatic expansion and then a sharp decline, which began the downward trend in farm numbers that has continued to this day. A slight aberration occurred in the long-range trend during the early 1930s when the government attempted an intensive rural rehabilitation effort, but it was impossible to reverse the inevitable.

The negative impact of the post–World War I era left a lasting impression on farmers into the post–World War II period, and the same mistakes were not repeated. A surprisingly strong world market helped to ease the transition from the wartime economy.

Telephones, radios, automobiles, electricity, tractors with associated machinery, trucks, and airplanes all were adopted by farmers during this period. The rate of adoption was cautious but steady during the early years, but with the economic recovery associated with World War II, it stepped up rapidly, causing a virtual explosion in production per acre and per worker. The second agricultural revolution was at hand as industrialization of agriculture continued.

The government and the global economy affected agriculture in ever-increasing ways during the era. The global economy caused serious gyrations in demand and prices, which gave boom and bust cycles to agriculture. The government's role was aimed at easing the impact of those gyrations, which somewhat tempered the unstoppable tide of farmers leaving the land. The government was much more direct in attempting to control agriculture than in the previous era when it strove to accelerate agricultural development.

Farmers no longer were the major sector in society, but they learned to effectively use the political process to win support from the government. The former governmental philosophy of drift toward some basic problems changed to one of easing the pain of those in farming. At the same time, governmental policy maintained its stance of assuring the consumers an ample supply of relatively low-cost food. The two aims were in conflict and required constant maneuvering to keep all parties involved pacified.

Continued Industrialization

The Impact of World War I

From 1897 to World War I, agriculture experienced one of its best periods. Prices rose slowly and steadily; land values rose gradually, giving farmers a more solid financial base; farming became a more stable business; living conditions improved; tenure and debt problems were satisfactory; and farmers maintained a strong voice in national affairs. Throughout the period, exports provided agriculture with an improved income and funds for the nation to balance its trade deficit. Even though agricultural exports increased in absolute terms from about $0.7 billion annually to $1.9 billion, in relative terms their share dropped from 71 percent to 45 percent. This was a reflection of the continuing rise of the industrial sector.

There was borderline hysteria prior to World War I because of a short crop. Before the war started, U.S. and European leaders acknowledged that the food problem "was serious and constructive action was necessary." The USDA led a campaign to reduce waste in agriculture and to increase production. A single example was the use of the refrigerator car for transporting eggs, which reduced annual losses by $50 million in that product alone. The South was encouraged to diversify so it could avoid having to import food.

The slogan "Food will win the war" encouraged a rapid increase in production. Wheat acreage increased 42 percent from 1913 to 1919, while overall wheat production increased 27 percent and the price per

Plowing and harrowing with Model "T" Fords converted to tractors to meet wartime demand for production, Michigan, 1917. (NAL)

bushel nearly tripled. Government propaganda for meatless and wheatless days, the discovery of vitamins, improvements in transportation and processing, and shifts in food prices all influenced changes in diets in the first four decades of the twentieth century.

World War I caused a permanent 10 percent increase in land in crops and a 26 percent increase in production. It accelerated agriculture's rate of growth and changed world trade channels. The war was a turning point in U.S. agriculture because it started an era of perpetual surpluses and hastened the trend toward an international economy. It brought about a sharp change in rural life because mechanization and industrialization became an integral part of farming. That permanently impacted farming, even though many farmers did not comprehend what had happened.

At the same time, there was no sharp increase in the rate of adoption of technology, as was experienced during the Civil War and World War II, because we were not in World War I that long, little new technology was available, and virgin lands were still being improved. In 1917 farmers planted 22 million acres more than in 1916, and in 1918 another 6 million. By 1919 farmers planted 33 million acres more than the 1910–1914 average, and livestock numbers were up nearly 29 million head over the same period.

The great irony was that production, land prices, and related input costs continued to increase during 1919 and 1920. The price of land in 1920 was five times what it was in 1900. This compounded the problem when a major price decline hit the industry in the summer of 1920. Farm purchasing power dropped sharply, which eventually helped to pull the rest of the economy into the 1929 depression.

Cotton, which traditionally was our leading export crop, did not share the early wartime boom. The cotton market was cut off from Germany, which was a major buyer in 1914 when a record crop was produced. Even with increased U.S. consumption, the total use did not equal prewar demand. Partly because of reduced cotton planting, conditions improved from 1917 through 1919. The 1919 cotton income was not equaled until 1948. In 1920 and 1921, cotton experienced the same drop in prices as the rest of agriculture, plus there was a severe boll weevil infestation.

Despite those negative aspects, the war served as the beginning for a New South. Farming diversified, farm income increased, and the rate of change accelerated. The movement that started in the 1920s provided the foundation for the dramatic change in the post–World War II era. The South and all of agriculture profited from legislative changes during and after World War I, such as the Federal Farm Loan Act, the Smith-Hughes Act, the Farm Credit Administration, the Marketing Acts, and soil conservation.

Writing in retrospect in 1928, the Secretary noted that farmers had answered the call to expand during the war. When world conditions changed, farmers took a hit that should not have been their total responsibility. He continued:

> The situation from which agriculture is still suffering has complex economic, social, and other roots. For these the nation cannot escape its just share of responsibility in that its officials advocated overwhelming expansion of production. . . . Failure to extend . . . help not only

would stamp the United States as ungrateful for the response of farmers to its appeals but would materially weaken the social and economic fabric of the nation.

The Impact of World War II

Historian Gilbert Fite stated that World War II sometimes has been looked upon as an event as important to agriculture as the closing of the frontier. It changed the face of agriculture by wiping out the surpluses of 1939 and 1940. Over 5 million people left the farm between 1940 and 1945, which enhanced the income of those remaining. Improved prices brought the greatest prosperity that farmers had ever experienced to that date. Unlike during the World War I period, farmers were reluctant to increase production unless the government gave them protection for the postwar period.

In 1938 good weather returned after the droughts of 1934 and 1936. The rains encouraged the farmers to start buying equipment after several years of virtually no purchases. Fortunately, 1940 was a very big year for farm machinery production. Dealers' lots were still full in 1942, when farmers somehow produced at what everyone believed to be full capacity.

In 1942 the government restricted machinery manufacturers to 83 percent of their 1940 production levels. Production of milk machines was allowed at 206 percent of 1940 to compensate for scarcity of labor. This proved to be a wise move, for in 1943 the greatest shortage faced by farmers was that of skilled workers. Farmers were encouraged to feed their dairy cows heavily because milk was in short supply. That was the best way to increase production without requiring more labor or equipment.

It was disturbing that the figure for annual milk production per worker on small farms was 25,000 pounds, while on large farms it was 80,600 pounds. Milk per cow on large farms was nearly double what it was on small farms. Worker equivalent in production was 1.4 on small farms and 3.6 on large farms. The urgency of the war would not permit small farms to be displaced, but clearly they required better management to reduce labor needs and still increase production.

In 1939, 35 percent of the farmers produced 84 percent of gross farm sales. Generally, they had the best and the largest equipment. The other 65 percent of the farmers, with 16 percent of gross sales, were encouraged to combine their holdings into units large enough to support a family. This would not only add more to the total production but would also make labor available for industry. Pooling equipment among small farms, particularly in the South, was advised. Farms enlarged as they mechanized, production costs decreased, yields increased, and living standards for farm workers improved.

In 1940 a farm economist wrote that if farming were done to gain maximum production at minimum cost, ample food could be produced with "half the existing farm population." The events of World War II proved that, for production steadily increased while farm population declined. An experience similar to that happened in Europe during the 1950s and the 1960s. This writer interviewed a farmer who was a draft administrator and who said publicly that his township, in an excellent farm community, had twice as many workers as needed. He was not popular for saying so, but events proved him correct. Prior to the war, 85 percent of the farmers in some industrial states worked off the farm. For them the transition was easy.

Farm productivity increased one-fourth to one-third, with a decrease in labor and little increase in machinery. From 1900 to 1939, personal income of those in agriculture suffered because there was no decrease in labor while productivity per worker increased. There was a surplus of labor on farms because the birth rate had not declined with the decreased need for labor.

In 1940 economist Theodore Schultz suggested that the United States could easily get by with 2 million fewer farms. Political leaders were reluctant to face this fact, but the war helped to reverse the program of keeping people on the land and preserving farm numbers. In 1942 a USDA economist suggested that the war would stimulate the growth of large farms and that the trend would persist after the war was over.

To combat the labor shortage caused by World War II, provisions were made to recruit, transport, and house Mexican nationals and individuals from the Bahamas, Jamaica, Newfoundland, and British Barbados. These people were kept mobile so they could be placed where needed. The Secretary of Agriculture was empowered to determine their wages.

World War II brought a decisive change in the role of the farm woman. Jobs were abundant, and she was expected to help wherever she could. One of the greatest benefits, in addition to the change in status, was that the woman finally had money to improve her home and living standard.

The food processing industry replaced much of the processing that was once done in the home. A dramatic change took place in the food industry because of the rise in institutional feeding involving hotels, restaurants, the military, factory canteens, schools, etc. The war started a great increase in women working outside the home, which they continue to do. This created a demand for convenience foods. The canning industry was the first to respond, followed by processors of instant, pre-cooked, heat-and-serve, frozen, and ready-mixed foods, all of which were designed to reduce meal preparation time. Many of these food products were developed to serve the military.

During the 1940s, over 1 million blacks left six plantation states. The South experienced an increase in labor cost for hand picking cotton of from 1¢ a pound in 1930 to 7¢ in the 1940s. From 1940 through 1944, the percentage of black males on southern farms fell from 41.2 to 28.0, and the percentage of black females from 16.0 to 8.1. The number of black sharecroppers fell from 392,897 in 1930 to 73,389 in 1959. At the same time, the number of white sharecroppers dropped from 383,381 to 47,650. Migration from southern farms during the war period was more than double the rate for the rest of the nation.

Prices that farmers received doubled from 1940 to 1945 and tripled by 1951. Fortunately, the cost of inputs did not rise nearly as rapidly, so farmers experienced new highs in net income. By 1947 net farm income had risen to $16.8 billion, about four times that of 1940. Total assets in agriculture rose from $53.7 billion in 1940 to $168.7 billion in 1952, much of which was caused by the rise in farm income rather than by an increase in the price of land.

Mortgage debt was only $7.1 billion in 1953 versus $10.8 billion in 1923. Non–real estate debt was slightly higher in 1953 than it was in 1921. Farmers remained cautious during World War II unlike during the previous war. After the war the distressed sales rate declined to 2 per 1,000 in contrast to 38.8 per 1,000 in 1932.

The big surprise that greatly benefited agriculture was that the postwar downturn experienced after World War I did not recur in the 1940s. In part, this was because the food aid programs of our government sustained the export demand for food. In the long run, this altered the market for agricultural products, because many goods that were shipped under those programs gained a permanent export market.

Lend-Lease, which was started during the war, was continued by the Marshall Plan of 1947–1951, followed by the Mutual Security Act of 1951–1961, Public Law 480, and the Trade Expansion Act of 1962. All of these focused on exporting agricultural products initially aimed at helping alleviate starvation in war-torn nations. Internally, they served to reduce our surpluses and, in the long run, created new markets.

At home both world wars changed long-term eating habits, which affected agriculture. In the first war, the big demand was for wheat, but in the second war, it was for meat products. During 1941 to 1945, the demand for meat products was 15 times greater than in 1936 to 1940.

The Price Decline in the 1920s and 1930s

By midsummer of 1920 prices of farm commodities were declining sharply. By December 1920 the crop was worth $3 billion less than the much smaller crop of 1919 and $1 billion less than the still smaller crop of 1918. As of November 1920 the prices of all crops were 48 percent below those of the previous year and 33 percent below those of spring planting time.

Prices rebounded to a four-year high in 1924–25. The cost of production dropped, while gross farm income rose to $12 billion, up from $9.55 billion for 1921–22. Production was still above the 10-year average, which worked against further efforts to increase prices. In 1925 Secretary Jardine wrote that for the first time since 1920 farm income allowed a "commercial return on invested capital and a fair reward for the farmer's labor, risk, and management."

A survey of farmers indicated that 42 percent felt low prices were the reason for financial problems, 17 percent high real estate taxes, 11 percent high labor costs, 10 percent high freight rates, 6 percent reckless spending, and 14 percent other causes. The Secretary indicated that surplus production was the greatest cause of depressed prices. The only way to face that problem was to control production or increase marketing. He added that the greatest challenge to individual farmers was to overcome the fact that so many of them were so inefficient that their cost of production was double that of the better farmers.

The price of wheat fell from $1.03 in 1929 to $0.36 in 1931, and hog prices fell from $11.36 to $6.14 per head. By 1932 the farmer's purchasing power was less than one-third of what it had been in 1914. Net income in 1932 was only $1.9 billion, compared to $5.7 billion in 1923–1929 and $9.3 billion in 1919. This resulted in a series of strikes.

The strikes of 1932 did not involve as many farmers as did earlier movements, but they were much more intense. Milo Reno was the legal mind in founding the Farmers' Holiday Association, an offshoot of the Farmers Union. In a two-month period the Association staged over 300 protests, chiefly in northwestern Iowa. The net result was that the strikes called national attention to the plight of agriculture.

The prolonged agricultural recession created a new economic and social philosophy, with the federal government playing a greater role in combating the excesses. But the problem of voluntarily reducing production was an impossibility for an industry as diversified as agriculture.

The federal government was the only institution with the power to control production.

Financial Problems

Secretary Henry A. Wallace, speaking in 1938, gave seven causes for the farm problem: (1) the wartime expansion of 40 million acres; (2) the change of the United States from a debtor to a creditor nation, which made it more difficult to export; (3) the displacement of horses by tractors, which eroded the value of horses and released more land for market production; (4) the fact that Europeans worked harder to produce their own food; (5) new competition from Argentina and Australia; (6) higher U.S. tariffs, which caused other nations to shut out our products; and (7) the growth of big businesses that could set the prices of what they bought or sold. Only the expansion in acreage and the switch to tractors were directly influenced by farmers; the other five causes were beyond their control.

Going back to August 1921, the year after farm prices started to decline, the Federal Reserve Board ordered country banks not to renew the notes of farmers and ranchers but to collect them in full at once. This smashed the farmers while the rest of the nation continued to prosper. At that time about half of the total agricultural output could not be consumed by humans, so it was basically wasted unless other uses could be found.

Surplus was the buzz word in agricultural circles. The next most discussed issue among the leaders was that most units were too small to provide a living. High fixed debts contracted during the boom period, followed by the two postwar slumps, one in the 1920s and the other in the 1930s, eroded land prices, which caused insolvency. Overlending, especially on less productive land or to inefficient managers, was an important mistake made by lenders. Another factor was that a large number of farms were located in areas of poor soils and climate or on cutover timberlands, which were not suited to agriculture.

A study was conducted by the USDA covering the period from January 1920 to March 1923 in 15 key midwestern agricultural states. It revealed that 8.5 percent of farmers lost their farms and another 14.5 percent were insolvent but not forced out of business. In addition, it showed that 14.5 percent of tenants were forced out of business, of which 20 percent were insolvent but left to operate farms they were on. The study revealed that 250 banks had failed in North Dakota, 150 each in South Dakota and Montana, and 100 each in Minnesota and Iowa.

The farmers' plight caused some of them to seek management help. In 1924 the Farm Bureau Management Service was organized to serve farmers in Illinois, but the work quickly spread to other states. The immediate results gained from better records were that farmers learned to concentrate on their most profitable enterprises and they enlarged to become more efficient. Next, they adopted the use of commercial fertilizer, followed soil conservation practices, and made better use of farm credit facilities.

The USDA closely watched the migration off the farm during the 1920s and by 1927 reported that the farm population was down by 3 million since 1919. To the government this was encouraging, because an off-the-farm migration was felt to be a partial solution to the problem. A 1929 study by the Department stated: "If the gradual reduction in the number of farms continues, then average individual income will continue to gain somewhat by reason of the fairly stable total [income] being divided among a steadily decreasing number."

Liberty Hyde Bailey advised that farmers should adopt the philosophy of the industrial society and solve their problem by underproducing. He said, "If there is consistent over-production of agricultural supplies, then there are too many farmers. . . . The situation will adjust itself in time, although the adjustment should be much safer if intelligently directed." Bailey did not advocate reducing output per acre or per worker, nor did he recommend that the problem be given to anyone running for political office because that would only complicate things.

In 1933 the Secretary wrote that the basic problem was that agriculture was overcapitalized. It was an overextended industry. He blamed the "outworn homestead policies" because they were partly responsible for bringing too much land into production, especially in the World War I era. "In many cases the area for the homestead was too small for efficient use. Many a homestead would not maintain a family."

Real estate taxes were a major source of agitation to the farmers. In 1913 real estate taxes were 1/10 of their net income, and in 1921 they had risen to 1/3. The actual amount collected doubled during that period. From 1914 to 1929 taxes tripled. Farmers were caught in a tax structure that looked upon real estate instead of income as its cornerstone. Property taxes produced nearly 80 percent of the state and local taxes. This put farmers in a bind. In 1923 only 29 out of every 1,000 farmers paid any federal income tax because their net incomes were so low, but real estate taxes were a burden.

In the early 1920s, the average farm family spent $1,504 annually, of which $634 came from the farm in the form of food, clothing, fuel, and housing. Incomes declined from that date and hit bottom in 1933, when the national average net was only $230 per farm after cash expenses. This allowed for no return to capital, labor, or management. At the same time, the capital in agriculture fell from $79 billion in 1919 to $38 billion in 1933, while the mortgage debt per acre was triple that of the prewar level.

In Iowa, one of the top farm states, taxes absorbed all the cash rent in some years. That was the worst example, but in most states taxes absorbed 50 percent or more of the real estate income. This caused the tax delinquency rate to rise above 50 percent in many good agricultural states. Mortgagees sometimes paid the taxes to keep title to the property. Eventually, real estate taxes proved to be a major cause of farm bankruptcy.

After the death of the Populist Party in 1908, there was no major agrarian protest until 1915, when the Nonpartisan League appeared. The League was founded in North Dakota and concentrated in 13 western states. Its major thrust was to improve farm income via state ownership of industrial enterprises. Without regard to the League, the federal government passed more far-reaching legislation favorable to farmers than had been passed up to that time. This encouraged the League, and it pushed even harder.

One of the major problems in agriculture was that once it became commercialized, it required more long-term capital than was available. Farmers had no choice but to borrow more-expensive short-term funds instead of obtaining less costly long-term credit. Sometimes this forced farmers to turn to questionable sources of credit, which often were frauds. This drained much-needed dollars away from the industry.

Perpetuating a Myth

Prior to the settlement of the United States, the image of farmers as self-reliant landowners was not well established. Even more unusual was the theory that such farmers were "essential to the realization of political democracy." The concept of the yeoman farmer was contested from the start, and during the last half of the nineteenth century it came under increasing pressure as industrialization opened up economic opportunities.

Probably no one did more to create the myth of the family farm than Thomas Jefferson. According to his thinking, such a farm had three basic traits: (1) it was a subsistence operation that bought and sold as little as possible; (2) the farmer did his own work and managing; and (3) he owned his land in fee simple. Those ideals were never followed because from the beginning the nation relied on commercial farmers who produced for the markets.

Three doctrines perpetuated agrarian idealism. First, agriculture was basic and was superior to other occupations because it supplied food and clothing for the nation. It was the original source of wealth and provided the raw materials that other industries needed. This doctrine was a combination of moral reasoning and economic ideas. As late as the 1930s, political, industrial, and labor leaders still agreed with the economic ideas, but the moral doctrine had lost its punch.

The second doctrine of agrarianism was that farming was a way of life and not a business. Because farmers were so close to nature and to God, they were morally superior to persons in other occupations. Individuals in other segments of society generally ignored this alleged moral superiority. This creed was strongest in the South, but it had long-lasting roots in rural states. Supposedly, farmers were more interested in the pursuit of happiness than in money. Most individuals and commercial farmers contended that farming was just as much a capitalistic business as any other enterprise.

The third doctrine was that the nation should remain rural to avoid the growth of cities and, hence, moral decay. Generally, commercial farmers were urbanized enough in attitude that they did not advocate this, but the traditionalists hung tough. Some segments of the media sided in with that doctrine, even though they realized that there were far more people on the land than were needed.

Many individuals espoused that the family farm was the cornerstone of society at the very time that large numbers of farm families were suffering from the tyranny of trying to exist on farms far too small to provide a living. Once the USDA started delving into the economics of farming, it saw the basic problem. A study of 4,100 Wisconsin farms of the 1880s determined that only 3 percent had net incomes of $2,000 over expenses. Only 12.5 percent of the 4,100 farms had net incomes of $1,000 or more. That left 87.5 percent of the farms with an average of $200 net income annually, "potentially a frugal living and nothing more." The bottom 3,600 farms had an average of 100 acres, while the 389 farms with $1,000 or more net income had 270 acres, and the 123 with a net of $2,000 averaged 356 acres. The study concluded that small farms were millstones around the farmers' necks.

The initial limit of 160 acres per homestead, regardless of the soil and climatic variation affecting productive potential, was not practical. But in 1862 the 160-acre limit was acceptable to public opinion. In the 1880s and 1890s, most of the Wisconsin farmers who moved west were those on the poorest land.

Historian James Schideler, in writing of the events from 1919 to 1923, concluded that those years were a "turning point in the great economic, political, and social trends of agriculture. . . . [It] shifted ideas concerning rural well-being, and established definite links of farm policy." A modern agricultural industry came into being. Technological efficiency became a reality, and the crisis of those years forced a need for improved business methods and production techniques.

During the 1920s, the tractor was adopted in farming faster than horse-operated machinery had been accepted earlier. The tractor brought about revolutionary changes in the farmers' manner of living and thinking. An economist for the USDA wrote in the 1920s that the tractor "would do more than any other factor to reduce the differences between farming and manufacturing as occupations and consequently to break down the differences . . . between farmers and urban workers." He added that the end of free land forced farmers to rely on credit more than previously. He saw no reason farmers should not operate on credit permanently just like other segments of industry.

During the 1920s and 1930s, various governmental agencies sought a solution to the persistent farm problem. While factory workers were employed about 2,000 hours annually, the farm work year, except in some areas of the South, was between 2,800 and 3,000 hours. This affected the members of the farm family, because in those decades only 15 percent of the farms hired labor and only 1.6 percent of them had two or more workers. One reason cited for the long working hours was that farmers were too far from their neighbors to visit, so they worked. If horses could have worked longer each day, the work year of farm laborers would have been even longer.

The study noted that farmers were reluctant to change. Agricultural colleges had tried to change them with little success. The farmers' attitudes about themselves and rural social institutions were reinforced by politicians, clergy, artists, poets, and others who stood to gain by appealing to the old, nostalgic ideas. Historian David Danbom added that that was true despite the fact that most of society knew that rural attitudes and institutions were no longer relevant, even though they persisted.

Historian James Malin suggested that one of the reasons there was such a strong myth about the farmer was that farm movement leaders and some farm historians had developed an underdog point of view of the farmer. USDA employees, along with farm organizations, did most of the writing about agriculture and supported the underdog attitude. Not until the 1940s was there a reversal of that trend. By then the industrialization of agriculture was in full progress.

Harold Breimyer, a highly respected agricultural economist, in writing of his early life on the farm, said that he did not enjoy farm work and took little satisfaction from it. He found most of the tasks uninteresting and "devoid of any artistic or spiritual uplift." To him, isolation was the most unattractive feature of farm life. It took its heaviest toll on the farm wife.

The lack of cash was a constant source of irritation between his parents. Farm payment dates caused a household crisis, and the children's savings banks were "borrowed from" many times and seldom repaid. When his father secured a job in a nearby foundry, family finances improved.

His father's job thrust Breimyer into the role as chief farm laborer. His real salvation was the 4-H Club, which opened a new world. "They were part of the cultural emancipation of rural America." One of the happiest days of his life was when he became 19 and could take a job in town.

Breimyer recalled that a good worker with good horses and equipment could farm 75 to 100 acres. However, by the late 1930s it no longer was possible for a family to survive on such a small farm. Farm economist Murray Benedict wrote that a farm too small for efficient production and not associated with off-farm income had little value to society.

Benedict saw no reason for a large farm population. He added, "There exists no proof that farm life is inherently superior to urban life." Necessary food products should come from units large enough to maximize efficiency and minimize cost. The commercial family farm of the early 1940s had that level of efficiency, and it would continue to grow as technology improved. Benedict concluded that extremely large farms probably had some social disadvantages, but the surplus farm population had to leave agriculture or else it would continue to live in substandard conditions.

By the 1940s, the USDA defined a family farm as one which required a full-time operator helped by the family, employed only a moderate amount of outside labor, and provided a satisfactory living for the family. The definition excluded part-time farms, residential farms, farms that did not employ the family, and farms that could not provide a reasonable living. By the late 1940s, the drop in farm numbers surpassed expectations. The USDA attempted to slow the increase in size of farms by limiting federal program benefits but with little success.

Despite the boom years of the 1940s, farmers still did not enjoy equal social and economic opportunities in society. A writer in the 1940s suggested that if farmers wanted to share in those benefits, they would have to enter other occupations. Some people were concerned that the disappearance of the small, owner-operated, family-size farm would pose a threat to democracy. To this a USDA economist answered:

> The romantic appeal of the symbol contrasts strangely with the economic fortunes of the reality.... There was an obvious discrepancy between the ideal and the real. The industrial revolution enabled a smaller farm population to produce more and better food ... releasing workers for relatively more productive employment in industry. Many moved with the trend; too many choose to resist it.

The Unstoppable Tide

The sharp rise in farm family living standards in the 1940s increased the need for additional income, which forced farm expansion or off-farm work. This enabled farm families to close the gap between rural and urban living standards. It was suggested that if we had not had a major off-the-farm migration, the rural United States could have been like the post-bellum cotton South. The irony of the early migration was that it was highest in the areas of greatest commercialization of farming and lowest in the poorest farm areas.

Between 1910 and 1925 the number of farms increased in the western half of the nation, except for the states of Nebraska, Kansas, Iowa, and Missouri. About one-half of the farms in the country in 1930 were in the southeastern quarter. They were very small, and nearly two-thirds of them grew cotton. Fortunately, a great migration of blacks from southern farms took place between 1916 and 1929 and lessened the plight of poverty-stricken small farms. Instead of diversifying, the remaining farmers learned to cope with the boll weevil by using better farming practices, and cotton acreage increased during the years of migration.

In 1921 the Catholic Rural Life Movement was started to benefit farmers and those living in rural districts. In 1923 this became the National Catholic Rural Life Conference (NCRLC). According to NCRLC historian Raymond P. Witte, its chief aims were to see "that farm life . . . [was] held in due honor; [to] endeavor to improve methods of agriculture . . . ; [and] to remove inequalities in the standard of living between city and country." Additional aims were "to care for underprivileged Catholics living on the land [and] to keep on the land the desirable Catholics . . . [then] living there. . . ." This movement was in part to offset the overwhelmingly urban Catholic Church, which was not reproducing itself. At the same time, the rural population was rearing children far in excess of the number required to maintain population. There was a back-to-the-land movement that accomplished little. Had there been a need to return people to the land, the prices for farm products and wages for farm labor would have reflected it.

The NCRLC movement had a precedence in an earlier colonization movement dating from 1879 to 1891. Witte suggested that its purpose was "not to settle Catholics on the land because the rural people reared more children. It was to remove poor Catholics from the evils of urban life." The Church chose the land and supervised the efforts of the settlers. The NCRLC viewed the migration of farm people as an "alarming danger to agriculture."

Secretary Wallace wrote that in 1922 an estimated 1.2 million people left the farm. The movement was uniform throughout the nation and continued in 1923 and 1924. It was composed of two major categories—successful farmers who could afford to retire and farmers in the 20 to 25 age group. A reversal in migration from the farm took place in 1932, when massive unemployment in the cities caused people to return to their former rural communities. Then the Secretary became concerned that such a movement would diminish hopes of making the adjustment necessary to reduce agricultural production to fit demand. The return to the land was most pronounced in areas of the South that had the poorest natural resources. At this point the administration faced the conflict of whether to encourage a back-to-the-farm movement or to encourage the long-range trend of declining farm numbers in an effort to continue to gain from increased efficiency in agriculture. The government's rural rehabilitation program reversed the trend temporarily, and family workers in agriculture increased from 7.7 million in 1930 to 10.7 million in 1935. Government programs worked in opposing directions. One program aided people in returning to the land, and the acreage reduction program encouraged a migration from the farm.

With the advent of World War II, the government's attempt at rural rehabilitation disappeared, and 11.4 million people left farming. In 1950 Calvin Beale stated, "A turning point in American rural life had been passed." Farm residents were no longer the majority of rural people. The rural nonfarm population remained quite strong because the automobile and interurban transportation enabled people to live in rural areas and work in urban areas.

According to Beale, the basic reasons people left the farm were (1) a reduction in labor hours because of technology; (2) the abandonment of the tenant system of row-crop farming in the South; (3) the consolidation of farms to achieve volume production because of low margins; (4) the movement of farm laborers to off-farm residences; (5) a decline in

Tabe 4–1. Trend of Total, Rural, and Farm Population

Year	Number (in thousands)			Farm Population as Percentage of	
	U.S.	Rural	Farm	U.S. Total	Rural Total
1840	17,100	15,218	9,000	69.0	59.1
1860	31,400	25,227	15,100	48.1	59.8
1880	50,189	36,059	21,973	43.8	60.9
1900	75,906	45,400	29,400	38.0	64.7
1920	106,089	51,708	31,974	30.1	61.8
1950	151,132	54,479	23,048	15.3	42.3
1960	180,007	54,054	15,635	8.7	28.9
1970	204,335	53,887	9,712	4.8	18.0
1980	226,545	59,495	6,051	2.6	10.2
1990	246,081	61,656	4,591	1.8	7.4
1994	260,000	N.A.	3,900	1.5	—

Sources: Calvin L. Beale, "Nonfarm Rural America," *Structures '80*, pp. 36–37.
Statistical Abstract of the United States 1993.
Historical Statistics, Colonial Times to 1970.

the birth rate of farm families; and (6) a strong nonagricultural job market with higher and steadier income and shorter hours.

Congress had declined to accept most of the recommendations of the Country Life Commission, but it did provide for the creation of the Rural Organization Service in the USDA. The Department's Rural Life Study gave rise to the discipline of rural sociology. In 1919 the Farm Life Studies Section was created to study rural population migration, tenancy, landlordism, rural town relations, and disability among farm people. A study by this section revealed that the most capable were the first to leave the farm. In 1925 the Purnell Act afforded funds for the experiment stations to research economic and social problems of agriculture. In 1935 the Bankhead-Jones Act provided for expanded agricultural research of social and economic problems. The Housing Act of 1949 allocated $280 million for rural housing, which resulted from research on the condition of rural housing extending as far back as the 1920s.

The chief reason generally given for rural housing lagging behind urban was that farmers historically financed housing and improvements out of current income but land and machinery by debt. When incomes were good, they paid off the debt rather than modernize the house. The saying that best expressed that practice was "A house never built a barn, but a barn will build a house."

Off-farm work has always been a factor in farming, but it has increased steadily since the 1920s. A USDA survey of the 1930s indicated that in one area, half of all small farmers who also worked off the farm received more cash income than the farm paid. "But even with these outside sources of income, the material standard of living of the operators on small farms was not more than half that on large farms of the nation." The conclusion of the USDA economist was that part-time farmers preferred rural living and worked off the farm by necessity. They raised a family on a modest income. They enjoyed the "proverbial freedom of the countryman." Governmental action was urged to make part-time rural work more secure. An alliance of rural and urban industries was needed to achieve this goal. Such a movement was deemed necessary to preserve rural life, which some felt was important to the nation.

One of the best ways for traditionalists to stop the off-the-farm migration was to slow down the consolidation of rural schools. The rural school was looked upon as one of the ways of holding on to the past even though the number of students was declining rapidly.

A study of Kansas in the late 1920s showed that there were 6 rural schools with no students, 15 with 1 student each, 34 with 2 students each, 68 with 3 students each, 132 with 4 students each, and 145 with 5 students each—a total of 400 schools with an average of 3.86 students. Changes had to be made and, unfortunately for the traditionalists, when the migration off the farm speeded up during and after World War II, the last hope to stop consolidation was lost. Improved roads and buses brought rural children into a new way of life.

Changing Roles

After the Homestead Act was passed, some women looked to the opportunity of going west to homestead as a way of shaking the traditions of the East. Whatever their purpose, women had a chance at a different life. Farm families traditionally were large, but after the Civil War, women became desirous of limiting family size and became interested in birth control. Women desperately sought to avoid conception, for in addition to the health risk, more children added to their household tasks. Sometimes women had to run the farms while their husbands worked elsewhere to earn additional income.

A USDA economist of the 1920s wrote, "For success in farming, health, strength, and ability for the wife are almost as important as for the farmer." Secretary Houston added:

> The woman on the farm is a most important economic factor in agriculture.... On her rests largely the moral and mental development of the children, and on her attitude depends in great part the important question of whether the succeeding generation will continue to farm or will seek allurements of life in the cities.

The study uncovered five major concerns of farm women in the 1920s: (1) the long workday; (2) too much manual labor; (3) the need for a better standard of comfort and beauty in the home; (4) more safeguards for family health; and (5) the need for a home industry to produce more income. The next two most mentioned concerns were better shopping facilities and better servicing of appliances.

Neighborhood farm women preparing wool, North Dakota, 1915. (June Tweten)

In spite of the awareness of the worth of farm women, their needs were largely overlooked by those attempting to improve conditions in farming. In the era of World War I, the USDA realized that home management directly related to the success of the farm. Labor-saving devices, sanitation, hygiene, and nutrition were key fields where extension could help the farm wife. A USDA survey of 55,000 farm women revealed that they wanted increased personal income, relief from burdens for them and their children, recognition of the monetary value of their work, and better education for their children. Farm women were portrayed as having freedom and happiness, but the fact that young farm women were leaving in record numbers told the true story. They may have believed in the agrarian ideal, but farm women were encouraging their children to leave the farm.

Studies indicated that farm women averaged 60 hours of work per week. About 80 to 86 percent of their time was spent doing housework, 11 to 18 percent doing farm work, and the balance doing other tasks. All women did farm work, but the smaller the farm, the more they were involved in those chores. In the central states, 89 percent of the women had poultry flocks, which averaged 102 birds. About 93 percent took part in dairy production, 33 percent made butter for cash, and 67 percent were solely responsible for the garden.

Home demonstration agent starting for a meeting, Montana, 1914. (NAL)

Typical scene in a farmhouse basement, with at least 17 different home-canned products, plus baskets of vegetables. It was not unusual for farm wives to can 500 or more quarts of fruits and vegetables. They also canned meat. 1930. (UMA)

Overall, farm women contributed to family income by producing as much as 70 percent of what was consumed at home. This was important, for on a small farm the expense of maintaining the family was one of the major costs of the farm. Studies of midwestern farm women indicated that their contributions were about 45 percent of the net farm income. A 1923 USDA report showed that income from poultry was exceeded only by income from dairy products, corn, cotton, and hogs. Women did nearly all the poultry chores and shared in dairy and hog production.

Women knew that they were crucial to the survival of the farm even though they seldom took part in financial decisions or shared in the profits. In 1920 only 16 percent of midwestern women retained the egg or butter money. This caused them to question the purchase of machinery that had limited use when they could not secure appliances that they used daily.

Technology eventually came to the aid of farm women. Those who started farming in the World War I era felt that electricity was the single greatest contribution to easing their lives. It changed both house and barn labor from hand labor to electricity, but at the same time it made a longer workday possible. The tractor was the next most named innovation by women of that era. Many had started with horses and then worked with the tractor.

Women who started farming in the pre–World War II era commented that getting rid of the chickens, hogs, and dairy cattle caused the greatest change and freed them from some of their most burdensome tasks. The little flock or the little herd could not compete with large-scale poultry, hog, or dairy enterprises. Women no longer had to make butter

Typical rural grocery store of 1910–1940 era. Farmer is trading eggs for groceries. Note the hanging sticky fly catchers and oil dispensers on the left. (UMA)

or cheese, collect eggs, wash the cream separator, or butcher. Commercialization of food production enabled the farm wife to work on the major enterprises or off the farm. Many women became the farm bookkeepers, "gofers," and marketers.

One of the most overlooked and yet most critical contributions of farm women has been that of serving as mediator between father and son (or sons) working together on the farm. Historically, fathers have been reluctant to give up management, while sons have been anxious to take over, even though the sons have relied on capital generated by their parents. Estate transfer has caused many bitter family feuds. It has taken much of the glamour out of the myth of the family farm, and the farm wife most often has been caught in the middle.

The irony of passing down the farm is that in spite of all the struggles, the success rate of a first-generation farm passed down is only 30 percent, and the success rate of a second-generation farm passed down is a mere 8 percent. This does not support the overworked adage that the only way to start farming is to inherit or to marry a farm.

Breaking the Isolation

Technology made it possible to break down isolation and bring the farm family closer to the rest of society. First came the telephone, then the automobile, the radio, and, finally, electricity. The gap between urban and rural residents narrowed. Instead of helping to hold people on the farm, the innovations caused rural youth to leave the farm at a faster rate than ever before.

The Telephone

An earlier discussion related how telephones first came to the rural areas after 1878, but it was not until the mid-1890s that they appeared in any sizeable number on farms. The census of 1920 indicated that 2.5 million farms, or 39 percent of the total, had phones. By 1925 that number had increased to 3 million. Telephone lines were relatively easy and inexpensive to construct, because farmers contributed poles and labor. The rate of expansion declined after the most densely settled areas had phones. Costs increased greatly in the more remote areas. It was not until 1949 that Congress provided federal funds to construct the more costly rural lines. Then, virtually all farms were able to have telephone service.

The Automobile

The automobile appeared in rural areas about 1910. The car fascinated farm people, and it greatly changed their life style. Cars were a financial burden, but bankers often loaned money for them before they did so for tractors. The car enabled farm people to expand their vision beyond the local neighborhood. They were able to go to larger communities for both business and pleasure.

Secretary Houston expressed satisfaction with the car because better roads and cars would encourage the flow of people to the cities. The one thing he did not want to happen was a back-to-the-land movement. A 1919 study indicated that the average grade school was 1.5 miles from the farm; church, 2.9 miles; market, 4.8 miles; doctor, 5.7 miles; high school, 5.9 miles; and hospital, 13.9 miles. Those distances were significant in the days of horses, but the higher speeds of cars greatly reduced travel time. It did not take farmers long to bypass the local general store for better shopping opportunities in larger cities.

Roads

By 1920 over 30 percent of the farmers had automobiles. This brought increased pressure for better roads. Local governments could not provide them, and it soon became obvious that a national network was necessary. In July 1916 the Federal Aid Road Act was passed, which became the backbone of the road development program. Under that act the Office of Road Inquiry was established within the USDA. Funds were provided for rural post roads on a matching basis with state and local governments.

In the early years, 79 percent of the roads had a sand, clay, earth, or gravel surface, and 21 percent had some form of hard surface. Each year the percentage of hard-surfaced roads grew. Starting in 1933 the federal government provided a bigger share of the total cost, and road building greatly increased. Road building lagged behind the demand, so the government advocated road construction as a way to put people back to work. In 1932 the USDA spent 69.33 percent of its budget for road building. This was justified on the basis that farm-to-market roads were critical to the entire society, not just farmers.

One of the interesting sidelights of the extended road system was discussion about reducing the number of counties to lower the cost of government. In 1918 when the first county merger took place, reduction in cost was realized. Counties platted in the days of horse transportation were no longer valid in the days of telephones, radios, and automobiles, but little progress was made in the movement to reduce their number.

The Radio In 1900 the Weather Bureau started broadcasting weather reports by wireless telegraph. Next, the Coast Guard started sending messages to ships at sea. In 1914 the Bureau sent wireless messages to operators throughout North Dakota, who then reported to farmers by telephone. By 1915 the USDA sent wireless market reports on some commodities to major markets. In 1917 the Local Market Reporting Service was established to send telegraphic news about markets and prices. The USDA felt that the more information available, the more confident people would feel about the market system.

On January 3, 1921, a radio station at the University of Wisconsin broadcast the first weather report. The following May 19, station KDKA in Pittsburgh gave the first market news broadcast. In June it announced regularly scheduled market reports by a USDA market reporter. In March 1923, KDKA employed the first full-time farm broadcaster. Radios caught on at once, even though many farm families struggled with batteries to power them. Early radios were expensive but were purchased as soon as money was available. The radio shattered the isolation farm families had long endured by providing both news and entertainment. It was more popular than the automobile.

In 1922 Iowa State College station WOI broadcast regular market reports, a station at Greeley reported the Denver market conditions, and the Chicago stockyards started daily market broadcasts. By 1924 about 500 stations were operating in the nation, of which several hundred emitted agricultural reports to farmers' 370,000 receiving sets. By 1926 the number of receiving sets had increased to nearly 1 million, of which 19 out of every 20 were battery powered. By 1929 there were 2 million farm radio receivers.

A banner program commenced in 1928 with encouragement from the USDA when the *National Farm and Home Hour* was first broadcast. It was aired six days a week until 1944, when it was discontinued because stations wanted to sell time and the USDA would not allow sponsors. The program resumed the following year but was no longer affiliated with the USDA. Allis-Chalmers sponsored the program in 1945 and continued to do so until 1960. It was the most popular farm program. *Aunt Sammy's Daily Housekeepers' Chat for Farm Women, Noontime Farm Flashes,* and the *U.S. Farm Radio School* were the other leading programs.

In 1944 about one-third of all radio stations had farm and ranch directors. They organized into a farm broadcasters' professional association. Much of their information came from state departments of agriculture, agricultural colleges, extension specialists, and the USDA.

In the 1950s the number of radio stations declined as the number of television stations increased. About 115 hours of farm programs were aired each week, but agricultural advertisers were slow to use TV until they learned how effectively large food advertisers were using it.

Electricity

From 1882, when the first central generating system was created, until 1935, less than 15 percent of the farms had been wired for electricity. Once it was established that electricity could be transmitted as far as 100 miles, the National Electric Light Association created a committee to determine if farms could be electrified. In 1906 the first known rural line was constructed at Hood River, Oregon. By 1912 Delco light plants were simplified enough that they were practical on farms. The Delco systems were driven by wind power or gasoline engines to charge the batteries and required constant supervision.

In 1914 the first known nonprofit electric cooperative power association opened at Granite Falls, Minnesota. Farmers built the line and purchased power from a local utility. By 1919 about 100,000 farms received electricity from power lines. By then other electric cooperatives had been founded near Webster City, Iowa, and in southern Idaho. The seeds of rural cooperative lines were sown.

In 1919 the USDA conducted a survey of users and determined that electricity would save farmers' wives at least 25 hours of labor per week. The savings were as follows: 2.75 hours in butter churning, 3.83 hours in separating cream, 4.46 hours in filling and cleaning lamps and lanterns, 3.52 hours in using the washing machine, 2.76 hours in ironing, and 7.32 hours in other applications. Another 10.5 hours could be saved in pumping water and 30.5 hours by using a milking machine. In all, it was estimated that electricity could save nearly 10 hours of labor per day. No credit was given for the better lighting and added comfort.

This prompted the USDA to establish a six-mile trial line at Red Wing, Minnesota, to determine the full benefit of an electrified farm and home. The good news spread rapidly, and all the neighbors wanted to get in on the experiment. Electricity was appealing, and farmers quickly applied for the service. In 1929 about 388,000 farms had power from central stations, in addition to those using private generating units.

The biggest handicap was the $2,000 to $3,000 cost per mile to establish power lines. Private companies were reluctant to take the risk. Between 1925 and 1927, Gifford Pinchot, a strong advocate of rural electricity, had a survey taken that indicated if everyone in the rural areas signed up for electricity, the cost factor could be reduced by $300 to $1,500 a mile over previous estimates.

To promote rural electrification, the Committee on the Relation of Electricity to Agriculture was organized in 1933 with support from the American Farm Bureau Federation and the Grange. The government responded positively, and in 1935, as part of the Emergency Relief

Appropriation Act, $100 million was set aside for rural electrification. The Rural Electrification Administration (REA) was created. Constructing rural power lines became part of a general program of unemployment relief.

Then came the big surprise. Many farmers resisted because they felt that they could not afford the service. A $5 share of stock in the cooperative was required. Many had difficulty raising the money, and others were reluctant to sign because of the government's involvement. Once construction started, the REA expanded rapidly from 11 cooperatives with 693 consumers in 1936 to 629 cooperatives with 549,238 patrons in 1940. Construction was delayed during World War II, but by 1945 there were 848 cooperatives with 1,409,000 subscribers.

Electricity was the biggest innovation in rural America since the automobile and the radio. The most popular appliance was the electric iron, and battery-operated radios were quickly replaced by electric models. REA teams traveled around the country demonstrating new appliances. A group wiring plan was implemented that brought the cost of wiring a farmstead down to only $55. By 1950 electrical service was available on 78 percent of farms, and by 1960 the figure reached 97 percent. At that time 1,000 REA cooperatives had 1.5 million miles of line serving 4.8 million customers. In 1973 the REA program was broadened to cover a loan program for Rural Telephone Cooperatives. Within a few years, everywhere but the most remote farms had electrical and telephone service.

Labor in Modern Agriculture

Labor Shortages of World Wars I and II

The historic shortage of skilled agricultural workers continued into the twentieth century. European immigrants filled the demand for workers in the North until shortly before World War I. Most blacks were denied a chance for an education, which limited their economic opportunity to do farm work in the South. By the end of the great migration of blacks between 1916 and 1929, nearly a million of them had left the former plantation states for jobs in the North.

Because of the excess of underemployed workers in agriculture, the migration from the farm did not handicap agricultural production during World War I. Bumper crops were produced with a labor supply that the USDA estimated to be about 37 percent short of needs. When World War I ended, many veterans did not return to the farm. Workers considered farm employment less desirable than urban jobs for several reasons: irregularity of farm employment; consistently lower wages; the fact that a vast majority of farm hands were expected to remain single; and limited chance for advancement. Most agricultural workers no longer had illusions about using the "agricultural ladder" to become farmers.

The percentage of farm laborers fell from 87.1 of the total labor force in 1820, to 47.5 in 1870, to less than 1.5 in 1994. Despite the extensive rural rehabilitation program by the government, which resulted in the creation of over 500,000 farms, a massive migration from the farms commenced in the late 1930s. Before that migration slowed down in the 1970s, more people left the farm than came to our shores in the 140 years from 1820 to 1960. The agricultural revolution worked miracles in freeing people from farming for other employment. Despite the exodus, underemployment persisted in the countryside.

During World War II the labor supply was still adequate on farms, but the quality was greatly reduced. Many of the remaining workers were un- or underemployed prior to the war. To keep workers on the farm, by 1942 wages were increased to 161 percent of the 1935–1939 average. However, wages were still well below industrial pay and less than what government leaders felt was satisfactory.

Japanese—Immigrant to Owner

The Japanese started coming to California in the 1890s. First, they became workers in industry, but eventually a large percentage of them migrated to fruit and vegetable farming. The Japanese always held agriculture in high esteem. By 1910 they were heavily involved in agriculture for the following reasons: (1) they were willing to work for

Hand picking potatoes in North Dakota, 1939. The average worker could pick 60 to 70 bushels per day; top workers, 100 bushels. Wage rates ranged from 3¢ a bushel in the 1890s to 15¢ by the 1940s. (NDIRS)

lower wages to establish the necessary contacts; (2) there was a labor shortage in agriculture in the early 1900s, and they were willing to take the available jobs; (3) most immigrants were single and were free to move from one job to another as the season required; (4) the organization of Issei workers helped both them and their employers; (5) Japanese labor agents provided a cooperative plan for the workers; (6) the agents did bookkeeping for the workers and provided a guaranteed supply of workers at a set price; (7) through the contract system many of the workers were able to secure leases on land and eventually become owners; (8) the Japanese willingly pooled assets and formed partnerships to start farms; (9) they were superior farmers and often outbid others for rented land; and (10) packers and commission firms frequently financed Japanese farmers because they delivered quality products.

By 1920 most Japanese had left the ranks of farm labor because either they were operating leased land or they owned farms. They improved their profits by packing and distributing fruits and vegetables. Before long they controlled about 25 percent of the wholesale and 50 percent of the retail outlets in some market areas. By 1941 the Japanese produced 30 to 35 percent of the California commercial truck crops and controlled much of the wholesale and retail distribution of fruits and vegetables.

Farm Workers on Larger Farms

By 1930, 56.1 percent of all farm workers were on farms with two or more employees. In some states that figure was as high as 82 percent. By 1935 one-third of all farm workers were on farms that had four or more workers.

Some states established labor exchanges or labor bureaus to standardize wage rates. This depersonalized labor relations and widened the gap between employer and employee, for the hiring was done on an industry-wide basis rather than by individual employers.

The farm labor picture changed sharply during the 1930s, because of the depression, government programs, the drought, and technological improvements. All of these tended to displace small- and medium-size farms, particularly tenant farms. Many of the displaced farmers were called "Okies" or "Arkies." They added to the surplus labor supply, causing widespread discontent that led to some intense strikes in agriculture.

Early Agricultural Strikes

Prior to the 1930s, unionism and strikes in agriculture were limited. Some collective action had taken place in areas of large-scale specialized farming, but in the 1930s nearly every state experienced some sort of a farm strike. Most were small, sporadic, and scattered. It was difficult to get agricultural workers to agree on any type of concerted action, because most were seasonal workers with an immediate need problem. The turnover rate was high, the workers were unskilled and very mobile, they represented diverse ethnic groups, and they included women and children as well as men.

The best-known early strikes were conducted by members of the Industrial Workers of the World (IWW), most of whom were floating seasonal workers. They were employed in mining, lumbering, railroading, and farming. The peak of their campaign came in 1915, when they held strikes and caused violence in the midwestern Wheat Belt and in some forestry areas. When World War I started, the federal government suppressed their activities. The IWW strikes were followed by at least 10 major strikes of central California farm workers between 1919 and 1921. Some of these were spontaneous, some were led by remnants of the IWW, and some were led by the American Federation of Labor (AFL).

In 1927–28, melon pickers and shed packers in California, beet workers in Colorado, and greenhouse and nursery workers in Illinois conducted strikes. By 1933 over 56,000 workers had been involved in 61 strikes in 17 states. By 1939 over 275 strikes involving 177,788 workers in 28 states had taken place. California alone had 140 strikes encompassing 127,176 workers. Some of these by the Cannery and Agricultural Workers Industrial Union (CAWIU) were the bloodiest strikes in American agricultural history.

One of the most significant strikes took place in 1933 by about 15,000 mostly Spanish cotton pickers. Public sympathy was for the strikers. The government became involved, and the strikers received a wage increase. By 1934 strike activity in general was reduced because wages had been raised, many workers disliked the fact that Communists were involved in the process, there was disagreement among the leadership, and many top leaders were arrested under the California Criminal Syndicalism Law.

Unionism

The first attempt to organize farm workers came in 1903, when immigrants from Mexico and Japan struck to obtain increased wages and to eliminate contractors from controlling their work assignments. The second effort came when the IWW attempted to organize farm workers nationally. Most of its members were workers on large wheat farms in the Midwest, workers on fruit and vegetable ranches in California, and lumberjacks in the Northwest. In 1922 a third effort to unionize was attempted when the grape and cantaloupe pickers of California banded together. That was followed in 1927 and 1928 when Mexican contract workers organized into the Mexican Mutual Aid Society (MMAS). They submitted demands to the growers and were answered by arrests and deportation, but they succeeded in forming a stable union. Within a year the MMAS had 20 locals and about 2,500 members, but then it quickly declined. It is significant that the MMAS is recognized as the beginning of trade unionism in agriculture.

In 1929 the Communist Party of the United States created the Trade Union Unity League (TUUL) to organize farm workers and others who were not union members. Unfortunately, leaders of the TUUL took control and the growers cried, "Bolshevism," which virtually was a death sentence to the movement. By 1934, after the truck crop and cannery workers were organized into unions, the TUUL went after the highly capitalized, industrialized, specialized dairy industry in the Los Angeles milkshed. The TUUL conducted a series of strikes in the dairy industry. Again the TUUL's connection with the Communist Party hurt its cause. Some dairies were forced to use armed guards to protect nonunion workers. In 1935 the TUUL dissolved and union activity shifted to the San Francisco milkshed under the AFL Brotherhood of Teamsters. By 1939 the Teamsters had 80 percent of the workers signed up.

In 1935 the AFL created a union of seasonal workers by associating them with the canning and packing unions. Many locals were formed, and by 1936 the union had an estimated membership of 7,500. In 1937 representatives from 52 locals in 11 states were chartered by the Congress of Industrial Organizations (CIO) into the United Cannery, Agricultural, Packing, and Allied Workers of America.

For most of the 1930s, agricultural labor disputes centered around the problem of which union should control California agriculture. The AFL eventually won, because the Teamsters were well organized and because they transported most of the commodities in the food chain.

From then on, unions have remained a part of the labor scene in California agriculture.

The Bracero Program

During World War I the government encouraged the importation of immigrants from Mexico to ease the labor shortage. Once the war was over, Mexicans continued to come. From 1920 to 1930 their numbers increased from 121,176 to 368,013. At the same time, 30,000 Filipinos were permitted to enter, and a large number of failed dust bowl farmers went to California. This created a surplus of workers, but farmer organizations and employer organizations opposed any attempt to restrict immigration from Mexico. Much of the reason was that both groups preferred Mexican to white workers.

In 1940 Texas cotton growers joined the clamor for more Mexican immigrants (braceros) because the migrants that they had employed were going north to higher-paying jobs in the sugar beet fields. In 1941, with the passage of the Selective Service Act, the fear of a labor shortage spread throughout farm areas of the southwestern states. Growers from Arizona, California, and Texas all appealed for relaxation of immigration laws to let braceros enter. Millions of Americans were still unemployed and on relief, so labor was still available. However, by April 1942 a labor shortage had developed. Organized labor opposed any importation but backed down rather than hamper the food production program. Even Mexican-Americans, many of whom had not bothered to become citizens, objected because they feared importation would keep wages low. Organized labor had one requirement—the braceros had to return home after each crop season.

In August 1942 an agreement was formalized between Mexico and the United States. The Mexican government remembered all the expense it had had in transporting destitute workers and their families back to Mexico during the 1930s. These immigrants had come at the beckon of the United States during World War I. Mexico had to be assured that exploitation would not take place and that no Mexicans would be returned penniless. A portion of every worker's paycheck had to be sent to the Agricultural Credit Bank of Mexico for a worker's savings fund. The exchange of individuals was highly regulated by both governments.

Some U.S. farmers did not like the contractual agreement because they felt that there were too many restrictions and regulations. On the other hand, labor leaders and Mexican officials felt it was too favorable to large farmers, enabling them to hold wages down. After World War II, the growers convinced Congress to extend the bracero program. They felt that the braceros would not unionize. However, in 1947 the National Farm Labor Union (NFLU) was founded with the help of the AFL. That year the NFLU brought a strike against a major California grower. This strike gained national attention but ended in 1950 without any gains for the NFLU. In 1949 the NFLU had a two-week strike against the cotton growers and won its demands. In 1952 it again won a wage increase and other benefits.

Much ground was lost with the start of the Korean War in 1950, when the number of braceros allowed to enter was greater than ever, causing the NFLU to lose much of its effectiveness. By 1957 over 436,000 workers had crossed the border under the bracero program. In 1965 the bracero program ended. Since then Mexicans have continued to cross the border on an individual basis.

The Global Economy

U.S. agriculture has always been part of the global economy. From our earliest times agricultural exports provided funds that were used to develop much of the rest of our economic base. As the world continues to shrink, foreign trade becomes an ever-increasing part of the economy. Each nation is subjected to gyrations in trade that it cannot control. Agricultural exports experienced a sharp increase from 1910 to 1920, an equally sharp decline from 1921 to 1936, a strong upswing from 1937 to 1951, and a sharp decline from 1952 to 1954.

When World War I ended, the United States was the world's leading exporter and importer of agricultural products. Our exports exceeded the total value of those from all other nations of the world. The United States had 4 percent of the world's farm workers but produced nearly 70 percent of the world's corn, 60 percent of the cotton, 50 percent of the tobacco, 25 percent of the hay and oats, and 20 percent of the wheat and flax seed.

Exports represented 15 percent of farm income in the five years prior to 1914, rose to 18 percent during 1914 to 1921, and then dropped to a low of 8.4 percent from 1934 to 1937. The immediate downturn in exports came when the United States discounted credits to European nations, which caused them to cut their imports. World War I brought a change in world purchasing patterns. This was caused in part by an increase in production in other countries; in part by our switch from a debtor nation to a creditor nation; and in part by a change in debt and tariff policy, which put our agricultural exports at a disadvantage in the world market.

In the 1920s the major farm organizations cooperated to secure important tariff legislation and obtained passage of the Emergency Tariff of 1921, the Fordney-McCumber Tariff of 1922, and the Hawley-Smoot Tariff of 1929–30. Foreign demand dropped sharply during this period.

The farm organizations wanted tariffs to prohibit imports that were produced by cheap foreign labor. (No thought was given to the productivity of that labor, so it may not have been cheap.) Farmers failed to realize that high tariffs blocked imports and thereby raised domestic prices.

The Hawley-Smoot Tariff of 1929–30 was the result of a campaign pledge in 1928 based on the belief that agriculture needed to rely on the domestic market and that it would profit from high tariffs. The assumption was wrong on both counts. Secretary Hyde noted that the share of agricultural exports had been dropping for 50 years but that in absolute terms, except for the depression years, they were still increasing.

In the 1930s world production grew faster than population, so it appeared that exports would not expand. The Hawley-Smoot Tariff, the

Table 4–2. Average Dollar Volume of Farm Exports and Farm Exports as a Percentage of All Exports

Years	Dollar Volume (in billions)	Percent of All Exports
1900–09	0.917	58.0
1910–19	1.90	45.0
1920–29	1.94	42.0
1930–39	0.765	32.0
1940–49	2.42	22.0
1950–59	3.53	22.0

Source: *A Chronology of American Agriculture, 1776–1976.*

highest in our history, sent a signal to other countries that they could not sell to us. They countered by reducing their purchases from us. Protectionist policies remained in force until 1934, when general trade liberalization policies were written by many nations. At that time Secretary Henry A. Wallace did not think that exports would increase because so much of the exports of the peak period were supported by U.S. grants. He stated that we had at least 40 million acres of surplus production and that we had to reduce our tariffs if we wanted to increase our exports. Wallace felt that exporting was vital to agriculture.

As World War II drew nearer in Europe, exports increased. In 1939 Congress eased its opposition to dumping, and exports rose rapidly. They declined briefly when the war broke out, because that sealed off the continental market. But soon there was a rapid rise in exports, which accelerated with our entry into the war. Demand continued strong after the war because of starvation in much of the world.

In 1947 the United States and 22 other nations created the General Agreement on Tariffs and Trade (GATT). Its general purpose was to lower tariffs and reduce trade restrictions among nations in an effort to avoid the mistakes of the 1920s. The Farm Bloc objected to GATT and did its best to undermine it. An escape clause on quantitative restrictions was permitted to appease our farmers. Because of our high support prices, export subsidies were necessary to maintain our export markets. At the same time, our high supports encouraged other countries to export their products to us. Our farm program was in conflict because price supports were used to control production, but without restrictions we created even greater surpluses. Fortunately, for most of the 1940–1951 period, the market outpaced production.

Government in Agriculture

The End of a Policy of Drift

Until the free land frontier came to an end, the chief theme in agricultural policy was exploitation and settlement by those who tilled the soil. After the Civil War, the emphasis began to shift to technology and education to make farming more scientific. Farmers were encouraged to produce for export, which was vital to the development of the rest of the economy. However, when the food-to-human ratio changed in the early 1900s, some of the more foresighted realized that a better long-range policy had to be developed.

The Country Life Commission of 1908 was created partly out of concern that food supplies might not always be in surplus. Congress was unconcerned as long as farmers had not made their troubles known to the legislature and retained the illusion that agriculture could handle its own problems.

In 1914 the USDA attempted to get farmers and rural communities organized for their own benefit. It suggested an organization for agriculture similar to the chamber of commerce. Again, the USDA learned that U.S. farmers were more individualistic and less organized than other sectors of our society. But the USDA did not give up, and in 1917 it called a meeting in St. Louis of all farm organizations to outline the food needs of the country. A by-product of that meeting was the creation of the National Board of Farm Organizations and an early attempt to establish a national rural policy.

Farm economists of the late 1800s and early 1900s had predicted that the nation would soon have to rely on industrial exports to purchase agricultural products from other nations. As late as 1921 the Secretary asked whether the nation should embark on extensive reclamation projects to keep up with consumption or whether it should depend upon imports. He seemed unaware that acreage and production had expanded because World War I had altered the outlook.

That all changed within months, and one thing was sure: farmers could not make the necessary readjustment once the export demand shrunk. Farmers were in financial distress and produced without regard for the demand. The USDA attempted to help farmers by forecasting demand. Greater emphasis was placed on the new science of farm management to determine cost of production, because the USDA recognized that the number one reason for farm failure was poor management.

Agriculture was in a twilight zone, passing from an era of three centuries of drift to one where measures had to be taken to control the gigantic productive potential that had developed. Not only was there a complete lack of policy, but there were no farmers in any position of national leadership. This was a reflection of the complete disorganization of agriculture.

Founding of the American Farm Bureau Federation

Farm prosperity evaporated when the agricultural price index dropped from 234 in June 1920 to 140 in December and to 115 by December 1921. Before that crisis occurred, a significant event in agriculture had taken place in 1919, when the American Farm Bureau Federation (AFBF) was created. One of its chief aims was to formulate a national farm policy.

Leaders were quick to sense that the AFBF could become an agent for uniting farmers and their organizations on behalf of a national program for the benefit of agriculture. In 1920 the AFBF called a grain marketing conference in an attempt to unite the grain interests into a single front. Aaron Sapiro, who had been involved with cooperative commodity marketing in California, encouraged a nationwide commodity cooperative. The United States Grain Growers was formed. Next followed attempts to consolidate livestock producers into a single cooperative. Then the American Cotton Association and the Burley Tobacco Growers Cooperative were proposed. In all cases, the AFBF provided the start-up money to found the cooperatives and some initial capital. Both marketing and supply cooperatives were formed.

The number of cooperative marketing agencies doubled in five years. The Capper-Volstead Act of 1922 gave legal status to cooperatives. This was followed by the American Institute of Cooperation in 1925, the Cooperative Marketing Act of 1926, and the National Cooperative Council in 1929. All were aimed at developing stronger marketing programs.

None of the attempts to improve prices by forming cooperatives were effective. Farmers signed restrictive contracts and agreed to market only through their association. However, in no case did enough farmers sign over sufficient production to affect the price. When the efforts of the AFBF in grain and livestock failed to accomplish the federation's objectives, the all-out campaign was lessened. The AFBF's marketing efforts peaked in 1923 and then subsided.

The first major effort to syndicate agriculture failed because the AFBF did not have the ability "to finance and the power to coerce." Not all was lost, however, for supporters of the cooperative movement learned that to succeed, local cooperatives needed large-scale cooperatives to compete with major corporate enterprises. Some farmers opposed that move because they believed they would lose their voice in the cooperatives.

Attempts to Forecast the Market

By 1921 most farm leaders realized that making agriculture more scientific and businesslike might not solve all the problems. It became clear to agricultural educators and extension people that farm problems were as much social and economic as they were technical. Farming was too overcapitalized, and farmers themselves were responsible for that problem. A land-grant college president suggested that some farmers were unable or unwilling to learn and to work with others. They were inefficient and should be eliminated from farming because they were doing no good to themselves or to society. They, however, were the ones who did the most complaining.

The problem of too many farmers did not disappear that easily, and land-grant leaders pointed out that having 25 percent of the population in agriculture was a calamity. Progress in agriculture could come only if the surplus farm population could be transferred to other activities. More economists and social scientists were employed by the experiment stations and the USDA to work with the problem.

A 1923 study of market reports stated: "[M]any [farmers] do not read crop reports or farm papers. . . . Many who read farm and city papers containing crop and market reports do not realize their significance and value and consequently are indifferent to them." The report continued that many farmers "do not know how to use such information in connection with their own business."

The above is remindful of a note that a longtime agricultural specialist had inside account books that he distributed to his clients: "There are those who do not know, they do not know that they do not know, they do not care that they do not know, so let them be, for they are happy."

Henry C. Taylor, the father of farm management, advised that forecasting was necessary to minimize the swings in production and in prices. In 1922 the first pig report was made, in cooperation with the rural mail carriers, who collected survey forms from 200,000 farmers. A 49 percent increase in intentions was shown, so the report was given wide publicity, and the increase was only 28 percent.

The USDA had given its first acreage estimates in 1910 for 13 crops, plus condition reports on 23 crops and 5 species of livestock. By 1920 reporting had increased to acreage estimates for 29 crops and condition reports on 44 crops and most livestock. The 1922 pig report mentioned above was the first one based on actual sampling. That kind of sampling increased, as did the commodities covered, so that by 1939 a total of 141 crops and 15 kinds of livestock were being reported on. Information gathered from the farmers was compared to census figures and data collected from stockyard companies, packer buyers, railroads, and brand inspectors.

Yielding to the demand for more information, President Harding gave permission to hold an agricultural outlook conference in early 1923. This conference had representation from industry, labor, the farm press, state governments, and farmers. The results of the conference were (1) it called attention to the plight of the farmer; (2) it improved the image of the USDA as a service bureau for farmers; (3) it broadened the scope of the work of the USDA; (4) it gave farmers an opportunity to air their grievances; (5) it affirmed that farmers expected the government to use its power to act in their behalf; and (6) it was the catalyst for the "Equality for Agriculture" movement. The conference leaders hoped that if farmers knew the market demands, they would cut production. The theory did not work in practice.

In the early 1920s, Cordell Hull, who later became Secretary of State, wrote that farmers would have to learn about tariffs or the industry would soon be in "a state of permanent decay." He continued that no tariff could help an industry that constantly produced a surplus that had to be sold abroad. Hull pointed out that existing tariffs hurt farmers by (1) increasing production costs; (2) increasing the cost of living; (3)

increasing transportation rates; (4) decreasing exports; and (5) decreasing property values. He led the reversal of the protectionist policies and helped to liberalize trade relations after 1934.

The Extension Service

The Smith-Lever Extension Act, sometimes called the Cooperative Extension Act of 1914, enabled the federal and state governments to cooperate in passing on information. The experiment stations were the major source of information for the extension service. Secretary Houston sensed that the population growth of the cities was far outstripping that of the rural areas. He hoped that the extension service might stem that tide. With the advent of World War I, Congress passed the Food Production Act aimed at all-out food production. The number of extension people was increased to 5,000. In addition, there were 1,300 home demonstration women. By the late 1930s, 191 black men and 268 black women were working in the South for extension. By 1953 there were 846 black extension employees.

In 1921, 15,000 students were enrolled in the 41 agricultural colleges. There was a big demand for agricultural graduates because the Smith-Hughes Act of 1917 created vocational agriculture mechanical arts courses for high schools. By 1926, 12 states required agriculture courses in high schools, and 19 states required agricultural education in elementary schools.

By the 1930s, the USDA was one of the largest governmental agencies in the world and was the most powerful one for a single occupational interest. In 1931 it had 25,000 employees, of which 15,000 were in the USDA and 10,000 were in the cooperative groups. The research and education programs were very beneficial to the top one-third of the farmers. On the other hand, they were of little value to the bottom half of the farmers. Secretary Wallace said, "Poor education and grinding poverty have closed their minds." The extension people could not reach those who did not want to be reached.

The 1930s produced the swiftest and most far reaching changes in agricultural policy viewpoints of any decade to that time. The decade was one of transition. It also was a period of great mechanization, and the innovators were quick to adapt to machinery. Others waited for the prosperity of World War II to mechanize. The 1930s saw a growing interrelationship between agriculture and the rest of the economy. The agricultural infrastructure was growing, but it was still not widely understood by farmers. The problems of surplus production, excess people, and overcapitalization forced the need for a long-range policy for agriculture and rural areas.

The editor of the 1940 *Yearbook* wrote that agricultural efficiency meant fewer people in agriculture but that inefficiency was tolerated through the medium of subsistence farming to take care of those displaced by technology. They were recognized as the disadvantaged farmers. He also noted that one-third to one-half of all farm families contributed little to the commercial supply of food and raw materials. "They cannot compete in the market and most of them live in poverty and produce relatively more children than other groups." We still had over 30 million people living on 6.1 million farms.

The Farm Bloc

By 1921, for the first time in its history, the USDA faced the problem of having to curtail agricultural production. This was a dramatic reversal for an agency that was developed to show farmers how to do a better job of producing. The USDA probed every aspect of agriculture to determine how to better the lot of farmers. The founding of the AFBF and the regulatory acts of the government during World War I were the beginning of an agrarian revolution as far as concepts of agriculture were concerned. Within one year of its founding, the AFBF had acquired a membership of over 1 million. That was far beyond what any previous organization had succeeded in doing.

In 1917 the National Farmers Union sent a lobbyist to Washington, who shared an office with members of the National Board of Farm Organizations. In 1919 the Grange sent a full-time lobbyist to Washington. The AFBF employed Grey Silver as its lobbyist. He was soon known by most members of Congress and became one of the most effective lobbyists there. All these groups joined hands, and during a special session of Congress in 1920, it became clear that an organized agriculture could exert strong nonpartisan influence on Congress. Due to the work of Silver and the other lobbyists, the Congressional Farm Bloc was formed. Its goal was to convince the public that agricultural welfare was important to the nation.

In rapid order, the bipartisan Farm Bloc was able to achieve passage of the Packers and Stockyards Act, the Futures Trading Act, the Agricultural Credits Act of 1921, amendments to the Federal Farm Loan Act, the Capper-Volstead Cooperative Marketing Act of 1922, and the Agricultural Credits Act of 1923. Then, members of Congress and representatives of the farm organizations split in the battle over the McNary-Haugen bill regarding how to improve farm income, and the Farm Bloc lost its effectiveness. However, it helped to mold the attitude of farmers, making them less resistant to government intervention in agriculture.

Henry A. Wallace, who later became Secretary of Agriculture, wrote in 1923 that farmers had a right to organize to reduce production just as did other sectors of our society. Many opposed that idea on the grounds that producing food was sacred and should not be controlled to reduce its supply. Wallace argued that farmers should not accept that view. He realized that farmers could not effectively organize to reduce production like city workers did because too many people were involved in the decision making for it to be possible on a voluntary basis.

Wallace explained that one-fourth of the farmers were beholden to their absentee landlord, banker, or merchant financier; another one-fourth were tenants who were pressed to make every dollar; another one-fourth owned their land but were "densely ignorant concerning the world supply and demand conditions and the forces which make farm product prices"; and the final one-fourth were "intelligent land-owning farmers who . . . [were] somewhat familiar with how the world market works and . . . [had] a sense of class solidarity to control their output."

Wallace added that the last group were strong individualists, who forged ahead on their own. They knew that they could profit under chaotic conditions at the expense "of those ignorant and poverty stricken farmers who don't know just what . . . [is] going on." He concluded his article with the thought that agriculture had never suffered enough to get all classes of farmers to work together.

Plans were considered in the early 1920s to control production, but it was felt that the taxpayers would not agree to any program of price supports unless there were restrictions on production. Price fixing was looked upon as a short-term remedy but appeared too costly and wasteful in the long run. It was suggested that if farmers did not agree to restrictions on production, they would have to take their chance on the world market. The philosophy of the nation was not prepared for a mandatory crop reduction program in the early 1920s, but the mentality had changed by the 1930s. Interestingly, the ideas of 1923 were the same ones used in 1933.

The Agricultural Marketing Acts of 1926 and 1929

After the price collapse of 1920, farmers questioned the marketing system because they saw that the rest of society was doing well while they were not. This helped agrarianism rise to new heights with the formation of the Farm Bloc and the cooperative movement of the 1920s. Some historians assert that cooperatives had their roots in the collapse of the Grange (1875–76) and the Farmers' Alliance (1890–91). That may be true, but the cooperative movement of the 1920s, which may have appeared spontaneous to outsiders, was very much part of the extension program.

The Sherman Anti-Trust Act of 1890 and the Clayton Act of 1914 aided cooperatives. After the extension service came into being, it encouraged and aided cooperative marketing and such cooperative ventures as rural telephone cooperatives, creameries, elevators, and shipping associations. The key legislation was the Capper-Volstead Act, which was specifically created to allow cooperatives to operate in restraint of trade.

Also known as Public Law 146, the Capper-Volstead Act had safeguards for the consumer, which made sure that farmers and cooperatives would stay within bounds. Government leaders clearly were more concerned about the high cost of living than they were about farm prices and markets. This was the reason for the safeguards, but everyone realized that something had to be done to help the farmer.

On July 2, 1926, Congress passed the Cooperative Marketing Act, and the USDA found itself involved with the cooperative movement. At this point about 800,000 farmers were members of various cooperatives. The USDA encouraged the formation of more cooperatives because it was hopeful that cooperative marketing could cope with the surplus problem. The Act gave farmers help through government research, service, and educational assistance.

In 1929 the Agricultural Marketing Act was passed. It gave cooperatives power and financial assistance in an attempt to stabilize agriculture by purchasing commodities and withholding them from the market. It established the Federal Farm Board, which attempted to control surpluses by orderly production and distribution. With the exception of using tariffs, this was the first effort by the federal government to directly increase prices.

A revolving fund of $500 million was allocated to the cooperatives to stockpile commodities. A huge stockpile of surpluses was created, but prices continued to decline because production remained high. Conditions worsened and the Federal Farm Board did not have the power or the funds to cope with the problem.

A strong undercurrent against the Federal Farm Board was based on the argument that China had used cooperatives for a thousand years and then abandoned the idea. Others were concerned about how to overcome the rugged individualism of the independent farmer. It was hoped that the voluntary nature of the Act would be socially acceptable to farmers who were not yet prepared to give up complete laissez faire. Over three years the Federal Farm Board lost $329 million and recognized that the farm problem could not be solved without the power to control production.

A major result of the Agricultural Marketing Act was the further expansion of cooperatives. A more significant result was the change in public attitude about what the federal government should do to help those in agriculture. After 12 years of agitation by farmers and their supporters, the government became directly involved in attempting to solve the problem through the creation of the Agricultural Adjustment Act of 1933.

Cooperative Shipping Associations

Shipping associations, elevators, creameries, and cheese plants were well established in the late 1800s and early 1900s. Natural forces, the USDA marketing research and education programs, World War I, and the agricultural marketing acts all influenced the cooperative movement. Between 1917 and 1920, shipping associations increased rapidly. By 1920, Iowa, the premier hog state, had 610 associations. That year they handled 25 percent of all livestock in the state. The number of associations in Minnesota increased from 115 in 1913 to 655 in 1919. Many of the smaller ones failed, and by 1926 there were only 577 associations, but the number of carloads of livestock shipped increased. That year Minnesota cooperatives handled 61.8 percent of all livestock marketed.

The Packers and Stockyards Act of 1921 brought most of the stockyard and commission charges at the larger livestock markets under regulation. This removed most of the reasons for cooperative shipping associations, but they continued to multiply. By 1923 cooperative

Milk cans being loaded onto a route truck, North Carolina, 1929. (NAL)

associations numbered 12,000, with over 2 million members, and did $2.2 billion worth of business. In addition, by 1925 there were 21 terminal cooperative livestock sales agencies founded by the AFBF.

County extension agents were actively involved in the movement, and the AFBF led in the formation in many of the key livestock states. Forming associations helped the extension agents' record. Associations were created for several reasons. In some cases, farmers had a grudge against a local buyer, or they were upset by a single transaction. Some associations were formed by hired zealous promoters of co-op evangelism. Others were organized to eliminate the profits of the private dealer, to reduce waste and loss, or to develop more efficient selling services.

Failures were experienced throughout the nation before the cooperative drive hit its peak. Too many associations had been created too rapidly. To combat the reversal, a field service program was established to keep members informed and to counteract competition. Most members seemed little interested in the affairs of their cooperative. One leader commented that a majority of the members would drive farther to see a circus, a ball game, or a horse race than to attend a co-op meeting. At least 80 percent never attended an annual meeting, 90 percent had never been visited by a field representative, and 80 percent could not name their district director.

Cooperative management disagreed as to how much information should be given to the members. Many associations were run by a retired farmer, by the elevator manager, or by the local banker. Some associations were too small to justify their existence. When trucks became a reality and went directly to the farms to pick up partial or full loads, there no longer was any need to pool for carloads.

Eventually county, state, and national associations were formed, which created more overhead and added to the cost of the local cooperative. This reduced the margin of net operating income and made marketing through the association no more attractive than selling through a local independent buyer. Disinterest or lack of loyalty of the members was probably as important as any single factor in the demise of many small associations.

Cooperative Oil Associations

As soon as automobiles, trucks, and tractors became common on farms, cooperative leaders sought to get farmers involved in cooperative oil associations. In Minnesota, one of the pioneer states in forming cooperatives, gasoline shipments rose from 7 million gallons in 1909 to 289 million in 1927. The first oil cooperative was organized in 1921, and by 1928 there were 51 in the state, with 27 service and 69 bulk stations. The major reasons for their formation were savings, dividends, and credit to tank-wagon patrons.

The Illinois Cooperative Act of 1923 paved the way for cooperatives in that state. In 1924 the first Illinois gasoline and oil cooperative was founded. In 1925 the Illinois Agricultural Association (IAA) started centralized purchase, in carload lots, of feed, seed, coal, binder twine, and other supplies at major savings. In 1926 other co-ops were formed under AFBF sponsorship. The Illinois Farm Supply Company (IFSC), founded in 1927, was one of 13 business units of the IAA, the general state farm organization.

In 1936 IAA did $23.6 million in business. Petroleum supplies were 87 percent of all business, with tires, paint, and insecticides making up the rest. Earnings totaled $1.1 million, about 10 times the original investment. That year the IFSC had 62 county service companies, with 161 bulk plants, 50 retail service stations, and 516 tank trucks. About 75 percent of all sales were delivered directly to the farm.

In 1937 the IFSC extended service to cooperative elevators, cooperative livestock shipping associations, and other cooperatives to handle feed, seed, fertilizer, fencing, twine, and other supplies. Illinois was clearly the leading state for supply cooperatives. At that time there were only 14 cooperative wholesale associations in the entire nation purchasing petroleum products for local cooperatives.

The Rugged Individualist Succumbs

Leaders of the Farm Bloc attempted to restore agriculture on a par with the industrial and commercial community. But in spite of the desperate conditions in agriculture, there was little success in achieving unity among farmers once they gained the early victories of the 1920s.

In 1922 President Harding expressed the then current philosophy, which endured for another decade:

> It cannot be too strongly urged that the farmer must be ready to help himself. . . . In the last analysis, legislation can do little more than give the farmer a chance to organize and help himself. . . . But when we have done this, the farmers must become responsible for doing the rest. They must learn organization and practical procedures of cooperation.

Self-help remained the basic philosophy of agricultural policy. The USDA could provide educational opportunities and service but not coerce. Sharply increased unemployment and reduced consumer demand caused the 1930 gross farm income to drop to the 1921 level. Prices continued to decline, but the experience of the Agricultural

Marketing Act of 1929 made it clear that no program would be effective without acreage and production controls. The leaders of the Farmers Union opposed such a program because there was no federal guarantee of the cost of production. In early 1932 the Farmers' Holiday Association was formed to get improved prices and to stop foreclosures against farmers. This became the height of agrarian radicalism (1932–1933) and made the nation aware of the plight in agriculture.

By 1933 the nation had experienced a change in philosophy and was prepared for legislation directed toward price supports and production

Striking farmers blocking truck hauling livestock to market, Iowa, 1931. (NAL)

controls. Such a policy was necessary to give a greatly overcapitalized agriculture time to downsize and to reduce the number of farms and people involved in production agriculture.

The Agricultural Adjustment Act (AAA) of 1933 was passed in an effort to achieve parity for agriculture. A sharp speculative rise in prices occurred in anticipation of the passage of the act. Other factors that aided the price increase were the abandonment of gold payments, a dollar devaluation, a drought with the prospect of a short wheat crop, and the expectation that the AAA and the National Industrial Recovery Act (NIRA) would mean higher prices. Prices rose 10 percent between March 3, 1933, and mid-July and then declined.

In June 1933, cotton farmers were informed that 10 million acres of cotton had to be destroyed to bring production in line with demand. On June 26, 1933, after several extensions of the deadline, over 1 million farmers signed up 10.5 million acres to be plowed down. Cotton farmers had virtually no alternative to protect against a further price decline, even though plowing under a crop was controversial.

By the spring of 1934 the trend toward more centralized control of agriculture became further apparent. Cotton farmers were taxed as a penalty if they sold more cotton than their quota. As one authority stated, "This tax was manifestly prohibitive and therefore coercive. It was intended to club into line those who had not signed the A.A.A. contracts . . . [T]he Federal Government simply took over the power of decision as to cotton production."

At the same time, a large majority of the nation's dairy farmers positioned themselves against government programs under which livestock farmers were paid to kill animals to reduce production just as crop farmers were paid to plow crops under. This was the first significant opposition against a government program and caused the AAA to withdraw from the dairy sector. Dairy farmers generally had hogs and poultry enterprises and were less reliant on the cash market for their livelihood.

Things were different for the wheat farmers who, in 1933, experienced the shortest crop since 1896. Plow-down was unnecessary for 1933, but wheat prices dropped from $1.04 in 1929 to $0.32 in 1933. When farmers signed up for the 1934 and 1935 crops, they were required to set aside 15 percent of their acreage and were paid benefits for their 1933 crop because of the drought. The severe drought of 1934 so reduced wheat production that the USDA had to do an about face on production curtailment.

To protect consumers a program was established under the Commodity Credit Corporation (CCC) to store crops to tide over years of crop failure. This also assured price stabilization, but it was a signal that with commodities in storage the government had the ability to control prices when crops became short. In later years supplies were released to hold down prices.

Secretary Wallace wrote in 1934 that the challenge of keeping the supply off the market to maintain prices was compounded because productivity was constantly climbing. He continued, "How can we reconcile increasing productivity in agriculture to a declining demand? This is the supreme question of our time."

In 1935 Wallace claimed that the first two years of production control were very effective. He hinted that the period of emergency adjustment was nearing an end. But he added that it was clear that "agriculture needs production control to prevent mass swings that lead to recurring cycles of over and under production." One of the worst droughts in our history had helped him ease the problem, but he understood the intensive productive capacity of agriculture that plagued farm programs from their conception.

Wallace added that no cooperative program to control production would work unless the government was strong enough to overcome human miscalculation. Later, in 1935, President Roosevelt made it clear that the program was not simply an emergency program:

> It had been laid out as a permanent system—the farm business to be governmentally planned and managed as part of a broader planned economy. . . . It never was the idea of the men who framed the act, of those in Congress who revised it, . . . that the Agricultural Adjustment Administration should be either a mere emergency operation or a static agency.

It was then decided to decentralize the program by establishing its administration closer to the farmers. Rexford Tugwell, who was one of the chief planners of the program, wrote in retrospect:

In no part of our social life is planning so carefully and so democratically done as in agriculture; in no part does it so nearly approach the necessary completeness for inclusion in a national plan. ... It is paradoxical too that in the industry which has been slowest to increase in unit-size, the scale of planning for it should be evident; but that is also true.

Tugwell continued:

Farmers may prefer to believe that they decide a good deal for themselves, and actually they may, for the moment, for the short run, but ... most of the issues about which they may regard themselves as free to choose, are decided in places and in ways which are far from local and far from individual.

Tugwell noted that by abstaining from voting in the referendums, the farmer was not so much of an individualist. He explained:

The farmer comes to understand the inescapable socialization of agriculture or he does not; but at least he feels himself more and more caught up in an on-going process. And he either participates in attempts to influence it or he submits to having others do it in his name. This is not very agreeable from the democratic point of view and the more independent-minded resent it. ... As a result, there has come into this most individualistic of occupations the most elaborate planning known to our economy, equipped with machinery for all the preplanning processes and with a bureaucracy which, like all bureaucracies, sometimes conceives itself as having divine rights, but which on the whole has given agriculture a far more secure place in that economy than it would otherwise have. And all of this has been in spite of professions by everyone concerned of unshakable loyalty to laissez-faire.

The AAA first sought to help farmers by reducing production. It experimented with control of production by long-range conservation programs, and it attempted to limit size of payments. Then the nutritionists looked at the need for more and better food for the entire population, but especially for the one-third who suffered from malnutrition. The conservationists became involved in controlling the direction of the programs.

With the drought of 1936, production of feed, livestock, and dairy products had declined more than was desirable, but the "administration took steps to mitigate the consequences of the drought to consumers as well as to producers." Wallace wisely repeated his fears of the latent potential of agriculture to overproduce. He suggested that instead of 360 to 365 million acres of production, there should be 285 to 290 million acres. He had little hope that exports would revive within the immediate future.

In 1936 the Soil Conservation and Domestic Allotment (SCDA) Act was passed as an amendment to the soil erosion control legislation of 1935. Part of the intent of the SCDA was to equalize income of farmers with that of nonfarm citizens. Parity became a goal of the program. At the same time, it was announced that there would be "protection of consumers by assuring adequate supplies of food and fiber now and in the future."

Farmers were fortunate, for until the mid-1950s they were over-represented in Congress, and they retained strong public moral support. Many farmers could have bettered themselves by leaving agriculture, but they stayed and turned to politics for their future. They became beneficiaries of government subsidies on an unprecedented scale. As one agricultural economist later suggested, "Plowmen in business suits have learned that no seed yields a better harvest than that planted in the fertile minds of Congress."

Except for the combined expenses of all the military organizations, farm subsidies were the largest item of expense for the federal government prior to World War II. In 1940 Wallace acknowledged that production control programs had not been successful and that the USDA was intensifying both the crop limitation program and the subsidization of exports. This sharpened the conflict between protecting agriculture and liberalizing trade.

Then came World War II, and to encourage farmers to increase production, the Steagall Amendment was passed. This amendment pledged government support to maintain prices for two years following the close of the war. By 1944 food production was 38 percent above the 1935–1939 average. It had increased every year for five years. The Agricultural Act of 1948 protected prices at 90 percent of parity until 1950, when price supports were to be reduced. The Korean War revived exports until 1953, when the USDA was again faced with the challenge of reducing surpluses.

At that time John Galbraith wrote that administrations "have promoted a liberal trade policy with one hand and in the farm policy have laid the foundation for economic nationalism with the other." Fortunately, the world demand held strong; but from that date on, monetary exchange, along with GATT and export subsidy programs, served as a major determinant, except when shortages appeared somewhere in the world.

Soil and Forest Conservation

In 1909 a gathering of people interested in soil science was held in Budapest. That was followed by other meetings in 1910, 1922, and 1924 in various European capitals. In 1924 in Rome informal plans were laid to create an International Society of Soil Science. An American was elected president, and in 1927 the First International Congress of Soil Science was held in Washington, D.C.

Since the early 1920s, the USDA had sought to get farmers to reduce production and to become more concerned about soil conservation practices. By then it was known that a loss of soil nutrients occurred with each crop. Some farmers were aware of the losses but did not know what to do, and others did not know or care.

Hugh H. Bennett, a USDA scientist, took it upon himself to make the nation aware of the need for soil conservation. In 1929 the first appropriations were made. The Soil Conservation Service (SCS) became the first agency in the world to do a soil survey of classifications of land based on the slope and on the degree and kind of erosion. By the end of 1930, over 800 million acres had been surveyed.

The Works Projects Administration (WPA) and the Civilian Conservation Corps (CCC) became involved in improving soil conditions on Indian reservations and in forests. By 1934 there were 161 CCC camps doing erosion work in nonforest areas and on privately owned agricultural land. Entire watersheds were targeted for demonstration projects, and eventually the personnel of 800 CCC camps, plus about 23,000 WPA workers, were engaged in areas of row-crop agriculture. They did contouring, made stock watering ponds, reseeded, repaired leaky irrigation ditches, controlled erosion on slopes, and did highway ditch erosion control. These were stopgap measures until farmers could be convinced to become involved. The USDA had a difficult time educating and selling conservation to many farmers.

In 1936 the original AAA was invalidated by the Supreme Court. Congress realized that to pass a similar bill, it would have to get farmers to "voluntarily" shift land out of production for conservation purposes. With the SCDA Act of 1936 the USDA gained the police power to enforce conservation and speed up the process of education. The first conservation practices were enticed by subsidies.

In 1937 cost-sharing grants were provided for carrying out certain conservation practices. The AAA of 1938 declared that its intent was to preserve soil fertility and to encourage proper land usage. Controlling production was secondary. The Secretary was empowered to make payments to producers based on their adoption of proper land practices. The conditions were pre-announced, and farmers were free to comply or not to comply.

Nearly 3,000 soil conservation districts were established. New England became the first region to go from forest to field and back to forest. Eventually, its forests became the basis for the tourist industry in that region. After the Watershed Protection and Flood Prevention Act was passed in 1954, communities throughout the nation became involved in conservation programs.

Originally, about 822 million of the 1.9 billion acres of land in the 48 states were forested. By 1922 we were the largest consumer of forest products in the world, and over 200 million acres of forest land had been cleared for farming. Our national timber harvest was twice the rate of forest growth. Conservation programs had started in the early 1900s, but it was with the work of the CCC in the 1930s that major progress was made in restoring our forests. Over 20 million acres of abandoned farm land were reforested.

In the 1940s the government changed its direction in forest management. Prior to that time its primary concern was custodial, and after that it became more interested in timber production. By the early 1950s forests were producing only about half their capacity. Much of the problem was in the hands of small owners, who did not give adequate care to the tree crop. With proper tree cultivation, the time needed to produce sawlogs or pulpwood in the South was reduced to 15 to 20 years. By 1952, in spite of the increase in demand, the rate of forest growth exceeded the demand. But one of the biggest remaining problems was that such a large portion of the timber cut was wasted.

Social Programs of the USDA

The USDA became one of the most far reaching branches of government. It expanded into many areas not previously thought of as being of concern to agriculture.

Nutrition

Nutrition first became an interest of the USDA in 1894. The Country Life Commission found that there was a vital need to do more about nutrition. With the passage of the Smith-Hughes Vocational Educational Act, the extension service became involved with the promotion of good nutrition.

In the 1920s the Food, Drug, and Insecticide Administration was organized to promote purity and truthful labeling in certain commodities essential to public health and to the economic welfare of the nation.

Law prohibited the shipping in interstate commerce of adulterated or misbranded manufactured or natural foods. It also protected farmers and others from buying and using insecticides and fungicides that were not what they claimed to be.

A 1935 amendment to the AAA was aimed at encouraging domestic consumption. Soon 12.7 million poor and/or unemployed were provided with food. Next came the school lunch program, and after 1946 the National School Lunch Act became permanent. It provided students with free or reduced-price lunches, depending upon need. In 1964 the Food Stamp Program became permanent. In 1966 the Child Nutrition Act created a school breakfast program and a special milk program.

Drought The drought of the 1930s was worse than the six previous nationwide droughts because it covered a much larger area. A total of 1,017 counties in 21 states were certified as having not more than 50 percent the normal crop of pasture, wheat, and feed grains. Employment was provided for farmers by shifting federal funds to the drought areas to build roads and carry out other public projects. Federal funds were disbursed to farmers in 17 states to purchase seed, feed, and fertilizer. The government provided feed to save dying cattle or to reimburse the farmers for dead animals. Despite the drought, which reduced supplies, the worldwide depression caused a sharp drop in export demand, and commodity prices fell to their lowest levels in more than 15 years. If the crop had been normal, prices would have been even lower.

Dust storms were so bad that people placed water-soaked sheets and towels over their windows at night to make it possible to breathe easier while attempting to sleep. Sometimes farmers had a difficult time sleeping because they could hear the lowing of hungry cattle. Bankers recalled finding notes under the bank door stating that the farm family had left and that all of their livestock and machinery were on the farm for the bankers to claim.

The droughts of 1934 and 1936 were so severe because the Plains had been drastically altered by overgrazing and tillage, which upset the balance of nature. There was no evidence of a permanent change in the climate that should have made the droughts more severe than previous ones. Great Plains cattle numbers declined nearly 2 million head from their peak in 1920, but in 1935 it was estimated that 100 percent of the Great Plains still remained overstocked.

In 1934 nearly 40 percent of the total cropland of the Plains was in wheat. The heavy emphasis on wheat was not only damaging to the soil, but it added to the surplus. Producing a crop wasted water that was needed for other purposes. In the drought of 1918 to 1922, about one-fourth of the farms were abandoned. Now, it was proposed that 20,000 farms should be purchased and consolidated into larger units. This would relieve the welfare budget of caring for those who had no chance to survive in farming.

From 1930 to 1936, an estimated 165,000 people left the Great Plains drought area. A national policy was needed to prevent people from returning to the area. The lack of underground water supplies and the fact that only about 5 percent of the rainfall ever reached the river systems gave little hope for extensive irrigated agriculture. To minimize the human and economic hazards, future farming operations had to be conducted with minimum population and maximum efficiency of mechanization.

The Resettlement Program

Drought and surplus production brought the issue of land conservation to a head. In the 1930s the public started to see land as a resource to be preserved. In 1935 at least 454,000 farms had land so poor that they were not capable of producing a living for the operating family. This represented a waste of land and of people. These farms averaged 164 acres, of which only 44 were in crops. According to M. L. Wilson, an administrator in the USDA, a program to purchase 5 million acres over 15 years was proposed. Other farms would be consolidated, and the surplus population would be resettled. The purchased farms were to be put into a soil conservation program.

The Federal Emergency Relief Administration (FERA) was created in 1935 in part to help rural families. The Resettlement Administration assumed those responsibilities. Its purpose was to aid the destitute farm families and to get marginal land out of production. Farmers were to relocate where they could make a living and earn enough to get themselves off relief.

The Bankhead-Jones Farm Tenant Act was created to help tenants establish farms of their own in areas with greater potential. Senator Bankhead hoped that the land relief measures would help to restore small yeoman-class farmers, who he said were the backbone of all former great civilizations. Bankhead wanted to remove people from the industrial centers.

In 1937 the Resettlement Administration became part of the USDA, and eventually it was renamed the Farm Security Administration (FSA). The chief initial task of the FSA was to finance and relocate tenant farmers by using the $25 million appropriated for that purpose. This massive rural rehabilitation program eventually affected 931,000 farmers. It was an effort to satisfy those who still believed that small-scale farming was vital to the security of the nation.

About 500,000 new farms were created by the resettlement programs. This caused an aberration in the decline in farm numbers, which had started in 1920. By 1940, under the European wartime stimulus, over 700,000 farms from the artificially created peak of 1935 and 1936 had ceased to exist. From that experience the USDA realized that the trends created by commercialization, mechanization, and the sheer force of economics, which it had generally supported, indicated the logical course to follow.

By the early 1940s the goal was to maintain "efficient family-size owner-operated farms." A 1941 directive stated that the USDA must attempt

> . . . to prevent large farms from becoming so large as to drive out family farming [and], . . . at the same time, do what it can to help make small farms large enough to provide each farm family with a reasonably

adequate minimum level of living. This objective is obviously inconsistent with a possible alternative objective of maintaining opportunity in agriculture for the maximum possible number of farm families.

The directive continued:

> The Department does not believe that agriculture should be made the dumping ground for the industrial unemployed. . . . [O]nly as many farm families should be permanently engaged in agriculture as can be afforded an opportunity to maintain a reasonably adequate level of living. There are at present more farm families attempting to gain a living from the land than can be provided with efficient family-size farms.

Any effort to maintain farm numbers ran counter to technology, which was continuously showing how to produce more with fewer farm laborers. The administration sought ways to handle the farm family displacement problem and still maintain family farms. Fortunately, the wartime demand for industrial workers solved the problem in the most logical way.

Unique Experiments

One of the most publicized rural rehabilitation efforts was the Matanuska Colony of Alaska. It had long been felt that Alaska had agricultural potential, so when it was decided to relocate farm families from impoverished areas in the states, Alaska appeared to be a logical choice. The goals of this effort were to reduce the number of people on relief in the states, to determine the viability of Alaska as a new frontier for a greater population, and to strengthen the Alaskan economy by producing more food locally, thereby reducing the need for imports.

About 75 percent of the families were selected from the northern, colder, cutover regions of Michigan, Minnesota, and Wisconsin. This "spur of the moment idea, carried out in the urgency of the times," was hampered by a lack of solid knowledge of the conditions in the settlement area. The political leaders often ignored the administrators on the job. Many directives came from California, where the leaders had little understanding of the local problem, and the program was carried out too fast to establish good foundations.

In the 20 years prior to 1934, Alaskans had homesteaded most of the better land in the Matanuska area. They claimed 23,000 acres, of which only 1,000 were cleared and cropped. By June 1935, 202 families with 903 people were on hand. The Alaska homesteaders were not willing to sell land at the price offered, so only about 175 acres of cropland were available for the colonists.

Of the families who were sent to Matanuska, most were not well versed in farming. The newcomers lived in tents while government-employed transient workers from California erected their homes. The buildings were poorly planned and constructed, and costs were twice the original estimates. By July 1939, 124 out of 202 original families had left, but replacements continued to come. Of those 124 families that left Matanuska, only 37 remained in Alaska. By 1948 only 12 of 282 transferred families were making all or part of their living from farming.

In addition to the poor planning and climatic conditions, the major problem was that the government had allocated only 40 acres of land per farm. Even in good farm states, this was not adequate for normal farming. The failure to provide economic units doomed the attempt from the start. The Matanuska effort persisted after the war, and from 1945 to 1950 another 110 homesteads were established. By 1955 only 11 of those homesteaders remained full-time farmers, 20 were part-time farmers, 22 were in other occupations, and 57 left the valley.

In 1930 the majority of the blacks still lived in the South and relied on agriculture. About one-fourth of them owned land. The rest were sharecroppers or day laborers. Most of them suffered from an overpopulation of farmers and farm workers and a shortage of credit. Between 1930 and 1940, about 200,000 black owners, tenants, or laborers were uprooted by the AAA programs that reduced the amount of land in production. Most of those displaced persons became day laborers at a time when fewer farm workers were needed. This created unfavorable publicity. The government reacted by applying the rural resettlement program to blacks who were facing local barriers (Jim Crow attitudes). The southern resettlement projects permitted only 9 out of 150 to be entirely for blacks. As in Matanuska, little was done to ease the poverty of persons involved in those resettlement projects.

The Power Revolution Booms

Harold Pinches, of the USDA, noted that the technological gains in agriculture had provided the nation with a more abundant and reliable supply of food. He continued:

> Only where agricultural production has advanced faster than a people's needs have the economic conditions been created necessary

Hillside combining with 24 horses and 4 men, Oregon, 1916. (Lou Kiene)

to release larger and larger segments of the population from limited production on the land and thereby enable more and more persons to advance in intellectual, cultural, and social development above static folkways.

By the late 1950s the rate of technological progress had slowed, and very little new land was being opened. At the same time, the population and industry continued to grow. Farms expanded and became specialized, which served to increase production and at the same time release labor for industry. According to Pinches, "The shift of human power out of agriculture to other sectors has been one of the great blessings of the agricultural revolution." The drain in labor from the farm caused a rise in wages, which hastened the adoption of technology. Agriculture amassed large accumulations of capital and capital facilities, which made it increasingly productive.

Many tasks once part of farming had shifted to urban areas. Processing and marketing operations were handled by specialized agencies. Buttermaking and livestock slaughtering are two of many examples. But the greatest shift of labor from farm to cities and towns was made possible by the adoption of mechanical power units. Tradition and inertia were being broken down, and farm families were less resistant to making change than in the past. There still were many farmers and farm youths who had strong feelings for the land and who thought they would not be happy elsewhere. But their number was diminishing.

Innovations did not benefit all farmers equally, for the early adopters most often made the greatest benefits from the improvements. The late adopters gained little, because prices were usually adjusted to lower costs from the improved methods by the time they began using the technology. But it was clear that the benefits of technological improvements ultimately went to the consumers. In addition to lower cost of food and fiber, all people enjoyed the increased leisure that technology had made possible.

Table 4–3. Percentage of Income Spent for Food

Years	Percent of Total Income
1780s	60
1850s	55
1880s	45
1930s–1960s	20–24

Years	Percent of Disposable Income
1970s	16–17
1980s	13–15
1991	11.6
1994	9.0

Sources: *The 1986 Yearbook of Agriculture* for early figures. Economic Research Service for 1991 and 1994.

The Tractor Replaces the Horse

In 1910 it took 72 million acres of cropland to feed horses and mules on farms and 16 million more to feed those in towns. In other words, 27 percent of the 325 million acres of all cropland was used to feed the work animals. In 1910 there were about 1,000 nonsteam tractors on farms. By the time horse numbers peaked at 26.4 million in 1919, tractor numbers had risen to 220,000. By the late 1920s, about the only market for work animals was the fox farm or the soap factory. Only on small cotton farms was there demand for "cotton mules." So many farms had adopted tractors that production peaked at 753,623 in 1948. By

Holt 120 crawler pulling Holt combine and grain wagon, two-person crew, Hardin, Montana, 1924. (J. R. Taylor)

1955 horse and mule numbers had fallen to 4.3 million, and tractor numbers had risen to 4.3 million.

The tractor industry had an amazing mushroom growth. A writer in 1916 commented, "It bled real hard money from the farmer when real hard money did not seem to exist." In the late 1920s an article in the *Yearbook* on declining horse production noted: "With improvements that are being made in tractors, it is difficult to foresee the extent to which tractors will eventually replace horses on American farms, but it is not likely that the horse will ever be entirely replaced. At least one team will be necessary on most farms."

From 1913 to 1919, only the larger tractors were promoted at tractor shows, but farmers learned that small tractors were also useful. Once it was realized that farmers would buy smaller, cheaper tractors, Henry Ford became interested. He had worked on developing a tractor since 1905, and in October 1917, he produced a commercial, light, inexpensive two-plow tractor—the Fordson. The Fordson was an immediate success, and in spite of some serious weaknesses, it captured 75 percent of the market share in the early 1920s.

In 1924 International Harvester Company introduced the Farmall—a row-crop tractor that was adapted to mounted equipment. The Farmall had a braking control device within the steering system and a power takeoff. In 1930 John Deere introduced the power lift. In 1932 Allis-Chalmers came out with pneumatic rubber tires, which speeded up most field operations, increased the drawbar power, and decreased fuel consumption. By 1938 over 68 percent of the tractors came equipped with rubber tires. In 1936 Harry Ferguson introduced the three-point hitch for a newly designed Ford tractor, which, among other features, provided for weight transfer.

The loss in asset value of millions of horses and mules had a negative financial effect on farmers. Shifting 25 percent of their cropland production to cash crops compounded the problem of an already glutted market. In the total transition from animal power to mechanical power, about 95 to 100 million acres of land were freed to produce for the cash market.

The immediate reaction of farmers who purchased tractors was that their farms were too small. The tractor forced farmers to become full-time farmers if they wanted to remain in farming. On the other hand, it was probably the most effective single machine in driving people from the land. USDA surveys of the 1930s indicated that a tractor displaced from three to five families, and in extreme cases as many as nine families. A Kansas study established that for each tractor added there was a decrease of one-half of a farm. The USDA determined that as early as 1938 the annual savings in labor by switching from horses was equal to the work of 440,000 laborers. Larger tractors and equipment made even greater labor reductions possible.

The greatest decrease in rural workers in absolute and relative terms came among those under 25 years of age. Although the USDA estimated that 350,000 fewer workers would be needed in 1950 as compared to 1940, over 2 million left the farm. Between those years tractor numbers increased from 1.6 to 3.4 million. Providing for the demands of World War II without the tractor, which freed land for human food production and released 2 million workers for industry, would have proven extremely difficult.

Plowing an acre of land with five horses and a gang plow required 1½ hours. By 1939, with the average tractor, that work was done in half an hour. In 1954, with a six-bottom tractor plow, it was done in 16 minutes. By 1972, using a 15-bottom plow, the time was reduced to 6 minutes for those still using the moldboard plow.

Machines Outgrew the Farms

From 1910 to 1957, agricultural output increased 85 percent while inputs increased 22 percent. By 1957 tractors, machinery, commercial fertilizer, chemicals, and commercial livestock feeds were the key inputs. Production per acre and per worker hour was greatly enhanced by the new inputs. Still, agricultural productivity was not increasing as rapidly as productivity in the general economy.

Small-scale beef operation feeding green chop, Pennsylvania, 1950s. (NAL)

Secretary Jardine warned in 1927 that rapid mechanization would force a painful adjustment on farmers not willing to adopt the newer ways. This was most obvious in the South, where much of the cotton crop required one worker for about 15 acres, while mechanized farmers could handle between 75 and 150 acres per worker, depending upon the size of the equipment used. The cost of production was reduced from $14.20 per acre with a mule and half-row equipment to $5.20 with a tractor and four-row machinery.

On Montana wheat farms, harvesting took 8.8 hours of labor per acre using threshing methods but only 0.75 hour per acre using the most

A 3,250-acre wheat field with 17 combines direct harvesting, pulled by 18-36 I.H.C. tractors, Hardin, Montana, 1928. (J. R. Taylor)

advanced combine and trucks. The cost per bushel of wheat produced dropped by 50 percent during the five-year transition period of the late 1920s. During those years, the farms studied increased from 627 to 1,284 acres, and the number of workers decreased on all the farms. The standard 6-plow tractor handled more than twice as many acres as a 16-horse team. Hauling a normal crop of wheat from 500 acres using a 6-horse team required 100 worker days, while a truck did the same work in 25 days.

Nationwide, farmers attributed their rapid adoption of tractors and larger machinery to (1) the need to reduce costs to survive; (2) the fact that farm programs took much of the risk out of investing in equipment; (3) the threat of unionization in some areas; (4) the exodus of workers during World War II for better paying occupations; and (5) greater economy plus improved timing of production.

Farmers who had not mechanized by 1951 gave the following reasons, in order of priority, for not having done so: (1) they were too old to change; (2) they were unable to get financing; (3) the farm was too small to justify the equipment; (4) they preferred working with animals; and (5) they were unable to hire skilled labor to run the machinery. By 1951 the 887,000 combines, 588,000 corn pickers, 686,000 milking machines, and 3.5 million tractors had saved millions of hours of labor and greatly reduced the sheer drudgery of farm work.

The increased efficiency kept farmers continually seeking to purchase the improved, refined, and enlarged machines. The larger machines worked to the disadvantage of the smaller farmers unless they secured custom operators to harvest, to plant, and to do a variety of other jobs.

Cotton Machinery

The production of cotton was mechanized more slowly than that of other major crops. From the Civil War to World War II the cotton states remained uniformly poor because they could not break the shackles of their traditions.

During World War I, it took nearly the same amount of labor to produce a pound of cotton as it had in 1860. The abundant supply of cheap, unskilled hand labor had hampered the progress of mechanization. By 1930 the only basic change was the use of power equipment in soil preparation, which reduced labor time per acre from 102 hours to 71. World War I did not have much of an impact on cotton production. In the South during that era, 87 to 99 percent of all black women and 50 to 66 percent of all white women worked in fields because of the lack of other opportunities. When the blacks started to move north in the 1920s, the labor supply was reduced slightly. This encouraged mechanization.

One of the major factors retarding mechanization was the fact that as late as 1940, 54 percent of the cotton farmers produced less than four bales of cotton annually. Custom, tradition, and small-farm mentality also were strong negative factors. But something had to be done to reduce the cost of production so cotton could compete with synthetics.

The Farmall tractor with mounted equipment made the first inroads into cotton production. By 1947 fully mechanized farms produced cotton by using only 39 hours of labor per acre, of which 33 was in weed control. Cotton farms had to enlarge and farm population had to decrease before the remaining farmers could mechanize. Many southern farmers dreaded mechanization because they realized the economic and social consequences.

The most sought after equipment in cotton production was a harvesting machine. The first cotton picker was patented in 1850 and a strip-type harvester in 1871, but the abundance of labor and problems at the cotton gin delayed their adoption. In 1926 the first strippers were used in cotton harvest. Unfortunately, the depression, a lack of cash, the supply of cheap labor, and a sharp decline in the cost of mules and mule feed delayed adoption of the strippers.

In 1941 the Rust brothers, together with International Harvester Company, marketed the first commercial spindle picker. Again, war intervened, and it was not until 1948 that 1,100 pickers were produced. The cost of mechanical harvesting was $7.38 per bale versus $37.76 for hand harvesting. Grade loss and field loss reduced the net saving to $4.36 per bale in favor of mechanical picking. However, the loss of crop from delayed harvest caused by the labor shortage made the timeliness of machine picking even more profitable.

Again, the cost of the equipment worked to the disadvantage of the small farmers. Unless they purchased a machine on a cooperative basis, smaller farmers could not compete with the larger, more mechanized farmers. Farms had to be restructured and labor needs drastically reduced before the overall cost of cotton was competitive with synthetics. During the war the labor shortage was so severe that almost 20 percent of the South's cropland was idle. This played a role in the re-entry of livestock, as pastures were seeded down and became a permanent part of southern agriculture.

After the war, farm consolidation and the reduced labor supply speeded mechanization. This decreased the time needed to produce a 400-pound bale of cotton from 155 hours to 10 or 12 hours by 1950. By then it was clear that mechanization was not the cause of the change. Most of the change was brought about by the factors that pulled workers away from the land. This was evidenced by pointing to areas where mechanization had not been fully developed but where tenant houses stood vacant because the workers had gone elsewhere.

Airplanes and Trucks

One of the greatest advances in the endless struggle to control insects came about almost accidentally in 1919 when airplanes were first used to search for violators of crop control laws in southern Texas. Those flights suggested the possibility of using planes in other ways. Before 1919 was over, planes were also used for forest fire patrol, aerial photography, and reconnaissance.

In 1921 planes were used at Troy, Ohio, to apply lead arsenate dust to a grove of nut trees infested with caterpillars. Six flights of less than six minutes total spraying time proved 99 percent effective. In 1922 planes were used to apply calcium arsenate to combat the boll weevil in Louisiana. Airplanes did the job more cheaply than other methods. In all, 1,100 farmers dusted 125,485 acres of cotton, which yielded an average increase of 339 pounds per acre. The application cost was less than the value of 100 pounds of cotton per acre.

The first planes specifically for crop dusting were produced in 1924. By 1925 a commercial aerial dusting company had 14 planes in operation, and aerial dusting became an established practice in cotton fields. Tests were made to determine the effectiveness of applying insecticides to control mosquitoes, tsetse flies, gypsy moths, and other crop, orchard, and forest insects.

The U.S. Air Service released fliers and mechanics to work with the commercial ventures. Because of an acute labor shortage in 1926, the above-mentioned aerial company was called on to dust several thousand acres of potatoes. Just two pilots, a mechanic, and one plane were able to accomplish what would have required 2,000 workers on the ground. By 1927 over 500,000 acres of cotton were treated by plane. In addition, planes were used to apply insecticides to peach, pecan, and walnut groves, vegetables, wheat, alfalfa, and other crops. The price per acre was no more than for application by traditional ground sprayers, and the procedure was much faster. The USDA saw aerial spraying as a means of relieving farm workers from being involved with the chemicals. Also, spraying could be done regardless of soil conditions. This was particularly important because the boll weevil liked wet conditions.

In 1925 planes were used in the barberry eradication program. In 1929 a California farmer used a biplane to seed 640 acres of rice. An average of 62.5 pounds of rice was seeded per acre at the rate of 8 acres every 10 minutes. The per acre cost was $1.97, including special insurance covering the pilot and the plane. The field yielded 15 bushels more than the national average because the pre-flooded field had an 85 percent germination rate.

After World War II the use of aircraft in agriculture accelerated. The effectiveness of DDT, produced during the war, was such that only a small application per acre was required. This enabled planes to spray

Loading rice for aerial seeding, California, 1950s. (NAL)

large acreages in relatively short periods of time. The airplane was so effective in the battle against insects that by 1952, 5,000 planes were in use for agricultural purposes. In 1954 planes proved their worth when they were used to control a serious outbreak of army worms at a cost of $1.50 to $2.50 an acre. By then 5,100 applicator aircraft were used to treat more than 61 million acres for insects, weeds, diseases, or brush.

In 1910 there were about 2,000 motor trucks and 50,000 cars on all farms. During World War I, railroad traffic was so congested that livestock shippers turned to using trucks. Trucks provided more comfort for the animals, and shrinkage was reduced. Between 1920 and 1923, trucks rapidly replaced railroads for hauling livestock. This had a great impact on the terminal livestock markets.

Trucks also made rapid inroads in the delivery of fruits and vegetables to markets in major cities. The producers quickly learned that the return per hour of labor by using trucks instead of railroads was much greater because of the redistribution time saved. By 1952 the 5.1 million farms had 2.4 million trucks (including pickups) and 4.35 million cars. Cars and trucks greatly speeded up farm work and reduced the isolation.

Productivity Booms

In 1927 Secretary Jardine wrote that he was excited to report that one worker with a tractor could take care of 250 acres of corn while the same worker using horse power could handle only 40 acres. Similar gains were given for other crops. He continued that U.S. farmers had led the world in efficiency in the previous century and realized that now they would make even more progress. Contrary to reports of fear expressed in the first decade of the century about having to import food, he now predicted that the nation would not have to rely on imports.

During the 1920s and 1930s, the more innovative farmers improved breeds of livestock and varieties of crops, and they adopted chemicals and commercial fertilizers to reduce the weed problem and increase yields. Total production increased despite acreage restrictions and the fact that only the more progressive farmers had made those changes.

From 1870 to 1925, the average acreage of cultivated land per worker increased from 32 to 49, chiefly because of mechanization. In constant dollars, the per worker machinery investment rose from $36 to $200. In total, this increase was from $270 million to $2.7 billion. But the real expansion was still to come. The output of agriculture rose 400 percent from 1870 to 1950. The labor force in agriculture in 1950 was 6.9 million, about the same as in 1870 but down from the peak of 11.6 million in 1910. During the same time, in constant dollars, farm output grew from $2.5 to $11.8 billion, and the rate of increase was accelerating.

Farmers learned that adopting technology was one of the best ways of reducing the cost per unit so that they could make a profit at lower market prices. They substituted capital for labor whenever they could. The mechanization movement gathered momentum in the 1930s. Productivity increased 11 percent in that decade, 25 percent in the 1940s, 20 percent in the 1950s, and 17 percent in the 1960s. The sharpest increase in productivity came during 1940 to 1945, at the time

A 6-foot combine developed in the 1930s and still in use on small farms in the 1950s and 1960s, North Carolina. (NAL)

The first Massey-Ferguson self-propelled combine, Model 20, being tested in western Canada by Tom Carroll, its inventor, 1939. (Massey-Ferguson)

when the most rapid migration off the farm was taking place. Farmers learned that the lowest-cost producer set the price. They had a choice of adapting or leaving the industry.

By the late 1950s U.S. agriculture had matured and was faced with new problems. Progressive farmers became interested in securing more specialized information. They wanted special regional and commodity magazines. They were less interested in general articles that appealed to a broader audience. The 1960s innovators started to adopt computers, which helped them refine their management techniques even more.

Technology and science had industrialized and mechanized farming and had reduced production costs to new lows. The USDA again confirmed that consumers were the ultimate gainers in the form of lower prices.

Tenant Farming

A USDA economist, writing in 1926, said:

Farm tenancy has been regarded by some as one of the outstanding evils of American agriculture—a foul monster dragging free American manhood down to the levels of European peasantry. But to others, more optimistic in temperament or more logical in their analysis of the situation, it has been looked upon as a normal element in the economic system.

The debate about the desirability or disadvantages of tenant farming continued during the first half of the twentieth century. The rate of tenancy decreased in the early part of the century in the older established areas but was on the increase in the newer western regions. The "agricultural ladder" still applied in the more recently settled states. Interestingly, more than half of the tenants were sons, sons-in-law, or close relatives of the landowners.

Henry C. Taylor suggested that the best way to keep farm tenancy down was to give rural youth a better education. This would not only facilitate the flow of surplus farm population into other occupations, but it would also improve the level of farming. Better education tended to reduce the birth rate in congested rural areas as people sought to gain a higher living standard. He suggested that the best way to reduce the rural birth rate would be to give every farm girl an opportunity for a high school education.

By the 1920s those who understood agriculture no longer considered tenant farming undesirable. Improved business management made large-scale farming possible and profitable. In 1925 the nation had a 38.6 percent tenancy rate. However, in the six best agricultural states, the tenancy rate averaged 50 percent, and in Iowa and Illinois, two of the premier farm states, it was 53 and 55 percent, respectively. In some of the very best counties the tenancy rate was 70 percent. With scientific methods, good managers had changed tenant farming from the weakest part of agriculture to the strongest. It was a boon to farmers, for it enabled them to use the capital of investors.

There will always be farmers who want to farm on a small scale. A USDA study expressed the hope that the pressure from large-scale farmers might at least make the smaller farmers strive to become more efficient and less wasteful of resources. The study's long-range social implications were that fewer farmers would lead to fewer but better rural communities, a more efficient use of resources, and a greater degree of stability in agriculture because of better leadership.

Since the Civil War the more commercially minded farmers had been the leaders in leasing land. They recognized the advantage of having someone else take the risks of ownership. Farmers realized that it was more profitable to invest funds in the operation of the farm and in equipment rather than in land. By renting, a farmer could operate a larger business with a far greater profit potential.

The USDA, which tracked tenancy annually, noted that the greatest increase came in areas of specialized cash-crop production, especially cotton, corn, wheat, and tobacco. In 1935 about 65 percent of cotton, 44 percent of corn and wheat, and 49 percent of tobacco farms were leased. As a sidelight, the Department had discovered that where birth rates were high, farms did not stay in the family for long.

Based on experience, USDA economists realized that if all farmers were given an equal number of acres, within a few years a large portion of the farms would not be owner operated. The history of farm inheritance has determined that, overall, 30 percent of the first generation of heirs will succeed. In the second generation the success rate is 8 percent. Century farms are indeed unusual.

In the prosperity of 1948, after a series of profitable years for agriculture, the nation probably experienced its highest portion of farm ownership. At that time there were 4 million full or part owners and 2 million full tenants. This was after legislation had been passed to aid farmers in acquiring land: (1) homestead tax exemptions; (2) graduated land tax on large or absentee holdings; (3) acreage limitations; and (4) a special farm tenant purchase program.

The 10-year experience of the Bankhead-Jones Farm Tenant Purchase Program indicated that the application rate was so low it would have taken nearly 400 years for all tenants to become owners. This was a generous act that had no limit as to the size of loan, and the borrower did not need to have a down payment. The farm had to be an efficient management unit for the family. Family and farm budgets were required, and the borrower could not rent or purchase additional land.

Progressive, business-oriented farmers have taken the same position as the rest of the business world. Those farmers realize that the basis of their welfare and independence is not land ownership but income. The restructuring of agriculture made it no different from other industries. It was more desirable to own sufficient equipment and manage a larger unit on rented land than to own and live from the proceeds of a small farm.

Farm Consolidation

The debate on farm size dates at least as far back as Aristotle, when everyone still worked the land by hand methods and nearly everyone was a farmer. In 133 B.C., within the Roman Empire, there were laws against farms of over 300 acres, because they were considered too large. In early New England farm size was discussed at town meetings. In the 1930s discussion ranged from "bittersweet invective to the most extravagant eulogy." Until then there was no effort to define a large-scale farm or why it was viewed with alarm or pointed to with pride. A land-grant college president, who had done extensive research in agricultural economics, told this writer his definition of a large farmer—"anyone who farms bigger than the farmer you are talking to."

Economist A. G. Black wrote: "In fact many persons, who by nature are inclined to look backward rather than forward, . . . seem to delight in crying `large-scale production' upon every occasion when a new technique gives promise of being generally adopted."

Agriculture lagged behind other industries in increasing its size of operation because a good power unit was not available. Horses and steam engines were equally inefficient and costly to operate. Large-scale farms, such as the bonanza farms of Minnesota and North Dakota, were simply multiple units of small farms. On the bonanzas, one worker and a full line of equipment and horses generally were used for every 250 acres. Up to 20 complete units operating under one field foreman made up one division of a bonanza, which might have several divisions. The bonanza farms gained their advantage because they had professional management, were specialized, had lower labor cost, and worked their equipment to capacity. They also profited from buying and selling large quantities of equipment or produce.

G. F. Warren concluded in 1913 that 320 acres were ample for peak efficiency but acknowledged that if a farm was well located, it might be as large as 640 acres. Warren judged his farm on the basis of the conditions in New York State. He determined that the larger the farm, the more likely farm boys were to stay at home. His data showed a range of 21 percent remaining at home if the farm was under 30 acres to 84 percent if the farm was over 200 acres.

One of the early USDA studies on large-scale farming indicated that consolidation was taking place on a gradual basis and that nearly all of it was being done by the "average family operator." The movement started in the Great Plains, where the distant horizon enabled the imagination to visualize a larger operation. Next, it took place in the Cotton Belt, where farms had become so fragmented after the Civil War that agriculture could not attain efficiency until there was consolidation. By the early 1930s the row-crop tractor encouraged consolidation in the Corn Belt.

A 1930 study of Great Plains wheat farms addressed the fact that consolidation was a response to economic forces and seemed inevitable. The economist who conducted the survey suggested that large-scale farms more satisfactorily met the needs of farm families.

The idea that farm consolidation would continue only within the limits of the family was still considered by British and American economists in the 1950s. It was the impression that the advantages of large-scale operations were less in agriculture than in other industries because the problems of management increased more with size in agriculture than in other industries.

By the 1920s it was understood that a farm ought to be large enough so that the operators could devote full time to managing the business of farming and not have to exhaust their energy in physical labor. The basic problem with the average farm was that at least 50 percent of the expenses of the farm were for family needs. By the 1940s the top 100,000 farms, 2.3 percent of the total, sold 22 percent of all farm products. Corn and wheat production per acre and milk production per cow on large-scale farms exceeded that on smaller farms. In all cases the smaller the average-size farm, the lower the average yields.

The need for some kind of rural development became obvious. In 1950, out of 5.4 million farms, 1.5 million had less than $1,000 net income from all sources. There were 2.85 million farms with less than $2,000 annual net income, and 3.3 million with less than $2,500 annual net income from on- and off-farm sources. Of those low-income farms, 2.7 million had a head of family over 65 years of age. Supportive studies indicated that the areas of greatest outmigration maintained the highest standard of living. The areas that had the highest percentage of original owners were those that had the greatest rate of consolidation.

Enter Agribusiness

After World War I farm numbers declined steadily, but that did not mean that agriculture was declining. The nation was building an agricultural infrastructure that would be unequaled and would do much to make farming more efficient. Agribusiness joined with production agriculture to give us the soundest food system in history. However, agribusiness has been a good target for attack by farmers who have not understood its contribution to a dynamic agriculture.

The Farm Credit System

One of the most serious handicaps to farmers was the lack of reasonably priced long-term credit. When the free land frontier ended and land values rose, farmers had to compete with the industrial sector for funds, and the need for money outpaced the supply. This caused acute stress for farmers. Bank panics in 1896 and again in 1907 revealed the weaknesses of our financial system and led to the creation of the Federal Reserve System. This served as a cornerstone for a farm credit system. The Country Life Commission realized the farmers' plight and recommended a cooperative agricultural credit system.

A commission was sent to Europe to study the agricultural credit system there. The result was the Federal Farm Loan Act of 1916, in which Congress provided capital stock for the Federal Land Bank. A system of 12 regional farm loan banks empowered to sell bonds to the public and reloan to farmers was created. The purpose of the Federal Land Bank was to make long-term money available to farmers. In the 1920s, when farm prices dropped and farmers faced financial difficulties, the government invested additional funds in the system. It was important to keep agriculture solvent, for in 1924 it represented more fixed capital than any other industry.

The Farm Credit System expanded with the passage of the Agricultural Credits Act of 1923, which provided for a system of 12 Federal Intermediate Credit Banks. The purpose of these banks was to provide farmers with intermediate-term loans for operating needs. In 1932 the Reconstruction Finance Corporation was established, in part to rediscount agricultural loans, make loans for feeding cattle, and buy stock in agricultural credit corporations. Congress also provided the Federal Land Bank with another $125 million for the purpose of extending the loans of hard-pressed borrowers.

In 1933 the Federal Farm Credit Act provided for the establishment of Production Credit Associations (PCAs) and the Bank for Cooperatives. Between 1933 and 1937, the government supplied funds under the Emergency Farm Mortgage Acts to keep the Farm Credit System afloat. This was critical to farmers, for during 1930 and part of 1931, a total of 2,232 rural banks went bankrupt.

One of the most significant early contributions of the Federal Land Bank was providing real estate loans of up to 40 years but with the freedom to prepay. Prior to that, farmers were limited to bank real estate loans of three to five years, and generally prepayment was not permitted. The Farm Credit System leadership felt that it was better for farmers to use their income for operating expenses and let the land serve as their lever. This was a basic change in the philosophy of farm financing. Historically, farmers felt that it was advantageous to get out of debt as quickly as possible.

A goal of the Farm Credit System was to maintain small-scale farming. However, the more astute farmers quickly grasped the advantages of long-term financing and increased the size of their operations.

After 1933 the new cooperative finance system brought profound changes to farmer cooperatives, and they experienced great growth in membership and volume of business. The Bank for Cooperatives not only financed those cooperatives, but it also provided financial advice. Cooperatives became involved in meat packing, sugar processing, milk drying, and rice milling and with feed mills, fertilizer plants, petroleum refineries, and a host of other agribusinesses.

Livestock Marketing

What is believed to be the first U.S. livestock auction was held in 1836 in Berlin, Ohio. In 1853 the second known auction was established in nearby London, Ohio, with sales held monthly in the public square. The London market remained important until the early 1900s. In the 1850s other auctions, called "court-day sales," appeared in central Kentucky. From those early events a system of auction markets spread slowly throughout the major livestock-producing areas.

In 1912 the Grand Island (Nebraska) Livestock Commission Company became the first livestock auction established in the West. It started as a horse and mule market and later expanded to all livestock. In 1937 it was still one of the largest in the nation. In 1917 farmers in Kern County, California, held their first sale. Within two years this market included a seven-county area. In the early 1920s it became part of the

California Farm Bureau Marketing Association. During the early 1920s auctions spread throughout the eastern and southern states.

Small-town merchants and chambers of commerce hoped that by holding auctions they could help Main Street. There was a quick rise and a sudden fall in the number of auction markets. However, many farmers desired to market close to home, partly because they could pick up the check on the same day. Attendance at the auctions ranged from 200 to 800. Auctions were semi-social events, and often entire families attended.

Low prices for livestock starting in 1929, plus the droughts of 1933, 1934, and 1936, gave added impetus to local marketing. This writer saw many cattle come to a major market that were so starved that they could not make the long haul. Nearly every carload contained dead animals. The dead ones were a cost to the farmer who shipped them, because the tankage value did not pay the freight charge.

Auction marketing expanded in the 1930s, and by 1937 there were 1,317 livestock auction markets in 37 states. Iowa was the leading state with 195, followed by Illinois and Kansas with 139 each and Missouri with 113. Traders and dealers were the chief promoters of livestock auctions, which had investments ranging from a few hundred dollars to $90,000, with an average of $7,345. Because of the low original capital requirements, 92 percent of the auctions were private and 8 percent were cooperative. Weekly auctions were usually held, with annual volumes ranging from less than 5,000 head to more than 75,000. Only 10 livestock auction markets handled more than 75,000 head annually. In 1936 the weekly auctions' dollar volume ranged from $10,000 to $4,437,000, with an average of $550,320 per auction.

Most livestock came from within 50 miles of the auction barn. Farmers purchased 36 percent of the livestock; packers, 22 percent; order buyers, who most likely represented distant farmers, 22 percent; and livestock traders, 20 percent. Traders supplied 26 percent of the cattle. Trader cattle were not always an asset, because many of them had traveled from one market to another. In addition to being run down, they also were carriers of disease. In 1937 only 29 percent of the auction markets carried bonds, over half of which were for less than $5,000. Farmers seemed willing to overlook the risk. The selling cost at local auctions was frequently higher than at the Chicago or Kansas City terminal markets. Despite many negatives, auction markets increased to about 2,500 in the early 1950s, and then they started to decline. The remaining firms increased in volume handled.

The cooperative marketing efforts of the AFBF made 1924 a pivotal year in livestock marketing. Many local livestock shipping associations were still being formed. On the other hand, the use of trucks and improved roads sealed the fate of many of those associations because farmers started bypassing the associations and the terminal markets and selling directly to packers. That was at the same time packer buyers began buying hogs in the country. This helped packers obtain a steady source of supply rather than having to rely on terminal markets. The results were clear, for in 1925 the major terminals handled 91 percent of the cattle; 85 percent of the sheep, lambs, and calves; and 78 percent of the hogs. By 1955 their share had fallen to 70, 40, and 38 percent, respectively.

One of the key changes in this process was that terminal markets lost their price-making influence. Larger livestock and grain producers began selling directly to the processors. The massive, flexible transportation and warehousing system made large-volume, efficient movement of goods possible. This was a positive factor for the larger producers, for it reduced their marketing costs.

Commercial Fertilizer

Commercial fertilizer was first used in the East and the South in the early 1800s for truck gardening, tobacco, and cotton production. However, most farmers in eastern and southern states and in the rest of the nation had never seen a sack of commercial fertilizer prior to 1940. In the mid-1920s a group of farmers had a carload of fertilizer shipped to Mankato, Minnesota, under a Farm Bureau cooperative arrangement.

These farmers were experimenting with some of the first corn planters with fertilizer boxes attached for row application. In 1929 these planters were introduced to the market.

When World War I started, commercial fertilizer was regulated because it conflicted with the needs of the military. Potash, which came from Germany, was not available. The United States had been a leader in production of superphosphate since the 1870s, but that was needed for the war effort. Nitrate of soda, which came from Chile, was also banned for agricultural use. When food supplies were threatened, fertilizers were released to increase production. Under the Food Productions Act, rock phosphate and nitrate of soda were made available to 300,000 farmers. This proved to be of great value to food production. After the war there was a surplus of nitrate, and in 1919 farmers from 38 states purchased it at salvage price.

The fertilizer industry realized the market potential in agriculture. In 1910 the nation's farmers had used about 844,000 tons of nitrogen, phosphate, and potash. During the 1920s and 1930s, fertilizer use rose slowly because of surplus farm commodities. However, by 1940 the volume increased to 1.7 million tons, and by 1953 to 5.3 million. During World War II and the Korean War additional nitrogen plants were built to supply the military. After those conflicts the nitrogen plants became available to supply agriculture. In the early 1950s commercial storage facilities for stockpiling and distributing anhydrous ammonia appeared in the Corn Belt. The USDA estimated that 55 percent of the increased production per acre between 1940 and 1955 came from greater fertilizer use.

Hybrid Corn

In November 1492, Columbus made a reference to corn that was grown in Cuba, Santo Domingo, Trinidad, and throughout the mainland of South America. In 1623 the Pilgrims mentioned how they might grow corn with greater efficiency. In the 1870s plant breeders started working on improving corn, but it was not until 1904 that work started on the first hybrids at the Connecticut Experiment Station. In 1920 inbreeding programs were commenced in earnest.

From 1870 to 1902 the real price of corn rose from 40¢ to 80¢ per bushel because of a rising demand. Gross production increased from 1.2 billion bushels to 2.8 billion because more acres were planted to corn. But the disturbing fact was that yields per acre did not improve in those decades. Because yields did not increase, the supply of corn per capita declined from 32 bushels in 1903 to 18 bushels in 1936. Yields remained stagnant until 1925, when corn acreage shifted away from the south central region to what became the Corn Belt. Greater yields were needed to meet the long-term demand.

In 1917 the Connecticut station produced the first hybrid from inbred lines that sold commercially. In 1918 the secret of the "double cross" was learned. This became the key to developing hybrid seed corn. In 1919 the method of practical hybrid corn seed production using the double cross was developed. In 1924 Henry A. Wallace produced the second hybrid, a single cross. Wallace went on to found the first seed company for the commercial production of hybrid corn.

Hybrid corn restricts its use to the first generation after crossing, so it is necessary to purchase new seed each year. This became the basis for a major agribusiness industry. By 1932 other commercial hybrid producers were established. In 1933 only 1 percent of the corn acreage was planted to hybrid seed, and in 1938 only 15 percent. By 1946 the total rose to 69 percent, and by 1960 it reached 96 percent.

The delay in farmers' switching to hybrids became the basis for some important sociological studies. Average corn yields were about 25 bushels per acre, but hybrids provided increases of from 9.0 to 15.2 bushels per acre. The great puzzle was why farmers did not adopt hybrids. The next 50 years saw more improvement in corn than had been seen in previous history. Hybrids were adapted to fertilizers, seed treatment prior to planting, insecticides, and pesticides. This increased yields to levels that were not previously thought possible.

One of the great benefits of hybrids was the fact that the corn could be bred to many traits, making it adaptable to mechanical production. By 1936 two-row tractor cultivators made up 59 percent and four-row tractor cultivators 7 percent of all cultivators sold. Tractor cultivation to control weeds during the growing season was a key to increased yields. The problem of root cutting by cultivation was greatly reduced when pesticides were adopted. Increased yields led to the purchase of mechanical corn pickers, which decreased harvest labor requirements by 50 percent by 1938. When corn heads were adopted for use on combines, harvest labor was reduced still further.

In 1946 the *Farm Journal* promoted a 300-bushel-yield contest. In 1947 the top yield among the 5 contestants was 115 bushels per acre. The following year 250 farmers entered the contest, and the top yield was 224 bushels. In 1955 a Mississippi farm boy produced 304.39 bushels of 13.93 percent moisture corn on land that, except for 1944, had been in continuous corn since 1926. This record stood until 1975, when an Illinois farmer produced 338 bushels per acre. Wheeler McMillen, editor of the *Farm Journal* who started the contest in 1946, told this writer in 1990 that in retrospect the contest was a mistake because there was no need to accelerate production that quickly.

As acres increased, corn became more susceptible to disease. In 1917 the European corn borer, which was imported with broom corn used in eastern factories, was discovered. By 1920 over 4,500 square miles were infected in the United States and another 3,000 square miles in Canada. Losses to the borer fluctuated, but each year it spread to more areas. A nationwide borer outbreak in 1954 caused severe losses. The outbreak of 1971 caused a major catastrophe to the corn industry. The corn borer remains a problem, but scientists are developing a borer-resistant seed.

Soybeans

The first reference to soybeans in the United States was in 1804. In 1907 the USDA started research on soybeans. In 1910 a Pacific Coast crushing plant imported beans from Manchuria to produce oil and meal. American-grown beans were first crushed at cottonseed oil mills in 1915, when there was a shortage of cottonseed. Soybeans were also grown for silage.

In 1917 soybean acreage expanded when mills contracted with growers to crush their beans. At that time we were importing soybeans for commercial uses. Soybeans yielded 15 to 40 bushels per acre and were complementary to corn in rotation. Soybean oil was used in the manufacture of soap and paint, as a substitute for butter and lard, and in the production of explosives, linoleum, varnish, and foodstuffs. The meal was used in foodstuffs, as a cattle feed, and in fertilizer. Soybean flour was used in foods of low starch content, such as bread, muffins, and pastries. Soybeans were packed as baked beans by canners and mixed with field or navy beans for home baking.

In 1921 the A. E. Staley Company offered its first contract to farmers for soybeans. By 1924 production had expanded to 1.8 million acres, of which 450,000 were grown for the beans and the rest for silage or hay. In 1929 improved soybean crushing plants were developed, and demands for the high-protein crop exploded.

Once a technique was developed to deodorize the soybean, it became acceptable for salad and cooking oils. During World War II soybean oil was used as a base for margarine, but domestic consumption did not expand until the late 1940s, when restrictions were removed on colored margarine.

After effective chemical weed control was developed in the 1960s, production grew rapidly. Up to the 1970s, soybeans were the most effective producer of protein per acre. Soybeans became the number one cash crop and the leading agricultural export commodity.

Animal Science

Historically, most farmers kept milk cows because milk was essential for the family. Any surplus milk provided the women and children

Dairy cow barn with straw ceiling for warmth. Note wooden stanchions and the horse harness on the left. This was typical of small-scale dairy barns prior to electricity, ca. 1915. (UMA)

a means to produce extra income. The milking herd was not considered a major enterprise, so little effort was made to improve production. In 1920 about 5 million farmers had 23 million cows. The average cow produced less than 4,000 pounds of milk and about 160 pounds of butterfat. In 1930 the USDA estimated that one-third of the dairy cows produced a profit, another third probably broke even, and one-third lost money.

Between 1905 and 1910 the first cow testing association was organized in Michigan. In its first year the average production for all cows enrolled was 5,354 pounds. Ten years later it was 6,637 pounds. During that same period, butterfat increased from 215 to 276 pounds. By 1925, 18,677 farmers with 307,073 cows were involved, but there was always a struggle to hold membership.

Farmers had to be convinced to keep records on each bull as well as on the cows in their herd because bulls often were sold before it was discovered that their offspring were good producers. This became the basis of the need for dairy herd improvement associations (DHIAs) and cooperative bull associations, which were founded during the same years as the cow testing groups.

In the 1780s Italy was the first nation to use artificial insemination (AI). Russia adopted AI on a large scale in the late 1800s. The United States first used AI in the mid-1930s, and by 1941, 237 bulls were bred to 70,751 cows. This was about 8 to 10 times what bulls would have been capable of breeding in the traditional way. By the 1940s each bull in an association provided semen for 644 cows, in contrast to 11.1 for a bull in a herd that still used direct breeding. AI was pushed because it was a good way to prevent disease and a very economical way to improve production.

As incomes increased in the 1940s, the demand for high-quality meats grew rapidly. Researchers sought ways for more efficient production. During the 1940s and 1950s, Streptomycin, Neomycin, Aureomycin, Terramycin, and Declomycin were developed to fight skin, intestinal, or general systemic infections. By 1948 vitamin B_{12} was discovered to be especially valuable for swine and poultry and also for humans. Growth rates for poultry and swine were outstanding. Additives lowered the death loss and reduced time and feed requirements to produce milk, eggs, and meat.

The antibiotics, which effectively helped to reduce outbreaks of "stress" diseases, enabled producers to safely concentrate larger numbers of poultry and livestock in a small area. This required specialized yards or barns but made for efficiencies in economies of scale by enabling skilled workers to care for large numbers of livestock or poultry. There was some criticism and opposition to continuous feeding of antibiotics because of the possibility of a Resistant Transfer Factor (RTF) in some bacteria that could transfer beyond the animals or poultry fed. In the first quarter century no negative results were determined in Europe or America, with one exception in the United Kingdom where animals definitely had been mishandled.

Chemurgy

Wheeler McMillen, a co-founder of the chemurgy movement, claimed that the idea of chemurgy first came to him in 1924 when he heard a remark that the stomach was inelastic. Experts were proclaiming that population had nearly reached its maximum. New uses had to be found for agricultural produce. McMillen wrote the first known article on chemurgy in October 1926, and in 1927 USDA scientists started looking for ways to increase the industrial use of farm products.

In 1935 the first nationwide conference on chemurgy was held under the sponsorship of Henry Ford and agribusiness and farm organizations. Chemurgy offered no quick solutions to the farm surplus problem. The chemurgists wanted private industry and the government to research farm commodities for new uses. In 1938 $4 million was allocated for building four regional laboratories to experiment with farm produce for industrial purposes.

One of the earliest attempts to find a nonfood use for farm products dates to 1906, when research was done on alcohol made from corn. During World War I, Henry Ford attempted to create an interest in producing alcohol from grain and garbage because of the threat of a gasoline shortage. In the Prohibition years producers of alcohol again sought ways to blend ethanol and methanol with gasoline.

In 1922 and 1923 Standard Oil attempted a 25 percent market blend. The program was discontinued after problems developed and customers complained. Other attempts in the 1920s also ended in failure. When farm prices dropped sharply in 1932, grain producers again looked to an alcohol motor fuel. In 1934 the Chemical Foundation pushed alcohol as a fuel. A plant was built in Kansas, and in 1938 Agrol was sold at 2,000 stations in 8 states. Unfortunately, the cost of production doomed the product to failure, and in 1939 the plant closed.

Prior to the 1920s furfural, a part of oat hulls, became a real problem for the Quaker Oats factory because it accumulated and had no market. Then it was discovered that furfural could be used in the production of a solvent, as an additive to lubrication oils, in germicides and fungicides, and for several other purposes. Soon there was a shortage of furfural, and a new source had to be found. The corncob proved to be the answer.

The first specific breakthrough under chemurgic research came in 1936 to 1939 when flax straw was used successfully to produce cigarette paper, fine text paper, airmail paper, and paper currency. During the 1940s a second chemurgic breakthrough came from the production of cornstarch. It was used for food, in the textile and paper industries, and for dynamite, insecticides, pharmaceuticals, paint, soap, and countless other needs.

The reasoning of the early chemurgic leaders' crusade to use farm products for industry was that crops took only 2 percent of their substance from the soil. The other 98 percent came from the air, rain, and sunshine. The minerals extracted from the soil could be replaced easily, and therefore crops could be continually reproduced. Chemurgic leaders foresaw chemurgy as a way of preserving oil, iron, coal, and other minerals, which they felt were finite.

The Food Chain

After 1900, urbanites no longer produced much of their own food, and farmers began to purchase more of their food in processed forms. The demand for processed food increased as improved methods of preserving and preparing were developed. By 1940 more than 65 percent of retail food was partly or fully processed. By 1960 that figure increased to over 80 percent, and food processing was one of the nation's largest industries. World War II greatly accelerated food processing, for the military needed to supply processed food throughout the world. That work created the convenience food industry. The increase of women in the labor force continued the demand for processed foods.

By 1920 grocery chains controlled 10,000 stores, with 22 percent of total sales. By 1928 the 315 chains had 32 percent of all grocery sales. In 1930 those chains had 30,453 stores, with an annual volume of $11,400 per store. By 1949 store numbers had fallen to 12,621, but the annual volume of each store had increased to $226,000.

Enter Agribusiness / 359

Farmers' market near Gamble-Robinson produce building, Minneapolis, 1914. (UMA)

A & P maintained its position as leader of the industry by eliminating credit and delivery and by reducing prices. Its stores remained small, one-person operations, with low rent and minimal fixtures, that sold at a narrow margin. By 1930, 15,737 A & P neighborhood stores having an average daily volume of $175 each sold a combined gross of over $1 billion.

In spite of the small size of the local stores, the chains were strong competition for the independents. The independent wholesalers who supplied the independent stores were bypassed by the chains at the central and terminal markets. Pricing once done by the terminals shifted to the exchanges and futures markets.

The independent stores, wholesalers, and terminal market interests went to Congress for help. The Federal Trade Commission prohibited trade pressure by the independents against the chains. In 1936 the

Robinson-Patman Act was passed specifically to regulate the chain store industry. Instead of harming the chains, the Act opened the door for them to buy directly from processors, cutting out the wholesaler and reducing the cost of their goods. Mass buying and scientific management, along with lower prices, wide selection, and quality, were keys to the success of the chains. Mass advertising, financial strength, and mass distribution also were helpful toward their success.

Initially, farmers feared the growth of chain stores, but in 1936, after a meeting of chain store and farm leaders, the attitude changed. Farmers learned that chains reduced distribution costs. This enabled lower prices to the consumer and, it was hoped, greater consumption, which would lead to better income for the farmer.

In 1916 the first Piggly Wiggly store featuring self-service opened. By 1928 there were 2,700 Piggly Wiggly stores or stores franchised by the chain. The self-service innovation was quickly adopted by both chains and independents.

The combination store for one-stop shopping appeared during the 1920s. This encouraged the advent of grocery store parking lots, which led to the development of the supermarket.

Henry Ford opened the first known supermarket in 1919. By 1926 he had 11 stores, with combined sales of $12 million. Most of the merchandise was prepackaged and price-marked. Because of public opposition, the stores were later restricted to Ford employees only.

In 1930 the first public supermarket that developed into a chain appeared. The King Kullen stores stressed low prices and concentrated on grocery, meat, bakery, and dairy products. Merchandise was available in large quantities. Concessionaires could rent space to sell produce, hardware, utensils, etc., in the stores. King Kullen stores had parking lots, and customers came from as far away as 30 miles. The stores averaged $10,000 a week in sales, while the average independent did $500.

The supermarket was heralded as being the most important retailing concept in food marketing history. Cash and carry, self-service, and wide variety were all combined. By 1937 there were 3,066 supermarkets, and the major chains had adopted their ideas. They featured prices on every shelf, parking lots, large refrigeration equipment, extensive lighting, displays, shopping carts, and checkout stands.

In 1925 Clarence Birdseye produced and marketed frozen fish. During the 1930s a complete line of meats, fish, fruits, and vegetables was successfully frozen, but sales developed slowly. However, wherever there were test markets, the sale of frozen food skyrocketed. Others experimented to determine the effectiveness of preservation of food by freezing. Frozen food locker plants were established in rural areas.

About 1933 General Foods rented freezer cabinets to retailers at a low monthly rate to entice them to stock frozen products and create a market. Refrigerated trucks came into use in the late 1930s, but railroads did not adopt mechanical refrigeration until 1949. By 1939 only 12,000 retail stores out of 600,000 had frozen food displays. Home refrigerators had freezing compartments, but that did not increase consumer acceptance. It was not until World War II that a mass market for frozen foods developed. By 1944 about 70 percent of the households had refrigerators, and frozen foods were widely accepted.

In 1947 marketing of fresh meat by self-service was done for the first time. This immediately spread to merchandising perishables in refrigerated display cases.

Processing, packaging, transporting, and marketing techniques all played a part in the explosion of the growth of the food industry. Canned goods, frozen foods, dehydrated foods, cake mixes, freeze-dried foods, and a steadily growing variety of innovations changed the food industry. The average number of products in a supermarket increased from 867 in 1928 to 3,000 by 1946, and stores continued to grow in size.

By 1952 supermarkets with over $375,000 in sales had 44 percent of the market, while superettes with sales of $75,000 to $375,000 had a 35 percent market share. The two groups made up 25 percent of all stores. Small stores with less than $75,000 in sales made up 75 percent of all stores and had 21 percent of the sales.

Summary

From 1914 to 1954 a mature agriculture started the long, slow process of reducing the number of farms, farmers, and farm laborers. Conditions improved for those who remained. Many of the changes can be directly attributed to the government. Government programs were designed to assure that there would be an ample supply of food at a reasonable price to the consumer and at the same time to improve the position of those who were below-average-cost producers.

It was too costly and wasteful of resources to maintain all who wanted to remain in farming. The gigantic rural rehabilitation program of the 1930s proved that point. Farm numbers could not be retained from either an economic or social standpoint. Each succeeding technological innovation made it possible to produce more with less land and less labor. Progressive farmers maximized the benefits of government programs for their personal gain and used them to expand their holdings to the disadvantage of the less creative operators.

While the progressive sector of production agriculture continued the process of commercialization and industrialization, another portion continued to look upon the farm as a way of life blended with off-farm employment. This gave the United States an increasingly diverse agriculture. Large-scale family farmers continued the business of farming as their sole source of income, while a growing number of part-timers derived a minor portion of their total income from the farm.

At the same time, the agriculture infrastructure went through the same process as the farmers. The number of service institutions declined while the total volume of business done by the survivors grew. This was true from the smallest rural cooperative elevator to the largest Main Street supermarket. All the time the relative cost of food continued to drop.

American farmers had adapted well. They provided for the ever-growing domestic market, and when called upon they met demands of the world market. Their major problem was that they continued to produce more than the market needed. In 1953, 24 million people, nearly 40 percent of the labor force, were involved in agriculture—about 6 million in farm supply; about 8 million on the farm; and about 10 million in processing and distribution.

References

Bailey, L. H. *The Harvest of the Year to the Tiller of the Soil.* New York: Macmillan Publishing Co., Inc., 1927.

Baker, John C. *Farm Broadcasting: The First Sixty Years.* Ames: Iowa State University Press, 1981.

Baker, O. E., Ralph Borsodi, and M. L. Wilson. *Agriculture in Modern Life.* New York: Harper & Brothers, 1939.

Banfield, Edward C. "Ten Years of the Farm Tenant Purchase Programs," *Journal of Farm Economics* 31, No. 3, August 1949, 469–486.

Bertrand, Alvin L. *Agricultural Mechanization & Social Change in Rural Louisiana.* Agricultural Experiment Station Bulletin 458. Baton Rouge: Louisiana State University & Agricultural & Mechanics College, June 1951.

Black, A. G. "Aspects of Large-Scale Farming in the Corn Belt," *Journal of Farm Economics* 13, No. 1, January 1931, 146–154.

Breimyer, Harold F. *Over-Fulfilled Expectations: A Life and an Era in Rural America.* Ames: Iowa State University Press, 1991.

Brewster, David E., Wayne D. Rasmussen, and Garth Youngberg, eds. *Farms in Transition: Interdisciplinary Perspectives in Farm Structure.* Ames: Iowa State University Press, 1983.

Busch, Lawrence, William B. Lacy, Jeffrey Burkhardt, and Laura R. Lacy. *Plants, Power, and Profit: Social, Economic, and Ethical Consequences of the New Biotechnologies.* Cambridge, MA: Basil Blackwell, Inc., 1989.

Ezekiel, Mordicai. "Schism in Agricultural Policy: The Shift in Agricultural Policy Toward Human Welfare," *Journal of Farm Economics* 24, No. 2, May 1942, 463–476.

Fink, Deborah. *Agrarian Women: Wives and Mothers in Rural Nebraska, 1880–1940.* Jack Temple Kirby, ed. Chapel Hill: University of North Carolina, Studies in Rural Culture, 1992.

Fink, Deborah. *Open Country Iowa: Rural Women, Tradition and Change.* Albany: State University of New York Press, 1986.

Fite, Gilbert C. "Recent Progress in the Mechanization of Cotton Production in the United States," *Agricultural History* 24, No. 1, January 1950, 19–28.

Genung, A. B. *A Brief Survey of 35 Years of Government Aid to Agriculture Beginning in 1920.* Ithaca, NY: Cornell University Press, Northeast Farm Foundation, 1960.

Griswold, A. Whitney. *Farming and Democracy.* New Haven, CT: Yale University Press, 1952.

Helms, Douglas. *Readings in the History of the Soil Conservation Service.* Washington, DC: USDA Soil Conservation Service, Economics & Soil Sciences Division, NHQ, 1992.

Hessor, Leon F., Raymond J. Doll, Glenn H. Miller, Jr., and Richard D. Hess. *Foreign Trade and American Agriculture.* Kansas City, MO: Federal Reserve Bank of Kansas City, Research Dept., August 1966.

Higbee, Edward. *Farms and Farmers in an Urban Age.* New York: Twentieth Century Fund, 1963.

Holley, Donald. "The Negro in the New Deal Resettlement Program," *Agricultural History* 45, No. 3, July 1971, 179–194.

Hopkins, John A. *Changing Technology and Employment in Agriculture.* Washington, DC: USDA Bureau of Agricultural Economics, May 1941. Reprinted, New York: Dacapo Press, 1973.

Iwata, Masakazu. "The Japanese Immigrants in California Agriculture," *Agricultural History* 36, No. 1, January 1962, 25–37.

Jones, Lawrence A., and David Durand. *Mortgage Lending Experience in Agriculture.* Princeton, NJ: Princeton University Press, 1954.

Knapp, Joseph G. *The Advance of American Cooperative Enterprise: 1920–1945.* Danville, IL: Interstate Publishers, Inc., 1973.

Lebhar, Godfrey M. *Chain Stores in America: 1859–1959.* New York: Chain Store Publishing Corp., 1959.

Lerch, Donald G., Jr. *Dissemination of Farm Information Through Newspapers, Magazines, Radio and Television*. Washington, DC: International Cooperation Administration, November 1959.

Macy, Loring K., Lloyd E. Arnold, and Eugene G. McKibben. *Changes in Technology and Labor Requirements in Crop Production: Corn*. Philadelphia: WPA National Research Project, Report A-5, June 1938.

Malin, James C. "Mobility and History: Reflections on the Agricultural Policies of the United States in Relation to the Mechanized World," *Agricultural History* 17, No. 4, October 1943, 177–191.

McMillen, Wheeler. *New Riches from the Soil: The Progress of Chemurgy*. New York: D. Van Nostrand Co., 1946.

Perkins, Van L. *Crisis in Agriculture: The Agricultural Adjustment Administration and the New Deal, 1933*. Berkeley: University of California Press, 1969.

Pinches, Harold E. "Revolution in Agriculture," *Yearbook of Agriculture, 1960*. Alfred Stefferud, ed. Washington, DC: USDA, 1960, 1–10.

Randell, G. G., and L. B. Mann. *Livestock Auction Sales in the United States*. Washington, DC: Farm Credit Administration Bulletin 35, May 1939.

Rau, Allan. *Agricultural Policy and Trade Liberalization in the United States 1934–1956: A Study of Conflicting Policies*. Paris: Librairie Minard, 1957.

Riley, Glenda. *The Female Frontier: A Comparative View of Women on the Prairie and the Plains*. Lawrence: University Press of Kansas, 1988.

Ross, Earle D. "The United States Department of Agriculture During the Commissionership: A Study in Politics, Administration, and Technology, 1862–1889," *Agricultural History* 20, No. 3, July 1946, 129–143.

Schideler, James H. *Farm Crisis, 1919–1923*. Berkeley: University of California Press, 1957.

Scholl, Kathleen K. "Household and Farm Task Participation of Women," *Family Economics Review*. Washington, DC: USDA ARS, Family Economics Research Group. Special Issue: Household Production, 1982, 3–9.

Scruggs, Otey M. "Evolution of the Mexican Farm Labor Agreement of 1942," *Agricultural History* 34, No. 3, July 1960, 140–149.

Spillman, W. J. "Factors of Efficiency in Farming," *Yearbook of Agriculture, 1913*. Jos. A. Arnold, ed. Washington, DC: USDA, 1913, 9–64.

Starch, E. A. "Experiments in the Use of Large-Scale Machinery Under Montana Conditions," *Journal of Farm Economics* 14, No. 2, April 1931, 335–340.

Stock, Catherine McNicol. *Main Street in Crisis: The Great Depression and the Old Middle Class on the Northern Plains.* Chapel Hill: University of North Carolina Press, 1992.

Tostlebe, Alvin S. *Capital in Agriculture: Its Formation and Financing Since 1870.* Princeton, NJ: Princeton University Press, 1957.

Tugwell, Rexford G. "A Planner's View of Agriculture's Future," *Journal of Farm Economics* 31, No. 1, February 1949, 29–38.

United States Senate Committee on Agriculture, Nutrition, and Forestry. *A Brief History of the Committee on Agriculture, Nutrition, and Forestry, United States Senate, and Landmark Agricultural Legislation, 1825–1986.* Washington, DC: Congressional Research Service, GPO, 1986.

Volti, Rudi. "How We Got Frozen Food," *Invention and Technology* 9, No. 4, Spring 1994, 47–56.

Welch, Frank J., and D. Gray Miley. "Mechanization of the Cotton Harvest," *Journal of Farm Economics* 27, No. 4, November 1945, 928–946.

Witte, Raymond P. *Twenty-five Years of Crusading: A History of the National Catholic Rural Life Conference.* Des Moines: National Catholic Rural Life Conference, 1948.

Unit V

1954 to 1994

Some Events and Technological Innovations That Affected Agriculture

1956 Soil Bank

1956 Food stamp program

1957 Four-wheel-drive tractors marketed

1960s Chemicals and fertilizers cause explosion in crop production

1985 USDA involved in biotechnology

1985 Export enhancement and the Conservation Reserve Program (CRP)

1985 Bovine somatotropin (BST) produced; approved for use in 1993

1993 North American Free Trade Agreement (NAFTA) adopted

1993 Global Positioning System (GPS) adapted to agriculture

1994 General Agreement on Tariffs and Trade (GATT) enhanced

Introduction

The period from 1954 to 1994 saw a continuation of past trends except that the rate of change was accelerated. The government was increasingly involved in efforts to maintain the delicate balance between demand and production. Its programs were aimed at attempting to ease the burdens for those in farming while protecting the nation's consumers with an adequate supply of relatively inexpensive food. The programs provided good insurance to low-cost producers who took the opportunity to expand their operations. At the same time, they worked to the disadvantage of less-efficient, higher-cost operators who were unable to make a profit at the price levels established. The inevitable trend in declining farm numbers continued, with the sharpest drop in the decade of the 1950s. A mature agriculture continued to reduce farm numbers and surplus labor in its struggle for efficiency.

The first major effort to reduce production and still aid disadvantaged farmers came through a massive soil conservation program dubbed the "Soil Bank." The Soil Bank enabled farmers to lease their land to the government for 10 years and gracefully depart from farming. The wisdom of the program was realized as the contracts expired and virtually 100 percent of the land was merged into existing operations. Very few of those who soil banked returned to farming.

The government simultaneously sought ways to reduce the surpluses through the food stamp program at home and Public Law 480 for countries in need of food but unable to pay for it in hard currency. The positive side of P.L. 480 is that many of those countries became customers for U.S. agricultural products once their economies developed.

The realities of the shrinking world impacted agriculture in several ways in recent decades. The devaluation of the dollar in 1971 spurted agricultural exports, causing a major improvement in the agricultural economy. Exports increased each year until 1981, when they reached

$43.8 billion, and then declined steadily until 1987. The years of decline meant years of recession for agriculture as commodity prices dropped and land values declined.

In the 1980s export enhancement programs were established to encourage other nations to purchase our surplus commodities. Domestically the Conservation Reserve Program (CRP) was created to take land subject to erosion out of production for 10 years. In addition to preserving land resources, the program was aimed at reducing the excess production that has plagued every Secretary of Agriculture in recent decades. Further attempts to improve long-range export potential involved the passage of the North American Free Trade Agreement (NAFTA) with Canada and Mexico. In 1994 the General Agreement on Tariffs and Trade (GATT) brought signatory nations another step nearer to free trade.

The technology parade that started slowly in the 1830s and gained speed with the development of the tractor accelerated in 1957 with the advent of the four-wheel-drive tractor. Labor costs per acre dropped to new lows. In the 1950s hand labor was virtually replaced for sugar beets, potatoes, tomatoes, and cherries by the development of harvesters and for milking by the introduction of herringbone milking parlors.

Chemicals and commercial fertilizers enhanced production per acre. Chemicals enabled innovators to adopt low-till and no-till programs, which reduced costs and erosion and minimized compaction. Strong environmental pressures were brought against agriculture starting in the 1960s.

While farmers innovated to become more efficient, the nation's consumers benefited from an ever-increasing variety of improved foods at steadily declining relative costs. Agriculture continued to do what it had been guided to do under the massive government programs. A decline in government research and development programs was offset by increased activity in the private sector.

Production Booms

A USDA study compiled in the 1980s indicated that from the Revolutionary War until 1870, productivity rose about 0.4 percent a year. In the horse-power era from 1870 to the 1920s, it increased 0.5 percent a year. During the mechanical power era from 1925 to 1945, the annual rate of growth was 1.2 percent, and in the science/power age from 1945 to the present, productivity rose 1.6 percent a year through 1994. The increases came in spite of the prolonged drought of the 1930s, the corn blight of 1954 and 1970, the sharp rise in input costs in 1975, and the droughts of 1980 and 1988. The number of persons supplied by one farm worker rose from 4.12 in 1820 to 6.95 in 1900 to 15.47 in 1950. In 1994, the number of persons supplied by one farm worker reached 128.

The United States remains one of the few net food exporting countries of the world. It usually is the number one food exporter. In the 1920s, 6.5 million farmers fed 106 million Americans; by the 1970s, 2.8 million farmers supplied 214 million Americans. In 1994, 1.9 million farmers were still counted, but about 350,000 supplied virtually all the produce to 260 million Americans and still exported about $44 billion in food.

Agriculture has supplied surplus farm labor to a steadily growing industrial sector. The agricultural labor force declined by one-third from 1950 to 1960. This made up nearly 40 percent of the new workers needed in other sectors. Total output continued to increase in spite of all efforts to control production.

Labor input needed in farming has been reduced each decade since 1920. Fewer people are employed on farms to supply a population of 260 million than were required in 1800 to supply a population of 5.3 million.

After agriculture proved that it could provide for the nation's needs, federal allocations dropped from 40 percent of all research and development (R&D) funds in the 1940s to 1.6 percent in 1966. In the 1980s public funds for agricultural R&D reached $2.1 billion annually, supplemented by private funds of $2.1 billion.

Historically, the first benefactors of research are the early adopters. As adoption by more farmers continues, production increases, prices fall, and the consumer becomes the major long-term benefactor. Ongoing technological change in agriculture has been a major restraint on the consumer's cost of living. At the same time, agriculture has released land for urban development, transportation, forestry, and recreation.

One of the greatest handicaps of many farmers has been their inability to manage the new technologies effectively. Specialization has also served to increase productivity as farmers have learned to maximize production in the commodity that provides the best comparative advantage. The more complex the production process, the more important it is to specialize.

During the last 40 years, crop production rose by more than 100 percent, with only a 3 percent increase in land and a decline of over 65 percent in labor inputs. The net result of agricultural research and technology has given consumers a greater variety of higher-quality products. It has also provided greater food safety and quality. Modern high-tech agriculture is the most environmentally friendly of any food system in history, and it has dramatically reduced human drudgery.

Chemicals

In the days of weed control via the hoe or hand pulling, hired workers often left the farm when such a job was scheduled. Farm children picked potato bugs for 5¢ a quart jar or spent days on end pulling mustard and other weeds. Norman Borlaug told this writer that anyone who is against the use of chemicals to control weeds has never spent time in the field getting blisters and thorns in their hands from thistles and then found themselves a few weeks later having to start the job over again. Until relatively recent times, insects and weeds plagued farmers everywhere. Fortunately, those days are over for modern agriculture.

Horse-powered sprayer requiring 5 workers to do 12 rows. Capable of spraying about 80 acres in 8 hours. California, 1921. (NAL)

In 1940 DDT was patented in Switzerland, and in 1943 it was available in the United States. It was so effective against beetles, flies, bugs, lice, mosquitoes, and other pests that the military immediately seized control of its production. DDT virtually replaced most other insecticides for agricultural use and was widely adopted throughout the world. Research in the late 1960s indicated that DDT is of a family of chemicals that has an extended toxic life. As a result, further production was banned in 1972.

In 1943 a chemical revolution in weed control took place with the discovery of 2,4-D. It was first released for public testing in 1945, and by 1950, because it was so economical in controlling weeds, its production reached 14 million pounds. By 1964 production rose to 53 million pounds, and crop losses to weeds dropped sharply. That year farmers spent over $270 million to treat 70 million acres of crops.

In the 1960s a movement was started to control the use of insecticides and pesticides. Sometimes their misuse endangered plants, fowl,

Aerial spraying Toxaphene on a cutworm-infested field, Arkansas. Spraying capability was 60 acres an hour or more with pilot and helper. (NAL)

animals, and humans. The issue of proper use has remained to the present. Because of criticism of the use of chemicals, research was directed to nonchemical methods of controlling insects. One of the first breakthroughs in that area was the sterilization of the male screwworm fly, which is a serious cattle pest. Another was the development of a sex attractant to control the gypsy moth and the Mediterranean fruit fly.

With constant refinement in how chemicals are used, the risk factor has decreased and the control rate in both weeds and insects has remained good. Farmers are the third largest users of chemicals and chemical products. Several industries, such as plastic, soap, cosmetic, and petroleum, all make extensive use of chemicals. Weed control is so critical to efficient farming that it is believed the herbicide market will exceed $6 billion by 2000. Fighting plant disease ranks high with farmers because billions of dollars are lost annually by direct crop losses or post-harvest spoilage. In 1989 producers and processors spent an estimated $2.3 billion for fungicides to prevent those losses.

Biotechnology

In the 1920s research concentrated on mechanical power and on the improvement of genetic stock of plants and animals. In the first 30 years the big gains were made in mechanical power. Next came the increased use of chemical fertilizers and feed additives, plus efforts to control pests and diseases. By the early 1960s a gigantic mixed-feed industry came about because of nutritional research aimed at increasing poultry and livestock production.

Serious research in animal growth factors had not taken place until the early 1900s, but by 1920 the value of vitamins was recognized. Then came knowledge of trace minerals, such as zinc, copper, and iodine. Vitamins and trace minerals helped to fight deficiencies and disease in

By using a totally mixed commercial feed, over 150,000 broilers are raised in this building each year under contract, Mississippi, 1977. (NAL)

livestock and poultry. Vitamin D enabled poultry growers to confine their flocks because they no longer were dependent on sunlight. Knowledge of vitamin B_{12}, the animal protein factor, led to a breakthrough in the discovery of antibiotics.

In 1963 the mixed-feed industry produced more than 57 million tons of concentrates and complete feeds. Over 85 percent of all prepared feeds had an additive or medication. Improved rations produced three major results: increased animal growth rate, improved feed conversion, and a reduction in subclinical diseases.

From 1943 to 1964, milk production per cow increased 71 percent. In 1962 over 7.7 million cows were bred by AI. AI was credited for about 35 percent of the increase in milk production from 1945 to 1975. During that period average per cow production increased from about 4,000 pounds to over 10,300 pounds because of AI, better rations, and improved management. Over 2.2 billion pounds of milk were added to our total production, while dairy cow numbers fell from 26.6 to 11.6 million head. World record-setting cows increased their production from 37,381

Milk tanker capable of hauling 50,000 pounds of milk directly from the farm bulk tank. Also, the first cow to produce 50,000 pounds of milk in one year. Pennsylvania, 1975. (NAL)

Herringbone milking parlor, where two people can milk six cows in six minutes by machine as opposed to one cow in seven minutes by hand. Milk flows directly to a cooling tank. Minnesota, 1977. (Drache)

Hogs in slatted-floor confinement barn with automatic feeders using a commercially prepared ration, Iowa, 1976. (NAL)

pounds in the late 1920s to 53,900 pounds in 1993. This was proof that greater average production was still possible.

AI in the hog industry helped to improve feed conversion from 3.5 to 2.5 pounds of feed per pound of gain. At the same time, litters per sow per year increased from 1.8 to 2.2, while pigs per sow marketed annually increased from 13 to 24.

Equal or better results were experienced in the poultry industry. Poultry was the first sector to industrialize, and by 1974, 16,500 farms controlled 90 percent of the production. The hog industry worked in the same direction, and by 1993 the top 30 hog farms produced 25 percent of the industry output. By the early 1990s the dairy industry was readjusting. Milking herds of 400 cows were becoming commonplace.

In the 1980s the United States entered the biotechnology and information technology era. Both technologies had a profound impact on agriculture as they accelerated the rate of change. Biotechnology focuses on living organisms to develop products to improve plants and animals or to develop microorganisms for specific purposes. Control over biological systems is one of the chief aims of biotechnology. Gene insertion, embryo transfer, and the use of bovine growth hormone were some of the first genetic engineering techniques. They give the ability to prevent, detect, and treat infectious and genetic diseases, and they increase production efficiency.

The American Association for the Advancement of Science has termed genetic engineering one of the four major scientific revolutions of this century. The Association compares it to the unlocking of the atom, the escape from Earth's gravity, and the computer revolution. Biotechnology in agriculture is potentially as important as it is in medicine. State and the federal governments allocate billions annually for biotechnology research, but most goes to biomedical research, not to research for agricultural and plant biology.

Information technology uses computers and electronic-based methods for gathering information on control and management of agricultural production and marketing. Detecting change in milk production or feed consumption is an example of what can be done by computers. Watching for disease, pests, or soil problems is typical of what computer-based technologies can do.

The Global Positioning System (GPS) has the ability to control chemical and fertilizer applications according to the needs of the soil as the machines pass over. This enables prescription farming according to soil capability. Cotton farmers using GPS reported that after seven years on the program, they had a net per acre increase of $36 to $54. The variance was based on the amount of input in the system. Grain farmers use sensors to monitor yield and moisture of grain as it is harvested and then can compare the results with their fertilizer and chemical applications.

Plant modification enables plants to develop more nutritious protein, to resist insects and disease, to grow in a more harsh environment, and to produce their own nitrogen. This will yield even greater results than the changes in animal agriculture.

Patenting of biological innovations and computer programs has accelerated the drive to technological changes. The U.S. Plant Variety Protection Act of 1970 gave some protection for new varieties developed. In 1980 the Supreme Court opened the way for patent rights on almost any genetic creation. Both the private and the public sector have greatly increased biotechnology research in recent years. Favorable legislation and court decisions have encouraged international biochemical companies to purchase about 1,000 traditional seed companies.

Integrated Pest Management (IPM) is a system of attempting to control plant pests via ecological and environmental methods. IPM uses biological and nonbiological techniques to control weeds, insects, and disease. A major goal is to reduce reliance on pesticides. IPM grew out of research started in the early 1970s. It was determined that damage from pests could be reduced by later application of pesticides, by using pest-resistant crop varieties, and by using natural enemies for biological control.

Both GPS and IPM are emerging technologies that have the potential to greatly increase plant and animal production. They are driven by producers seeking new ways to lower per unit cost. Both technologies are expected to reduce land and water requirements, making agriculture more environmentally friendly. They will further reduce the need for workers in production agriculture, but those remaining will need to be more skilled.

Many innovations in biotechnology are opposed by environmental groups because of the fear of toxicity or the possibility of causing some weeds to become resistant. In spite of major governmental mechanisms for safeguards, environmental groups have sought and secured injunctions to suspend some research. Whatever the eventual outcome of those struggles, at this point in history, agriculture supports about 5.6 billion people. It is estimated that only about 10 million people lived in the pre-agricultural world. The limits of our agriculture are still unknown. To date, only about 250 plants have been domesticated out of more than a quarter of a million plants in existence.

Energy and Irrigation Use

In the days when farmers did not keep records and used human or animal power to operate their farms, little thought was given to the cost of energy required to produce a crop. Previous discussion pointed out that animal power was estimated to be about 3 percent efficient and consumed about one acre of every four that it produced. With the adoption of the tractor, farmers became more aware of the cost of energy. However, they were never more efficient in its use.

From 1950 to 1980, productivity in agriculture more than doubled, labor was reduced by half, and purchased energy input quadrupled. When the energy crisis struck in the early 1970s, farmers, like all other sectors of the economy that required power to operate, became very aware of their energy bill. In the late 1970s agricultural exports paid for over half the foreign oil imports. Those exports required a very small fraction of U.S. energy consumption.

The total food and fiber production chain requires about 16.5 percent of the total energy used. Of this, food processing and manufacturing requires 29.1 percent; home preparation, 26.0 percent; commercial preparation (all public and institutional establishments), 17.0 percent; retailing, 4.9 percent; wholesaling, 3.0 percent; on-farm production, 17.6 percent; and transportation, 2.4 percent. Of the nation's total energy usage, about 4 percent is needed for home preparation, of which refrigeration and freezing rank high. By comparison, all farm energy usage is only 3 percent of the total. The natural fiber and forest industry requires only 5.5 percent of the total energy needs. Nitrogen fertilizer is responsible for about one-third of our total crop production. Nitrogen and other fertilizers that are mined and delivered to farms make up about 33 percent of the total energy used in crop production.

As the nation has become more aware of conservation of natural resources, more attention has been given to the use of water for irrigation. On western flat lands, where irrigation was first used, water was delivered by flood irrigation. This was extremely labor intensive. On rolling land, contour grading was used to construct ditches to carry the water to various levels. Siphon tubes sucked the water into furrows and spread it throughout the fields.

Siphon irrigation of a cotton field using 2-inch tubes to draw water from a cemented ditch, Texas, 1962. (NAL)

In the 1930s sprinkler irrigation was first used in the Pacific Northwest and in California. Sprinklers were expensive to purchase, but they were water efficient and could be used on land too uneven for older systems.

Eleven quarter sections under center-pivot irrigation in which 1,397 acres of the 1,760 are irrigated, Nebraska, 1967. (NAL)

In 1952 a patent was granted for the first successful center-pivot irrigator. In 1953 the first commercial systems were marketed. The center pivot enabled automatic irrigation of large tracts of land with a minimum of labor. With just one worker, a large-scale irrigated farm can operate at least 15 to 20 center-pivot systems covering 133 acres each. The operation includes applying fertilizers and chemicals along with the water and having the worker walk the fields to check for problems.

If all water use, including rainfall, is figured, agriculture consumes about 84 percent of the total water needs of the nation. If only groundwater sources are calculated, agriculture ranks second after industry in water use. About 90 percent of the use is for irrigation, which also requires 13 percent of the energy of production agriculture. Most of the irrigated land is in high-value crops and would not be productive without added water. Potatoes, vegetables, cotton, rice, corn, hay, and wheat, plus orchards and pastures, are the major users of irrigation water. Corn and hay are grown on nearly 40 percent of the total acres under irrigation. Irrigation has enabled Nebraska to nearly equal Illinois in corn yields. Open-range Nebraska pasture land, which once produced an annual average of 27 pounds of beef per acre, has increased production to 700 to 900 pounds under irrigation.

Total irrigated land increased from 3.6 million acres in 1889 to a high of 50.3 million in 1978—roughly 5 percent of our land in crops. Irrigated acreage has since declined because of conservation efforts, participation in farm programs, declining net income from irrigation, and lowering of the water table in some areas. California has about 17 percent of the nation's irrigated acreage, which covers about 90 percent of the state's cropland.

A 7-foot-long shank being pulled by four Caterpillars (only one is visible in the photo) to break the hardpan and stir the soil, California, 1968. (NAL)

Trickle-irrigation tubing buried under tomato plants at the root level to make maximum use of water, California, 1981. (NAL)

In recent years trickle or drip irrigation has been adopted on high-value crops with great success. The capital cost of laying the system is offset by the reduction in water cost. This is a critical factor in areas where rainfall is limited and water is scarce, hence, high priced.

Alternative Products

Greenhouse, Floriculture, and Nursery Products

As the urban population increased, the demand for greenhouse and nursery products expanded. A previous discussion told how the industry was started and the problems of developing greenhouses. In 1885 the Society of American Florists held its first meeting. By then, 24 million roses and 120 million carnations were being produced annually by the greenhouse industry. They were grown on 12,000 acres of farm land. Several thousand additional acres were needed to produce the seed.

Ten acres of crops under glass, West Virginia, 1972. (NAL)

In 1899 the industry was composed of over 10,000 firms, with at least 619 acres under glass. The firms grew vegetables for the off season, along with potted plants and flowers. That year wholesale vegetable sales were valued at $2.25 million and flowers at about $11.25 million. In all, 275 million roses, carnations, and violets were sold, in addition to about 100 million potted plants. The greenhouse, floriculture, and nursery sectors had become a major industry.

Nearness to the market is an important factor in the business of cut flowers and cut greens, which explains why commercial production of those commodities is found in 42 states. Potted plants are grown commercially in 46 states, with California supplying 18 percent of the national production. The demand tripled from 1980 to 1989, with total sales of potted plants reaching $661 million in the latter year. In 1989 the bedding and garden plant category, which includes vegetable plants sold in various forms, flowering plants, and leafy plants, had sales of $964 million. At the same time, nursery products, which represent the largest subsector of the greenhouse/nursery industry, produced $4 billion in sales.

As transportation improved, the amount of imported fresh and frozen fruits and vegetables rose sharply. This was due partially to the increased demand from Asian and Hispanic immigrants, who wanted products that they had had in their homelands. It was also due to the fact that U.S.-based multinational firms were establishing processing facilities and otherwise encouraging production for exporting to the United States.

By the late 1980s the three green sectors were one of the fastest growing alternative enterprises in U.S. agriculture. In 1989 the nursery industry alone grossed more than $7.1 billion, excluding Christmas trees, seeds, and food crops grown under cover. In the 1990s it is outpacing all other major farm sectors. In 1990 cash farm gate sales were $8.1 billion and were equal to 10 percent of all farm cash crop sales. The three green sectors ranked seventh among all agricultural commodity groups and in 21 states was among the top 5 commodity groups. That year total retail expenditures for greenhouse and nursery products reached $38 billion ($150 per capita). This contrasted to sales of $49 billion for the fresh produce industry. When value-added services and employment were included, the floral and nursery industry outranked most traditional agricultural sectors.

From 1970 the greenhouse, nursery, and turfgrass industry grew about 10 percent per annum, including crop production, wholesaling, retailing, landscaping, and related activities. By 1991 it had an $8.8 billion farm value and represented 11 percent of all farm receipts, ranking sixth after beef, dairy, corn, hogs, and soybeans. Its sales were greater than the combined farm gate sales of tobacco, sugar crops, peanuts, grain sorghum, and food grains.

By 1994 the greenhouse/nursery industry had expanded to all 50 states, and in 20 it was in the top 4 commodity groups. Greenhouse/nursery crops were first in five states; second in five of the major agricultural states, including California (the first-ranking agricultural state) and Florida; and third or fourth in seven states, including Texas (the second-ranking agricultural state). Over 60,000 producers of greenhouse/nursery crops (including producers of Christmas trees) received more than 50 percent of their income from those crops.

In 1990 the average net income for farms that specialized in green crops was $53,589, the highest of all production specialties and four times the average farm net of $13,458. The farm gate proceeds for 1994 were expected to be $10.4 billion ($700 million more than in 1993), again outpacing the proceeds of all other segments of agriculture. Total retail sales were expected to reach $40.4 billion ($155 per capita). At the growth rate experienced since 1970 the greenhouse/nursery industry will be the fifth-ranking agricultural commodity group in 1995 and either the third-or fourth-ranking by 2000.

Aquaculture Products

Fishing was a major industry in the colonial United States and provided supplementary income for a large portion of pioneer farmers. As the rest of the economy expanded, fishing became relatively less important, but prior to World War II the United States was second only to Japan in total fish catch. That changed rapidly, and by 1960 the United States ranked behind Japan, China, the U.S.S.R., and Peru. In the 1960s we had about 80,000 fishing boats, of which 12,000 were vessels of 5 tons or more that harvested most of the fish. Over half our vessels dated prior to World War II, so our fleet was falling behind those of the world's leaders. The fishing industry employed about 225,000 people, of whom 60 percent were involved in open-sea fishing; most of the others worked in canneries.

In the 1960s the United States, with just over 5 percent of the world's population, consumed about 11 percent of the fish catch, of which about 70 percent was imported. By 1990 seafood was our third-ranking import, after petroleum and automobiles, and totaled about $9 billion annually. Starting in the mid-1970s, the world fish catch from the oceans leveled off. Even with improved technology, fishing is becoming more costly.

The first known effort to raise fish for commercial purposes came in 1927 at Auburn University in Alabama. During the 1930s experi-

Fish-feeding ponds where small channel fish are placed and fed a commercially prepared feed. They grow to 2 pounds. Texas, 1962. (NAL)

ments proved that fertilizing the ponds significantly increased their carrying capacity. During the 1940s feeding fish with commercially prepared feeds became a major success. After World War II other nations came to Auburn to learn about aquaculture to increase their food supplies. By 1969 the United States had 2.2 million artificially made ponds. It was estimated that nearly one-fourth of all the nation's fishing was done at them. By 1973, 43,000 acres of catfish ponds alone produced 38 million pounds of fish.

Aquaculture had to expand to make up for the demand for fish products. In 1980 aquaculture production reached 203 million pounds, with a farm gate value of $192 million. In the 1980s the industry was recognized as the fastest growing sector of agriculture. In 1986 production reached 620 million pounds; and in 1990 it reached 860 million pounds, with a farm value of $760 million. Feed mills, processing plants, equipment manufacturers, and fish farmers employed 300,000 workers, with a total economic impact of $8 billion.

Most of the production originally came from Mississippi and other southeastern states, but aquaculture has spread throughout the nation, even to the areas of colder climate. The United States ranks tenth in world aquaculture, producing 9 percent of our domestic consumption. This is encouraging to those in aquaculture because the world seafood industry is being challenged by more harvesting restrictions and the possibility of a diminishing resource base.

Harvesting catfish at a fish farm, Mississippi, 1975. (NAL)

The consumption of farm-raised fish has risen steadily. The trend is for increasing commercial availability of major freshwater fish, such as striped bass, walleye, yellow perch, and shellfish species, which are being shifted almost entirely to aquaculture. Currently, about 25 percent of salmon and shrimp sales and nearly 100 percent of catfish, trout, and hybrid striped bass sales are from farm-raised fish. Predictions are that by 2000 more than 25 percent of all freshwater fish and shellfish will be farm raised. This is good news for fish farmers, who have operating costs of up to $2,000 per acre and need to increase their volume to make fish competitive with beef, pork, and poultry products.

Industrialized Agriculture

Thomas Urban, a leader in one of the nation's largest agribusiness firms, expressed his views on trends in our dynamic agriculture as follows:

> The industrialization of agriculture is with us. It's driven by consumer and processor needs, supported by new and useful technology, and augmented by the severe agricultural recession of the 1980s, which changed attitude toward risk. The consequences for farm policy and rural development are significant, and should be favorable.

Urban explained that industrialization is the key to the new structure in agriculture. The production segment is rapidly becoming part of the industrialized food system. He predicted that by 2000, 25 percent of grain corn will be used for energy, sweeteners, starch, proteins, and oils. Because processors have special needs, they want a special type of product that they can receive directly from the producer on a contract basis. The processors need identity preservation because that is important for processing requirements.

Industrialized agriculture has little use for the small-town blacksmith or for the local foundries, flour mills, butcher shops, harness makers, and wagon manufacturers. All of the above either have faded out of existence or have been replaced by large-scale firms. The need to achieve economy of scale or the inability to meet government regulations has driven many small firms out of business.

Agribusiness Infrastructure

In pioneer days, off-farm purchases consisted of bars of iron, log chains, leather for harnesses, and salt. The rest of the inputs came from the land and the family. By 1900 off-farm purchases had risen to 24 percent of inputs, and by 1976 they had increased to 62 percent. In the 1990s they are 75 percent or more. Those purchased inputs provide one of the reasons for the success of U.S. agriculture. They are supplied by the world's leading agricultural infrastructure system.

Farming belatedly followed the trends of the agribusinesses that make up our food chain. Economy of scale has enabled the farm machinery, seed, feed, chemical, fertilizer, and farmstead equipment industries to establish large dealer networks to service farmers. The four largest tractor manufacturers have over 80 percent of the market, and the four largest combine manufacturers nearly 90 percent. Dominance of the market by a few large corporations holds true also for seed, chemical, fertilizer, and exporting firms.

Components of farm production expenses

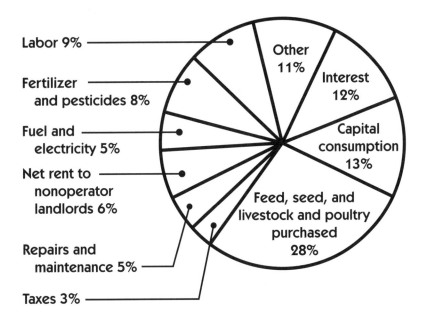

1988 data. Includes operator farm dwellings. Other expenses include machine hire and customwork; marketing, storage, and transportation; and miscellaneous.

(*1990 Chartbook of Agriculture*, USDA)

Farming represents only 1.4 percent of the Gross National Product (GNP) and employs 1.5 percent of the labor force. But it is one of the largest sectors of our national economy and part of the food chain that makes up nearly 16 percent of the GNP. Of that amount, the farm sector represents 1.4 percent of the GNP, input industries make up 5.2 percent, and the processing and distribution sector accounts for 9.2 percent. In the early 1990s, agricultural exports created 1.06 million additional jobs. Each $1.00 in exports generates another $1.52 of output.

In 1989 the food and fiber system contributed $820.6 billion to the GNP. This was allocated as follows: $71.4 billion to farming (8.7 percent), $272.8 billion to the input sector (33.2 percent), and $476.3 billion to manufacturing and distribution (58.0 percent). Out of the 21 million workers in the system that year, farm workers represented 2.459 million, the input industry 5.356 million, and the processing and distribution firms 13.246 million. In addition, another 2.6 million

The food and fiber system

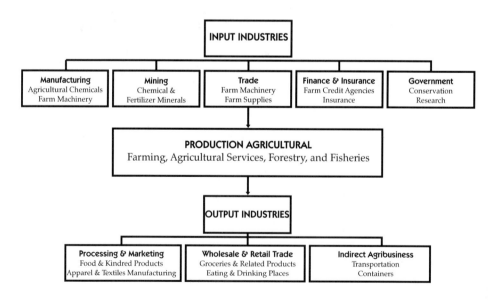

(*1990 Yearbook of Agriculture*, USDA)

employees were in indirect agribusiness firms. By 1994, 16.2 million of the total 23 million agri-related workers were in metropolitan areas, and the other 6.7 million were in non-metropolitan areas.

From 1945 to 1990 the number of farm workers declined from 9.8 million to 2.4 million. The number of hours worked on farms in 1990 was less than 20 percent of what it was in 1945. The number of migrant workers declined from 420,000 in 1949 to 159,000 in 1985. Between 1945 and 1990 the total number of hired workers in farming dropped 45 percent, even though the amount of work done by those persons increased from 22 to 31 percent. Mechanization increased labor productivity ninefold during those years. The gain came chiefly from the use of larger machines on fewer farms and from a 75 percent increase in the use of fuels and lubricants.

In 1987 farm earnings represented only 8 percent of rural earnings, and in all but the most rural states farmers made up only about 5 percent of the population. Only 514 out of our 3,042 counties receive 10 percent or more of labor and proprietor's income from agriculture. In the late 1980s, the agribusiness sector—farming, input, processing, and marketing—accounted for one-third or more of the employment in 785 rural counties.

A writer in the 1960 *Yearbook* explained what was taking place in agriculture: "This revolution is not finished. . . . It is continuing without any foreseeable end. We must expect it to accelerate." As more businesses saw the opportunities in agriculture, they placed technically trained people to the task of developing more services and equipment. Agriculture was industrializing like the rest of the economy had already done. The USDA, which had observed much of the trend, had 135,000 employees in 1994—an employee for every 14 farmers. If only the commercial farmers were counted, the USDA had an employee for every 3.08 farmers who grossed over $50,000 that year.

Professional Management

"Management" was basically an unknown word to pioneer farmers. Nearly all of their emphasis was on having the family work as hard as possible so that they could produce enough to survive. That style of farming changed slowly after the 1840s. After the 1920s, farmers were forced to change from thinking about farming as a way of life to thinking about it as a business. Each innovation forced farmers a step closer to business management. Each change doomed more of those who had no desire to change or saw no need to change.

The rapid rise and then decline in the farm economy between 1973 and 1981 forced farmers to become more astute money managers. Commercial farmers had to spend more time managing, relying to a greater degree on skilled labor and leased land than on family labor and family-owned land. Farm managers had to refine their management techniques regarding crop input, harvesting, and especially marketing.

The reasons most often given for farm expansion were (1) the need to gain economy of size; (2) better use of equipment; (3) the ability to spread fixed costs over more units; (4) lower input costs; (5) the stable prices and benefits based on volume that government programs provided; (6) better use of management resources; and (7) the fact that net income improved with volume if unit cost relationships were favorable. On the other hand, if unit cost relationships were not favorable, the chance of failure increased.

The Office of Technology Assessment (OTA) made an in-depth study of agriculture in 1986 that built upon major studies made in 1979 and 1980. The OTA study identified five classes of farmers: (1) Farmers with less than $20,000 income who were "living in poverty or who used the farm as a source of recreation." (2) Part-time farmers with $20,000 to $99,999 farm income who likely produced a significant net income but also worked off the farm. (3) Moderate-size farmers with $100,000 to $199,999 farm income who were large enough to have farming as a primary source of income and who could still rely on one operator. Most of them also had off-farm income. (4) Large farmers with $200,000 to $499,999 gross income, whose farms were chiefly family owned and operated and had one or more full-time operators. (5) Very large farmers who grossed $500,000 or more annually. Their farms were chiefly family-oriented operations, of which 5 percent were family corporations.

These farmers tended to contract at least part of their production and often were involved with vertical integration.

The above groups, in the order given, made up 60.5, 26, 8.1, 4.2, and 1.2 percent of all farmers. Their respective portion of national net farm income was –3.8, 5.4, 14.6, 20.4, and 63.5 percent. In other words, the smallest farms made up 60.5 percent of all farms and averaged 3.8 percent net loss, while the largest 1.2 percent of all farms had 63.5 percent of the total net farm income. The OTA study pointed out that "while the lowest average cost of production may be attainable on a moderate sized farm, average cost tends to remain relatively constant over a wide range of farm sizes. Thus farmers have a strong incentive to expand the sizes of their farms to increase total profits."

Agricultural economist Leonard Kyle studied large-size farms in 1970 and documented that these operations were nearly all family businesses but that they were industrialized farms. He concluded that the traditional two-worker farm was in "for a tremendous competitive struggle in the years ahead." Size was a relative concept for farmers, which changed over time and had a "relationship to each person's image of a benchmark. . . . Most people accept all types of large scale organization . . . , but agricultural tradition emphasizes smaller economic organization as being more desirable."

Kyle pointed out that using fixed standards, the number of so-called large-scale farms increased 16-fold between 1929 and 1964—from 7,875 to 125,000. The pace did not let up, for by 1992 those farms were called "full-time commercial farms," and their number had increased to 497,567. They produced over 85 percent of our total food and fiber. Kyle determined that economy of scale in most cases was reached with two workers, but there were still opportunities to profit on a larger scale by buying inputs and improving marketing. The real key to the larger units was in their efficiency in the use of hired labor and management. Professional managers, who devoted their total energies to managing and not doing manual work, gave their farms a strong edge over the operations of farmers who attempted to manage while doing physical labor.

One of the tests that the OTA applied to determine why some farms grew and others did not was the rate of adoption of new technologies in both animal and plant agriculture. Generally, the larger the farm, the more rapidly innovations were adopted. The OTA study pointed out that U.S. farmers had to continue to adopt the new technologies to produce more efficiently and compete with other countries in the export market. If that market could not be maintained, it would cause serious structural change and adjustment among farmers and others in rural communities.

The 1980s brought about long-range changes in the financial markets that forced farmers to be more business management oriented. This new stance was brought on, in part, by the regulatory and competitive changes in the financial markets that affected the financial framework of farming. Economics forced financial institutions to concentrate on larger-size loans. The service cost of small loans made them prohibitive to accommodate. This worked a disadvantage to the small-scale borrower. Said the OTA: "The farm of the future will be treated financially like any other business—it will have to demonstrate profitability before a bank will finance its operation. Managing a farm efficiently and profitably, which will necessitate keeping technologically up-to-date, will be a key to access credit."

One of the forces changing the attitude of farm financiers about financing farmers was that a greater percentage of purchased inputs were required as technology use increased. Farm management studies pointed out that in 1945 it took an investment of $4.73 to generate $1.00 in income. By 1965 the required investment had risen to $15.00 for each dollar of generated income. This was one of the reasons that astute managers turned to renting instead of owning land. That enabled them to use the landowner's capital and save their own for operating.

Capital-Intense Farming

A study of full-time commercial farms indicated that agriculture is among the most capital intense industries in our nation. From 1870 to

1920 the capital invested per worker increased 10 percent per decade. Then, because of the depressed agricultural economy, the investment slowed down until 1940. From then until the 1960s, the rate of capital invested per worker increased between 31 and 34 percent per decade. Beginning in the 1960s, the investment slowed down to about 20 percent per decade, except for a period in the 1980s. This has brought the average capital investment on commercial farms to well over $1 million, averaging better than $500,000 per worker. This compares with the petroleum industry, which is the next most capital intense, with about $320,000 invested per worker.

Farmers have always complained about the high cost of starting to farm. But by the 1960s the growing capital cost became a very real threat to the traditional family farm. The cost-of-entry argument was a popular complaint of farmers as soon as the free land frontier came to an end. Programs for beginning farmers have been put in place in agricultural states to aid those individuals. However, the greatest change has come about in the young farmers themselves, who have approached farming with enlightened attitudes toward financial management.

To meet the needs of commercial agriculture and supplement capital from the established financial institutions, farmers are able to secure credit from dealers, merchants, processors, and other middlemen. Such credit is made available on open account, on extended open account, and under a formal contract. Dealers often extend credit because it helps them make sales. Some dealers make money by financing. Sometimes either there are no finance charges or they are hidden in the sale price. Processors extend credit to secure a source of supplies. At least 60 percent of the farmers use their suppliers for credit. One of the undesirable features of scattered credit is that it is often difficult for financial institutions to determine the financial status of their borrowers.

A 1992 USDA study concluded that to generate an income comparable to the average household income, a farm needed at least $750,000 in assets. As farms became more capitalized, their variable costs increased significantly, making them more vulnerable to gyrations in the market. Part of the financial problem was caused by the fact that farm consolidation had not taken place fast enough to maintain net income per farm.

The amount invested for each dollar of farm income varies with size of farm. In 1989, for all farms nationwide, the average investment for each dollar of sale was $5.67. This ranged from $45.55 for farms grossing less than $10,000, to $6.22 for farms having sales of $40,000 to $99,999, to $2.38 for farms grossing over $500,000 in sales. This explains why larger farms are able to support a higher percentage of debt load.

Farming historically has maintained a low debt load in relation to assets. That is what enabled many way-of-life farmers of the past to survive. Their chief concern was having cash for payment on the mortgage and for taxes and small consumption needs. The debt-to-asset ratio has varied from a low of 11.8 percent in 1960 to a high of 23 percent in the mid-1980s; by 1993 it was back to 16 percent. The quick recouping in the 1980s was aided in part by major write-downs of debts by financial institutions, particularly by the Farmers Home Administration (FMHA). Total assets in the industry exceeded $1 trillion for the first time in the early 1980s. After dimming in the mid-1980s, the financial picture has improved and again is nearing the $1 trillion dollar level.

Integrated and Corporate Farming

Former Secretary Bob Bergland noted that a survey of rural electric customers indicated that contracting dominated perishable agriculture. He added that more perishable agriculture would become part of a totally integrated system. Vertical integration is beneficial to both the producer and the consumer because the operations can be coordinated to take advantage of new production methods. In some sectors processors have taken over the basic production of a food product to maintain a steady supply and to control quality.

Enhanced biological and informational technologies helped to change the vertical market structures. Contracting became a means of coordinating a commodity within the system, but it avoided integration even though some strict contracts resembled integration. There are several reasons for contracting: to increase efficiency; to gain market advantage; to reduce risk or uncertainty; and to obtain financing. Contracting will accelerate along with changing technologies because informational technology and biotechnology have the potential to change production methods faster than ever. Contracting serves as a means of shifting the risk during the period of change.

The trend toward increasing growth in farm size fosters specialization, which causes greater income variation—hence, the desire for the stability of contracting. Labor laws have encouraged contracting, as have economic development bonds, which have provided for larger farms. To date there is little reason to believe that the government will become involved in farmer contract operations. By the mid-1980s about 10 percent of crop and 35 percent of livestock production was sold under contract. Nearly all of the fluid milk, most of the broilers, 40 percent of the turkeys, virtually all of the sugar cane and sugar beets, and 75 percent of fruits and vegetables for processing are contracted. Overall, about 30 percent of all farm products are delivered under some form of contract.

Integration came about in the poultry industry when feed manufacturers financed growers. The manufacturers developed a contract system rather than open-account financing and were paid when the poultry was sold. Gradually contractors assumed an increasingly greater managerial role along with added production and price risk. New technologies enabled skilled management to extend over larger operations. Because financing came with production contracts, many underemployed farmers had an opportunity to expand in farming.

The growers' obligation was to furnish land, buildings, equipment, water, electricity, and labor. The contractors provided the chicks, feed, and other inputs, including management. The producers received a per pound fee plus incentives for efficiency. Broilers were the initial contract operations, followed by large-scale egg units. Most of the early large broiler producers were feed companies and meat packers located in the southeastern states.

Between 1950 and 1964, the share of cattle purchased by packers at terminal markets dropped from 75 to 36.5 percent, and the share of hogs dropped from 40 to 24 percent. Packers went directly to large feeders to reduce acquisition costs and to better determine the availability and flow of supply. Packers integrated with beef and hog feeders when they found it necessary to maintain a regular source of supply. At times they purchased cattle ahead and held them with the feeder or moved them into their lots until the animals were needed for slaughter. However, as recently as the 1980s, feed manufacturers represented the greatest direct involvement in farming. Cooperatives had about 20 percent of the total integrated business involving poultry, beef, and hogs.

The large beef, hog, dairy, and poultry complexes were economy driven because of the better efficiency in feeding, management, financing, breeding, nutrition, disease control, buying, and selling. To further reduce price risk, the larger operations made more use of the futures market, where professional speculators were an important factor in the overall marketing process. Farming is increasingly bimodal, with a limited number of large farms producing an ever greater share of the produce, such as meat, eggs, milk, vegetables, and fruits. The smaller-scale operations continue to dominate production in grains, oilseeds, and cow-calf operations, which continue to have easy market access.

One of the most discussed issues in agricultural circles has been that of corporate farming. Some factions have long felt that corporate farming had to be tightly restricted. In the late 1960s and early 1970s there was a rash of campaigning to limit corporate farming, and some farm states passed laws against nonfamily operations. The basic problem was whether the nation wanted industrialized agriculture or whether it preferred the traditional system. Some people felt that corporate farming would "erode the social and economic strength of rural communities." Others feared "monopoly control of food production if company farms" became dominant.

As late as 1969 few concrete studies were available as to how extensive corporate farming really was. Opponents testified that the "social virtues and social values of family farming" were no longer being discussed. They believed that corporate farming was an evil. Others argued that there was no evidence that there would be a great social impact. In 1965, 680,000 of the 3 million farm income tax returns showed farm losses offsetting nonfarm income. This established that the decline of the family farm and of small farm communities existed prior to the rise of corporate farming, as did rural-to-urban migration. It pointed out that a complete halt to involvement of corporate farms would neither remedy the country's farm problem nor bring about a return of the family farm.

A Senate committee discussion concluded that "the future of small business in smaller towns and cities . . . is closely tied to the solutions of our farm problems. [Corporate farming was only part of these problems.] The advantages and disadvantages of industrialization . . . [and] business management techniques should also be studied."

A 1964 USDA study pointed out that the greatest reduction in farm numbers came in farms with under $2,500 annual gross sales. There was little to indicate that large-scale corporations or factory farms were taking over agriculture. Most of the growth in farm size was in larger commercial family farms. There was no evidence of increasing predominance in agriculture of larger-than-family units.

A later study by the USDA indicated that 13,313 farming corporations made up 1 percent of all commercial farming operations, farmed 7 percent of all farm land, and made 8 percent of all sales. Nearly two-thirds of those farms were family owned. The owners were involved in the management, and farming was their only business. Of the remaining one-third, 15 percent had farm-related businesses and 18 percent had businesses not directly related to farming. Most of those businesses were local firms, such as automobile dealerships, grocery stores, and real estate firms.

Most of the employees of the farming corporations were stockholders or family members. Corporate farms owned or rented an average of

4,480 acres, in contrast to an average of 491 acres for all farms nationwide or 702 acres for the commercial farms. The largest corporate farms were in the mountain states, where they averaged 11,423 acres and were primarily cow-calf operations. A 1989 USDA study indicated that family-owned farms made up 86.6 percent of all farms; partnerships, 9.6 percent; family corporations, 2.9 percent; nonfamily corporations, 0.3 percent; and cooperatives, estates, and trusts, 0.6 percent.

The corporate farm issue has proven to be a very emotional one in some rural areas. It is chiefly a matter of accepting the changing needs of agriculture and recognizing its commercialization and industrialization. By the 1990s some states that had had strong anti-corporation movements in the 1960s and 1970s were reconsidering their position.

Farm Tenancy—A Vital Part of Modern Agriculture

When land was free and easily acquired from others, farm ownership was "regarded as normal, and tenancy as abnormal." Tenancy increased because those who owned the land did not desire to sell it. A 1923 USDA study stated that 39 percent of farm homes were rented versus 60 percent of urban homes. Most rented farms were owned by former farmers or by children of farmers who had left the farms but were interested in retaining them as investments. The most numerous landlords were retired farmers who had moved to nearby towns.

Forty years later USDA studies indicated that larger commercial farms were much more likely to be rented, in whole or in part, than smaller family operations. Progressive farmers recognized the advantages of renting over owning. They looked upon borrowed capital and custom-hired machinery in the same way as rented land. In spite of many efforts on the part of the government to favor totally owned and operated family farms, more land was rented as farms increased in size. The trend to part-owner, part-renter changed the attitude toward tenancy. Tenancy was often highest in the best farming areas and among the most successful farmers.

Renting is no longer considered an inferior way of farming but a means of freeing up operating capital. In some areas of specialty crops, renters as well as crops are rotated on certain lands. At the same time, renting machinery and hiring custom operators are commonplace for the larger farmers. Some farmers hire much of their work done on a contract basis, and others have leased livestock. In the 1990s large milk processing facilities lease milk cows so they can maintain their supplies to run their plants. Cows are leased to large-scale operators, who are then free to provide better milking facilities. Other farmers contract to manage the dry cows and raise the heifers.

The trend toward leasing land was so pronounced that by the 1980s full ownership almost signified part-time farming. In those cases, the farms were rural residences with the owners employed off the farm or otherwise living on nonfarm income. Farms averaging under $10,000 in annual sales indicated an 88.5 percent ownership, while farms grossing over $500,000 in sales had a 54.9 percent ownership. As the trend increases to larger-size farms, the percentage of leased land will also be greater.

In a 1990 study by the USDA, nearly 55 percent of all farms were completely owned farms that averaged 354 acres and controlled 33 percent of all land. On the other hand, 8 percent of all farms were completely rented farms that averaged 715 acres and controlled 13 percent of all land. The renters tended to be younger farmers who were more risk oriented and who realized that they had to manage larger farms if they wanted to survive as farmers. Part-owner, part-tenant farmers made up 37 percent of all farmers and controlled 54 percent of all land, for an average-size farm of 905 acres. These farmers knew that it was not practical to own all the land necessary to make the proper business volume to support their families.

Surplus Farm Labor

Historically, agricultural laborers generally received 30 to 40 percent lower income than persons in comparable nonfarm occupations. There was some hope for improvement when in the 25 years starting with

World War II, 25 million persons transferred out of farming. But technology improved faster than the labor force declined, and wages remained low. One reason for the delayed exodus from farming was that farm labor did not have the skills to work in urban industry. Another was the attitude that farming was a superior occupation; thus, sons and daughters were encouraged to remain on the farm. Far more stayed than could make a living from farming.

In the past too many farm youth had limited opportunities in nonagricultural vocational training. As recently as the 1960s, most still took vocational agriculture courses. Even worse was the fact that of 600,000 rural farm males of high school age, less than 90 percent were enrolled in high school and 85 percent were working in agriculture. Equally unfortunate was that 40 percent of all male farm youth hoped to remain in production agriculture—a percentage far in excess of what the industry needed. There had never been a long-range program to train or care for the excess labor that was generated on the farms and, hence, a policy of drift continued.

Unionism

In Unit IV reference was made to some attempts at unionization in the 1930s in specialized areas of California. With the advent of World War II and the decades of high labor-force growth, agricultural labor was overlooked by the unions. In the early 1960s, Cesar Chavez organized the Farm Workers Association (FWA), which became self-supporting by 1963. In 1965 the FWA became involved in a strike started by another union at Delano, California. The FWA received the support of national leaders and financial help from the United Auto Workers. Chavez called for a consumer boycott and a march. The strike was settled in 1966 and proved a major victory for the FWA. Then the FWA merged with the Agricultural Workers of California and formed the United Farm Workers Organizing Committee (UFWOC). The UFWOC immediately went to battle with the Teamsters, who also sought to unionize the field workers.

The UFWOC won the battle and commenced to unionize the agricultural workers.

In 1968 the UFWOC became involved with the issue of pesticide use in fruit and vegetable production. By 1970 the UFWOC had won concessions from the growers on how to use pesticides. The favorite tactic of the UFWOC was the boycott, which was effective because the union's members were chiefly involved in the production of perishable commodities.

Unions approached farm workers in other parts of the country, especially on the larger farms or where processing facilities were part of the business. But the lack of concentration of farm workers made it expensive and impractical from the unions' standpoint to attempt to organize farm workers.

Communal Farming

The Amish are in some respects the best known of the unique religious groups that have remained in agriculture. The soil holds spiritual significance for the Amish, and their goal is to blend farming into a preferred way of life.

From the start the Amish were diversified farmers with intensive cropping and also with large livestock or poultry enterprises. In recent years some of the communities, especially in Illinois, have tended to break away from diversification and have specialized to gain economy of scale and remain competitive. However, because of the reluctance of the Amish to mechanize, they have not done as well financially as highly mechanized intensive crop farmers.

In recent decades the Amish have accepted diesel-powered bulk milk coolers because without them they would not have been permitted to sell milk for public consumption. Gasoline engines are allowed to power the hay mowers, balers, corn pickers, and similar machines as long as they are horse drawn. Tractors may be used for mixer-grinders and silo fillers but are otherwise shunned. Generally, government payments are refused.

The Amish were always expected to farm, and prior to 1954 those who sought other occupations were excommunicated. However, that has changed, and by 1980 less than half of all Amish households were engaged in farming. The majority of those not farming work in independent trades associated with agriculture.

There are three alternatives to dealing with the excess Amish population. One alternative is to establish new settlements. This is becoming more difficult as land values rise and little land is for sale in large blocks. The second choice is to subdivide existing farms and intensify poultry, dairy, or hog enterprises. This presents a problem because the concentration of manure is so great that it has caused environmental problems. The third and least satisfactory alternative is to seek nonagricultural occupations while maintaining the traditional life style.

The Amish are dependent on commercial markets for their major income from dairy, livestock, poultry, vegetable, tobacco, and other products. They purchase staple foods, such as sugar, salt, and flour, and basic groceries but not canned goods. Most of their food and clothing are still made in the household. The Amish have settlements in Ohio, Pennsylvania, Indiana, Illinois, and Iowa. There are about 600 Amish districts, with nearly half of them in Lancaster County, Pennsylvania; Holmes County, Ohio; and Elkhart County, Indiana.

Tomorrow's Challenge

The shrinking world makes agriculture increasingly vulnerable to the gyrations of the world market. Most of the changes are beyond the control of agriculture or the U.S. government. Within our nation, conditions are no more stable. Chemurgy researchers are constantly seeking new industrial uses for farm products because agriculture's productive capacity continues to outpace the demand. Cooperatives have become revitalized and have taken on a more businesslike attitude, which has enabled them to play an increasing role in the value-added sector of agriculture. The food chain continues to grow to meet the expanding needs of the ever-more-fickle consumers.

At the same time, government programs continue to attempt to maintain the delicate balance between supply and demand. Often this hampers agriculture's freedom to compete effectively in the world market. The continued aim of the programs is to assure the public of an abundant supply of reasonably priced food. Agriculture remains overcapitalized and farm numbers continue to decline in the struggle to put the industry in line with long-range demand.

Agriculture in a World Economy

It was not until the 1930s that attaché posts given status as agricultural representatives were established in 10 countries. By 1938 the number of posts had increased and the Foreign Agricultural Service (FAS) was established. FAS officers were to report on commodities, attend conferences, and be involved with embassies or legations in gathering agricultural data and informing foreigners about the availability of U.S. agricultural products. The FAS grew rapidly after World War II, and by 1970 there were 61 attaché posts.

World agricultural trade increased after 1934 when new trade agreements were written. Since the late 1940s exports have ranged from a low of 20 percent of agricultural sales to a high of 40 percent. Our agricultural exports have exceeded agricultural imports every year and have helped combat our trade deficit.

Except for the war years, the United States was not an active participant in world trade. After a post–World War II low of $2.8 billion, agricultural exports rose to $7 billion by 1970. They hit $43.8 billion by 1981, then fell to $26.3 billion in 1986 and bounced back to $43 billion by 1994.

After World War II, farmers in other nations had become more productive and also sought to export. Western Europe changed from being a net importer of 26 million tons of grain during 1960 to 1962 to a net exporter of 58 million tons during 1989 to 1990. The U.S.S.R., on the other hand, attempted to become agriculturally self-sufficient, but in 1962 it admitted to the world that it had failed to do so and became a leading importer.

World trade reached a pivotal point after Europe and Japan recovered from the wartime destruction of their economies. This changed U.S. agriculture because the total volume of world grain trade increased, several net exporting countries became net importers, and the United States became the leading agricultural exporting nation. We became involved in world trade policies and helped to stabilize world grain prices and demand.

Our position as the largest exporter of agricultural products was aided by the rapid industrialization of Western Europe and Japan. Unfortunately, by the 1960s the European Community (EC) countries were beginning to object to freedom in agricultural trade because they wanted to protect their farm economies by subsidy and import restrictions. The United States continued to battle for more freedom.

In the 1970s the value of agricultural exports increased fivefold. This was caused by the devaluation of the dollar, aided by improved trade relations with China, the U.S.S.R., and other nations, and by the industrial economic boom in the Pacific rim countries. A key event to the changing export picture took place in July 1972, when the U.S.S.R. purchased $1.1 billion of farm products from the United States in one year. This caused much talk of a possible world food shortage. The result was the 1974 World Food Conference at Rome.

Farmers expanded production in anticipation of increased demand. This put them in a volatile position, partly because domestic consumers became disturbed by rising prices. Also, railroads and export facilities greatly increased their ability to handle grains. In the years since the great export boom of the 1970s, capacity for exporting has been about double the existing need.

World food trade rose from 10 to about 15 percent of total production. In 1979, at the height of the export boom, the United States commanded 82 percent of the soybean, 67 percent of the coarse grain, 41 percent of the wheat, and 37 percent of the world's total agricultural exports. We accounted for 20 percent of the total world agricultural trade. The basic reasons for the export growth were that many countries were more open to world trade; total world trade was expanding about 20 percent annually; population continued to grow at 1.8 percent annually; and real incomes increased 3 to 4 percent yearly.

Grain elevator, only 68 feet less than a half mile long, with a capacity of 18 million bushels, where grain is accumulated for export, Kansas, 1975. (NAL)

One of the contradicting factors in the great expansion of world agricultural trade was that the tariff rate on agricultural goods rose from 21 percent in the mid-1960s to 40 percent by 1991. During those years tariffs on manufactured goods dropped from 40 percent to 4 to 6 percent. Another contradiction was that in the late 1980s the United States exported only $7.6 billion in consumer-oriented value-added agricultural products, while the rest of the world sold $133.7 billion. In 1988 we exported $9.4 billion in intermediate agricultural products, compared to $43.5 billion for the rest of the world. We did 35.3 percent of the world trade in bulk agricultural commodities but only 17.8 percent in intermediate agricultural products and less than 6 percent in consumer-oriented value-added agricultural products.

After NAFTA was implemented, we reversed our position, and by 1993 consumer-oriented exports of snack foods and vegetables experienced their seventh year of increase—to $15 billion. Value-added exports climbed to one-third of our agricultural exports. In 1993 NAFTA countries took 20 percent of our agricultural exports.

American agriculture has to overcome certain obstacles in exporting. We are looked upon as a supplier of last resort. Countries buy from the United States only when no other supplies are available. The exchange rate has often been against U.S. commodities because the dollar has frequently been overvalued. Much of the rest of the world suffers from a shortage of dollars and saves them for other needs.

Retiring land via the Soil Bank Program or the CRP keeps our production down and our prices above world market levels. School lunch programs, food stamp programs, and export subsidies were all put into

Over 800,000 bushels of corn piled at a country elevator, Illinois, 1975. (NAL)

effect to keep agricultural prices up. Again, the programs were at conflict, so Congress empowered the Commodity Credit Corporation (CCC) with export provisions to barter with needy countries as well as to provide transportation for the exports.

The most significant early act of such nature came in 1954 with the passage of P.L. 480. This allowed for the sale of surplus farm commodities for foreign currencies not to exceed an annual loss of $700 million to the CCC. Surpluses could also be used for overseas charity or foreign aid or for disaster relief as long as they did not interfere with normal marketing. By 1966 P.L. 480 and other giveaway programs to reinforce "food for peace" had exported $15.033 billion. By 1986 those programs had provided $43 billion in food to over 100 countries. Unfortunately, trade policies often hinder comparative advantage from working, and much of the world's agricultural output is produced inefficiently at a higher cost to consumers. To correct that problem the United States was active in establishing GATT. Agriculture was somewhat bypassed because several industrialized countries wanted to protect their farmers. Much of the cause lay in the memories of food shortages during the two world wars.

The 1979 GATT round brought a number of tariff reductions on agricultural commodities, and regulations were established on countervailing duties. The variable levy of the EC caused a reduction of U.S. exports to Europe. The levy protection stimulated European production but caused higher prices, which resulted in reduced consumption. The 1994 position of the USDA was to attempt to reduce agricultural trade barriers, to improve market access, and to minimize trade-distorting internal supports, export subsidies, and pseudosanitary regulations.

Secretary Espy stated that the USDA wanted to increase agricultural trade because the U.S. budget support for agriculture will continue to decline. In 1993 exports took 20 percent of all farm products and accounted for the production of one out of every three acres. The Secretary's pronouncement was recognition that exports are critical to prosperity in agriculture.

Since 1973 there have been several disturbances to agricultural exports. In 1973 a total export embargo was placed on high-protein feedstuffs. This eventually affected 41 agricultural commodities of the oil seed family. The immediate impact was the reduction of soybean prices by nearly half. This gave Brazil an opportunity to increase its exports. In 1974 the government caused a partial suspension of sales of wheat and corn to the U.S.S.R. because of tight supplies in the United States and in the world. The markets anticipated that act, so it had little impact on prices.

In 1975 a second partial embargo was placed on exports to the U.S.S.R. when its crop was 75 million tons short. This came after it had purchased 9.8 million tons of grain and caused an increase in U.S. prices. That year the International Longshoremen's Association announced that it would boycott the shipment of grain to the U.S.S.R. On January 4, 1980, a third partial embargo was placed against the U.S.S.R. when that nation invaded Afghanistan. The embargo canceled sales of 17 million tons of grain, which were in excess of our 8 million ton grain agreement of 1975. The embargoes hurt long-term trade relations between the two nations and were not totally effective because there was sufficient leakage by other nations to fill the Soviet demand.

As a nation the United States was no longer totally capable of controlling the destiny of agriculture. We had to export at least 30 percent of our annual production to keep agriculture prosperous. Our ability to overproduce was as powerful as ever.

Chemurgy

Some agricultural leaders have suggested that food production could be limited in the future because it is increasingly reliant on mineral and fossil fuel energy. This could lead to using farm products to produce energy.

Because of the finite supply of fossil fuels, raw material prices for them will continue to rise. But the real price of renewable resources, such as corn, will continue to fall because technology has shown less expensive ways of producing those products. If the rate of increase since World War II continues, the price of crude oil could double relative to the price of corn every 17 years.

The Chemurgic Council continues to seek ways to develop nonfood uses for farm crops and profitable uses for agricultural wastes and residues. In the 1970s biodegradable cornstarch derivatives partially replaced petrochemicals in the production of plastics. Soy ink from soybeans supplanted petroleum-based ink products. In addition to current ethanol work, research is also being done on using alfalfa and other forages to produce electricity. Many other areas of research in the field of chemurgy indicate a brighter future for the use of agricultural products in industry.

Cooperatives

In the previous discussion on cooperative financing, the development of the Farm Credit System into the 1950s was explained. Basic additions since have involved establishing the Future Payment Fund, which allowed farmers to make pre-payment into a reserve fund. After 1961, PCAs were authorized to make crop and livestock loans for as long as seven years. Intermediate-term loans avoided the necessity of having to reopen long-term mortgage agreements. They could be used effectively to finance breeding stock and machinery, which could not always be paid for in a shorter time.

The additional need for credit put added burdens on the agencies financing farmers and farmer cooperatives. To resolve the problem, the Farm Credit Act of 1971 removed many restrictions on lending to cooperatives. This gave them greater authority, which enabled them to broaden their services. Farm Credit System cooperatives were encouraged to pay off their government loans. When those loans were repaid, farmers were in complete control of their cooperatives. The greatest disadvantage was that the associations became responsible for income tax on their earnings. On the other hand, associations were free to appoint their own leaders without government approval.

After the Bank for Cooperatives was established, money and financial advice were more readily available for borrowing associations. The financial advice greatly improved the success rate of cooperatives. The Bank for Cooperatives raised money by selling debentures to the public at only a fraction of a percent higher interest than the federal government had to pay. Cooperative debentures were neither insured nor guaranteed by the government, but the opinion in financial circles was that the government would not allow the Bank for Cooperatives or the Farm Credit System to fail. This was proven correct in the 1980s when the government advanced several billion dollars to tide them through the agricultural recession.

The other benefit of the Bank for Cooperatives was that once money and better management were available, many local associations grew rapidly. This made them more competitive with privately funded businesses. In 1976 seven farmer cooperatives were on the *Fortune* 500" list. Four were listed in the top 200. Five others were listed on the second 500 list.

The largest cooperative was Kraftco, which was number 33, with $4.9 billion in sales. Others on the *Fortune* 500" list were Farmland Industries, number 135, with $1.51 billion in sales; Associated Milk Producers, 141, with $1.48 billion; Agway, 154, with $1.33 billion; Land O' Lakes, 180, with $1.12 billion; Gold Kist, 239, with $815 million; CF Industries, 359, with $468 million; and Dairylea Cooperative, 438, with $355 million.

In 1994 agri-food had 34 firms listed in the top 290 on the *Fortune* 500." Of those top 290 firms, 7 were cooperatives, led by Farmland Industries, which ranked 109 with sales of $4.7 billion. Next was Agway, 154, with $3.1 billion in sales; Land O' Lakes, 172, with $2.7 billion; Cenex, 215, with $2 billion; Central Soya, 224, with $2 billion; Del Monte, 271, with $1.6 billion; and Gold Kist, 289 with $1.4 billion. Several other well-known cooperatives were on the list of 500. In recent years cooperatives have become increasingly active in the marketing and processing of agricultural commodities. The long-range trend of decreasing the government's involvement in agriculture and the uncertainty of markets may aid cooperatives.

One of the early large-scale food processing co-ops was American Crystal Sugar, which was formed in 1972 to take over an independent company that was having difficulties. By 1993 its revenues were $542 million, and its stock had increased 20-fold in value. In 1994 American Crystal and another beet sugar co-op pledged to invest $51 million to help start Northern Corn Processors. Farmers will invest a 49 percent equity to found a $250 million high-fructose corn syrup processing plant.

In 1982, 1,000 farmers formed the Minnesota Corn Processors Co-op, which has become one of the world's largest producers of ethanol. A second plant was built in Nebraska, and the combined plants have the capacity to use 70 million bushels of corn yearly. By 1993 the investment per share had increased by 10-fold, in addition to paying a dividend of nearly 30 percent that year.

The above accounts show the possibilities of cooperatives in the processing sector. Such firms will increase in years ahead. Cooperatives have been forced to grow to reach economy of scale. The number of marketing and purchasing cooperatives declined steadily from over 25,000 in the 1920s to 5,782 in 1986. Of those remaining, 60 percent are involved in marketing. From 1950 to the mid-1980s the cooperative share of marketing rose from 20 to 30 percent, and input sales increased from 12 to 27 percent. The cooperative system has been aided by fewer, larger, and more specialized farms, by fewer and larger marketing firms, and by a restructuring of the tax system.

The Food Chain

In previous discussion, reference was made to the emergence of chain stores and the first supermarkets, but those stores were found only in larger centers. As recently as 1924, 20 million rural people were still

Typical rural store. Note caps, alarm clock, socks, handkerchiefs, and shoes at left. Note also 100-pound sacks of flour on the ice cream freezer, post office with 54 boxes, cookie display, and onions and apples in boxes. Baker, Minnesota, 1920. (Fran Iverson)

dependent upon 39,000 small towns, villages, and hamlets for their supplies. Because of the lack of competition and small volume, it was common that retail prices there were as much as 25 percent higher than in larger centers. The fault was not with the local merchant. It was simply that the merchant and the farmer remained locked in the grip of an antiquated and wasteful system. The big change in that marketing system took place between the two world wars and into the 1950s. However, in the 1990s some of the most remote rural areas still experience the impact of that system.

In the 1940s and 1950s, frozen food technology was developed. It kept food fresher, and it reduced swings in seasonal prices for the producers. Consumption of frozen orange juice concentrate rose from

226,000 gallons in 1945 to 84 million gallons in 1960. Frozen dinners did not sell well initially, but in 1955 the Campbell Soup Company started a large-scale advertising campaign of "TV dinners" and the product boomed. In the first year 70 million dinners were sold. By 1960 sales hit 214 million, and in 1993 over 2 billion frozen dinners were marketed. The instant potato flake of the 1950s opened an expanded market for potatoes via the dehydration process. Chemical finishing made cotton more wrinkle resistant, which enabled it to compete with polyester and nylon.

Improved packaging was one of the most important breakthroughs in the food industry. One of the earliest significant developments in food packaging came when the National Biscuit Company introduced the Uneida Biscuit package in the late 1800s. This replaced the traditional cracker barrel for transporting and marketing crackers. In the 1960s a series of packaging innovations commenced that helped to fill a niche when half of all women were in the labor market and did not have time to prepare traditional foods. In 1960 boil-in-bag foods entered the market. The foods were prepared and placed in bags that could be immersed in hot water and boiled until the foods were ready to eat. Tear-off aluminum tabs on juice cans became popular. In 1962 aluminum beer cans and aerosol cans were introduced. In 1963 steel coffee cans with plastic reseal lids were an instant hit. Shrink-wrapped corrugated fiberboard trays were adopted in 1964 for ease in the rapid handling of canned goods. That was followed in 1967 by the introduction of tamper-resistant closures for milk jugs. The plastic foam egg carton was adopted in 1968. Large bottles for soft drinks were introduced in 1970, followed by polyester bottles in 1977. In the early 1980s aseptic squeezable ketchup dispensers replaced glass bottles, which were difficult to pour from. Tamper-evident closures and heat sealing of rigid plastic all became part of convenience packaging.

In 1967 an event took place at Hunts Point within New York City that illustrates the immensity of the food chain. That year the world's largest wholesale produce market, located on 126 acres, was dedicated. The complex handled 15 tons of fresh produce every minute of every daylight hour of each working day. One of every eight carloads of fresh produce and vegetables grown in the United States was sold at that market, in addition to products from 35 other countries. It was the price-making market of the nation. Because of the efficiencies in handling fresh fruits and vegetables, marketing costs were reduced $15 million annually on just those items. Later the complex was enlarged by an additional 350 acres to handle all foods for New York City.

A technological revolution in the 1960s in the meat packing industry caused a shift away from the huge terminal markets into areas where livestock was being fed. The terminal markets had declined steadily since the 1920s. Good transportation and cheaper labor fostered the construction of newer, technologically advanced packing plants in livestock-producing areas. Older, massive plants in terminal market centers with high labor costs could not compete with the decentralized modern plants. Revised federal beef grades made it easier for new companies to compete with the mainline companies. At one time the beef trust controlled 85 percent of the sales of dressed beef. Legislation reversed the trend, and by 1970 the four major packers controlled only 20 percent of the national cattle slaughter. By 1987, with the adoption of boxed beef, they again increased their share to 67 percent. By 1988 boxed beef, which greatly reduced the labor cost for the retailer, made up 82 percent

Federal meat inspector checking hog carcasses, Iowa, 1975. (NAL)

of the market. By 1993 the top five beef packers controlled 89 percent of the cattle slaughter.

In the 1980s the food processing and manufacturing industry made up 13 percent of all manufacturing, with 1.1 million workers in 23,000 plants. Of the $276 billion in food and beverage production, meat packing led with $66 billion, followed by dairy products with $37 billion, all beverages with $34.7 billion, processed fruits and vegetables with $29 billion, and bakery goods with $18 billion. Food wholesaling and retailing, including food service, employed 8.2 million and added another $155 billion, plus $15 billion for transportation. The fiber and apparel goods made from agricultural products added another $74 billion and provided employment for 520,000 workers.

A privately owned (one-farmer) potato processing plant that uses over 1 million pounds of potatoes daily. Nearly all the potatoes are grown by the plant owner. Minnesota, 1994. (Larry Monico)

Interior of a giant supermarket, Maryland, 1982. (NAL)

Supermarkets, which were born out of the depression of the 1930s, peaked in 1960 with 33,300 stores. Their sales volume had plateaued, and by 1963 the number of stores decreased to 28,140. They made up 12.3 percent of all grocery stores but controlled 69 percent of the total business, with an average volume per store of $1.424 million annually.

While the supermarkets increased their market share, the independent stores' share declined from 30.2 percent in 1948 to 9.1 percent in 1963. The number of independent grocery stores declined from 378,000 to 245,000. Most of the decline in numbers came in stores that had done less than $100,000 in annual volume. To defend themselves against the supermarkets the chain stores (a chain is any firm that has 11 or more stores) and the independents adopted the techniques of the supermarkets. They formed associations to pool their buying.

The large supermarket chains had several competitive advantages: they had superior access to capital via the securities market; they were able to buy more cheaply; they had lower transportation costs because they purchased in carload lots and had their own truck fleets; they were able to make more extensive use of private label and integrated processing; they had a better-coordinated supply organization, which gave them lower costs; and they were more systematic in their training, which enabled them to increase their dollar volume per employee.

In the late 1950s convenience stores were developed in the South and in the West. To a degree they replaced the neighborhood "mom and pop" grocery stores found in the older suburbs. The initial business of the convenience stores was primarily in milk sales, because home delivery of dairy products was declining. Later, they added self-serve gasoline, carry-out foods, and hot sandwiches. Their variety was not large, and their prices were higher than those of the larger-volume retailer. Their advantage, as the name implies, was convenience. These stores sold about 3.5 percent of the total grocery volume in the 1980s.

In the 1970s the superwarehouse stores entered the retail grocery market, and in the 1980s the superstores and hypermarkets developed. The latter aimed at one-stop marketing of a large variety of products. Hypermarkets are four to five times larger than the typical supermarket. In recent years warehouse and club stores have cut into supermarket sales. From 1990 to 1993 their share of the market increased from 3 percent to 15 percent of total sales. Smaller grocery stores still sold about 11 percent, but their share continued to decline. Specialty stores sold about 4.5 percent, and other retail places about 5 percent. Thirty-five percent of food sales took place at restaurants (14 percent), fast-food places (10 percent), and institutions, vending machines, places of entertainment, or malls (11 percent).

By 1963 the number of items stocked by the average supermarket had climbed to 6,800, and by 1986 it had grown to 11,000 for the smallest stores and to 39,000 for the largest. In 1992 the average supermarket carried 30,000 items, a mere fraction of the 300,000 packaged grocery items or over 700,000 labeled food products available.

Since the 1970s the growth rate in the introduction of new items has averaged about 11 percent annually. The number of new items that

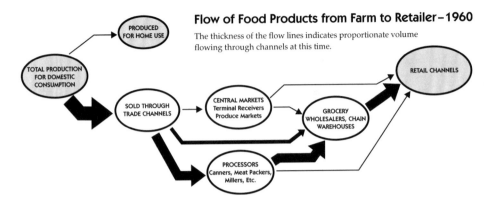

(Lifeline of America)

enter the market each year varies from as few as 3,000 to as many as 20,000. The failure rate of new products varies from 80 to 99 percent; normally only 1 to 3 percent are successful. A 1992 study indicated that in their first year, only 5.8 percent of new products had attained $1 million in sales. Risky and costly as the introduction of new items may seem, the consumer is still the winner, for each dollar spent on research and promotion of a new product yields a reduction of $30 to $50 in food cost or an improvement of that amount to the producers.

Entering a new product is a highly competitive and costly venture. First, the retailer has to be convinced to make shelf space available. Then, the product must be accepted by the consumer. Pre-sales advertising, which is extremely costly, gives the large firms the marketing edge in introducing new products. In 1972 the four leading companies controlled 90 percent of the sales in soft drinks, 84 percent in breakfast cereals, 84 percent in chewing gum, and 95 percent in canned soup.

In the 1970s there were about 20,000 food manufacturing companies, but the largest 200 firms held 81 percent of all the assets. They accounted for 64 percent of the total shipments, 85 percent of all the media advertising, and 100 percent of the network television advertising by food manufacturing firms. In 1982 the top four firms had 58.3 percent of total retail grocery sales, and in 1992 they accounted for about 65 percent of all the food industry sales. Concentration on the retail level of the four largest chains varied from a low of 27 percent to a high of 90.6 percent in the 318 standard metropolitan statistical areas. In 1992 the margin on sales of the top eight supermarket chains varied from 0.10 to 2.65 percent, with an average of 0.87 percent.

More compact equipment has enabled a reduction in size of most new supermarkets to about 43,000 square feet. By 1990 the average opening cost was $3.8 million. Most retailers relied on wholesalers for financing, which continued the process of consolidation in the retail sector. Supermarkets have added bakeries, delicatessens, carry-out foods, beauty and health-care items, and pharmacies. In 1992 supermarket pharmacies filled 16 percent of all prescriptions.

Consumer pressure for one-stop shopping has caused the rise of discount warehouse clubs. These clubs sell everything from cars to cereals and have cut into the supermarket share of the market. In 1991 over 500 club stores had $28 billion in sales. This was equal to the total of the convenience store industry less gasoline sales.

As technology in the food chain increased and food was further processed and delivered to the consumer, the farm gate percentage of the food dollar decreased. As recently as 1950 farmers received 47¢ of each food dollar if the food was purchased at a grocery store and prepared at home. By 1970 that figure had fallen to 37¢, and in 1992 it was 26¢. If the food was purchased at a commercial eating establishment, the farmer's share was 41¢, 32¢, and 22¢, respectively. The reason for the difference is that retailing takes only about 23 percent of the food dollar, whereas food service gets 61 percent.

Percentagewise, the food dollar is spent in the following manner: farm value, 22; labor after the farm, 35; packaging, 8.5; intercity transportation, 4.5; depreciation, 4.5; advertising, 4.0; fuel and electricity, 3.5; before-tax profits, 3.5; rent, 3.5; interest, 3.0; repairs, 1.5; and all other costs, 6.5.

Although the percentage of people in farming has been steadily declining from the days when the figure was 95 percent of the work force, agriculture is still the nation's largest industry. In 1989 agriculture employed 23,184,796 persons. From 1975 to 1989 the number of farm workers declined by 780,000, while the number of persons in farm-related work increased by 5.1 million. Of the nearly 23.2 million agricultural workers, 12.993 million were in wholesale or retail trade, 3.2 million were in processing and marketing, 3.2 million were in farm production, 2.59 million were in indirect businesses, 842,000 were in agricultural services, and 426,000 made up the agricultural input sector. Another 843,000 were in forestry and fisheries. By 1994 the number of farm workers had declined to 2.4 million, or about 10.4 percent of the agricultural sector.

Distribution of food and fiber system employment

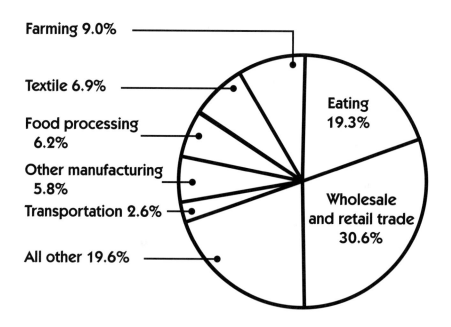

(*1990 Chartbook of Agriculture*, USDA)

In 1992 those 23.2 million jobs represented 17.3 percent of total U.S. employment and earned 16 percent of the GNP. Farmers made up less than 10 percent of all persons employed in the food chain. In only two states, North Dakota and South Dakota, did farm production jobs exceed jobs in the wholesale and retail sectors of agriculture. Our most urban states have the highest number of people employed in the food chain, most of whom are in wholesale or retail trade or in machinery production. California has 2.5 million; New York, 1.4 million; Texas, 1.4 million; Florida, 1.1 million; Pennsylvania, 1.1 million; and Illinois, 1.0 million.

The highest percentage of farm and farm-related jobs that deal directly with farm production are in the most rural states. Those states are Iowa, with 431,380 employed, who make up 27.2 percent of that state's work force; South Dakota, 100,503 and 26.7 percent; North Dakota, 95,247 and 26.4 percent; Nebraska, 243,833 and 25.0 percent; Idaho, 124,581 and 24.8 percent; and Arkansas, 271,959 and 23.6 percent. The total agricultural work force of those states is less than half the agricultural work force of California, which has 11 percent of all farm and farm-related jobs.

Of the 23.2 million agricultural workers, 16,439,334 (over 70 percent) work in metropolitan areas, and the other 6,723,231 are employed in non-metro areas. In 1990 the supermarket industry alone employed 3 million, while 2.5 million worked on farms. By 1999 it is expected that supermarkets will need 3.6 million workers.

One of the changes brought about because of greater mobility of population has been the growth of food service catering to those away from home or to two-worker families. In 1925 Howard Johnson opened his first ice-cream parlor / drugstore / restaurant. Except for local independent restaurants, there was little additional activity in the food service industry until the 1940s. Then, White Castle pioneered a fast-food hamburger chain. Others followed in the 1950s, and by the 1960s home delivery of prepared food commenced.

The share of two-earner families increased from 39 percent in 1950 to 60 percent in 1992. At the same time, the ratio of single-person households rose from 11 to 25 percent. The share of food dollars spent away from home rose from 24 percent in 1950 to 46 percent by 1990. The market share of the fast-food industry expanded from 8 percent in 1948 to over 40 percent by 1990. Sales at eating places with fixed menus made up over 80 percent of the sales of the food service industry. The fast-food industry accounted for 87 percent of the sales of frozen french fries.

The share of disposable income spent for food would have dropped faster if it had not been for the shift from eating at home to eating out. In 1929 consumers spent 23.8 percent of their disposable income for food, of which 20.6 percent was spent on food eaten at home and 3.2 percent on food eaten in commercial establishments. By 1959 consumers spent 17.7 percent for food, of which 14.2 percent was on home-consumed food and 3.5 percent on food eaten away from home. By 1992 the national food bill represented 11.3 percent of disposable income, of which 7.1 was for home-prepared meals and 4.2 percent was for food eaten at commercial places. It is not uncommon for the tip at luxurious dining establishments to be greater than the actual cost of the food served. Our food system is not only very responsive to consumer demands, but it is also efficient.

The total spent for food in 1992 was $605 billion, plus $87 billion for alcoholic beverages. Because beverages and tobacco are derived from farm products, they always are figured in the agricultural sector. The food and fiber sector generated a volume of $950 billion. Of that total, 79 percent was paid for directly by those who consumed the food either at home or away. Another 20 percent was paid for by the government for the military, by institutions, or by businesses. Only 1 percent of the food bill today is for food produced in private gardens.

The spread in the farmer's share for various commodities depends upon the degree that the raw food must be processed or its degree of perishability. For example, the farmer's share of the food dollar is different for each of the following: corn syrup, 4¢; oatmeal, 5¢; frozen french-fried potatoes, 10¢; margarine, 19¢; sugar, 39¢; canned peas, 21¢; frozen orange juice concentrate, 40¢; lettuce, 18¢; fresh potatoes,

21¢; lemons, 23¢; milk, 43¢; broiler chickens, 51¢; eggs, 54¢; and choice beef, 57¢. The spread on some products can vary greatly with a change in farm value, but in most cases the percentage spent for the raw product is not great enough to make a major difference in the retail price.

Farm Programs

For innovative farmers six decades of agricultural programs have provided incentives to industrialize, while for many others the programs have merely delayed the inevitable. In 1900, of the 38 percent of our people who still lived on farms, many were underemployed. At the same time, nonfarm industries needed workers. Food prices were rising, causing consumers to complain. The rest of society realized that agriculture had to become more productive and efficient to reduce food prices and to free farmers and their families for work in factories and offices.

The government initially encouraged agriculture to increase its productivity. After the 1930s and perpetual surpluses, the programs were designed to provide a safety net for farmers, which lowered the risks and enabled them to adopt new equipment, technologies, and facilities. Farmers reduced their reliance on diversified farming because of the insurance provided by price supports. The more they specialized, the more their productivity increased and the more they expanded.

The programs worked to the benefit of large-scale producers who could deliver directly to large-scale buyers, but small-scale producers had to rely on the declining terminal markets. Farm credit programs encouraged substitution of capital and management for labor. This hurt way-of-life–oriented farmers who were not accustomed to managing and helped those who managed, borrowed, and expanded the most. Progressive farmers often purchased additional land just to offset diverted acres required by the programs. This practice caused a significant structural change in farming.

Our early policy of disbursing land made food and fiber abundantly available and hastened development of the nation. That policy was supplemented by the creation of land-grant colleges. These institutions fostered the trend of making agricultural products available at low prices. The colleges increased knowledge about agriculture, which made it more productive. Agricultural education was initially neutral as to farm size, but after the advantages of scale were better understood, the colleges tended to be pro large farm. The educated farmers were quicker to adopt the new techniques.

Each step of the process favored the innovative farmer, whether it was guaranteed prices, direct payments, or commodity loans. The larger the unit, the greater the potential gain. Technologies increased the economies of scale, which aided the larger enterprises. This was better understood after the work of the National Food and Fiber Commission of 1967. The Commission concluded that federal and state governments had been "prime movers in the creation and promotion of new agricultural technology." Large-scale farms had profited most, and farm numbers continued to fall. The Commission stated: "This excess manpower and the excess crop acres are the heart of the U.S. agricultural adjustment program." Farm policy was not solely responsible, for farm expansion was consistent with national growth, as was the case in all other industrialized nations.

The only lament the Commission had was that "People are leaving farming for other occupations, but not as fast as [the] manpower requirement [in agriculture] decreases." It did not feel that the growth of agricultural productivity should be decreased but that the United States should take advantage of expanding world demand.

Our agriculture had to be ready to adapt to changing needs, and it could not do so if it were denied access to technology and markets. On that basis the Commission recommended a more market-oriented agriculture. This was the first step in a long-term change of policy.

The Commission was very foresighted in its recommendation, but changing the political course was difficult. The Commission was aware that the necessary changes in farm employment, farm size, and production patterns would not be accomplished as fast as necessary unless political efforts were made to bring about the inevitable adjustments. It continued: "No farm policy consistent with national growth can preserve farming opportunities for all manpower currently on farms. . . . [W]hether we follow a policy of production restraint or even a policy of all-out production, we would still need approximately one-third less labor in agriculture in 1980 than we need now." The Commission suggested that less than 10 percent of farm boys could expect to enter farming as their chosen occupation.

A Congressional study in the 1980s concluded that price support programs had become less effective in holding down surpluses and costs. It was decided that the Food Security Act of 1985 should make corrections and start reducing the government's role in commodity programs. This would save tax dollars and move agriculture toward the free market, as had been recommended in the 1967 study. Agriculture had matured, and downsizing had to continue. This meant a further reduction in farm numbers and in the number of people employed on farms.

The problem, as the USDA well understood, was that production continued to outpace demand except for brief periods, such as the early 1970s. The 1980s study pointed out that the time to produce a bushel of corn had dropped from 20.4 minutes in 1919 to 1.8 minutes in 1986; the time to produce a bushel of soybeans from 12 to 1.9 minutes; the time to raise 100 pounds of beef from 275 to 50 minutes; and the time to raise 100 pounds of pork from 220 to 20 minutes. The labor reduction needs forced a sharp drop in the number of farms and the growth of part-time farming. By the 1980s the government's effort to have cheap food for everyone and to free farmers for work in other industries had been accomplished.

Now the objective was to determine how industrialized agriculture should continue. The Congressional study hoped that by getting farmers to produce for the market it would reduce the stigma that programs were merely "rural welfare." New directions were recommended, which included protecting the quality of food, developing overseas markets, helping the needy in the United States and abroad, protecting the natural resources, and providing education and information assistance to stabilize rural America.

Secretary Henry A. Wallace said in 1933: "Having conquered the fear of famine with the aid of science, having been brought into an age of abundance, we have now to learn how to live with abundance." He prophesied at the time of writing the first farm program that it would hasten the demise of the small family farm, because he understood the economic implications of the support system. In 1955 farm economist John D. Black wrote to Secretary Benson: "Price supports mean very little to our low-income farmers. The truth of this statement is most clearly obvious in the case of the families that simply do not have enough land even though it were developed to the optimum point."

The truth of Wallace's statement was proven many times over when it was recognized that the programs had little value to the bottom three-fourths of farmers. The bottom 50 percent of farmers of the following commodities received the percentage of benefits indicated for their crops: wheat, 10.9 percent; cotton, 6.2 percent; rice, 7.0 percent; and feed grains, 13.3 percent. These farmers received 9.7 percent of all payments to farmers. On the other hand, the top 10 percent of farmers of the same commodities received the percentage of benefits indicated for their crops: wheat, 50.5 percent; cotton, 53.3 percent; rice, 39.8 percent, and feed grains, 39.5 percent. These farmers received 46 percent of all payments to farmers.

The 1967 Commission knew that out of approximately 3 million participating farmers, 2.4 million collected from $40 to $3,750 in annual farm payments while the top 16,000 collected from $19,000 to $100,000. By the 1990s it was concluded that the $144 billion loaned out by the FMHA, a lender of last resort, had "played a more important role in moving farmers out of agriculture than in keeping them in." The initial purpose of the FMHA was to help young farmers get started.

The wide swings in income, assets, and equity between 1974 and 1987 did not have a major impact on the long-range trend of declining farm numbers. Farmers' net worth dropped $235 billion. Their nominal net incomes declined from $25 billion in 1975 to $12.7 billion in 1982 but increased to $58.9 billion in 1989. Government payments jumped from $1 billion in 1974 to $11.9 billion in 1988, while off-farm income remained at about $20 billion and in 1988 exceeded 50 percent of net farm income.

When it was clear that price support programs were not effective in keeping people on the farm, public policies changed toward ways of improving the rural nonfarm economy. New programs encouraged the creation of jobs in rural areas, and in recent years they have been more beneficial than farm commodity programs. During much of the 1970s and since 1985, the average farm family has earned more than the average U.S. family. In addition to improved income, the median net worth for farm-operator households in 1988 was $167,000 (over $600,000 for commercial producers) compared to $36,000 for all U.S. households.

Probably two of the most far reaching programs to provide farm families with a graceful exit from farming were the Soil Bank Act of 1956 and the CRP of 1985. The Soil Bank Act was the government's response to coping with surpluses after the sudden drop in export demand following the Korean War. The government rented entire farms on 3- or 10-year contracts to take land out of production. In all, 28.6 million acres were removed from production. However, technology grew so rapidly that in spite of massive land idling, production did not decline. The positive benefit was that farm numbers dropped 1.7 million (31.1 percent) during the decade of the 1950s and farm population declined by about 8 million.

The CRP, which is still in effect, took 37 million acres of erodible land from cultivation for 10 years. It was not as effective as the Soil Bank in retiring farmers because it did not require entire farms. However, many older farmers used the CRP to reduce their farming operations, or they rented their remaining cropland to expanding farmers. The future of the CRP is not in the hands of farmers. Budget constraints, international competition toward production controls, and environmental pressures will decide its future.

After the 1950s and 1960s witnessed declining acreages in production from post–World War II levels, the 1970s brought about an increase in acres cropped from 294 million to 363 million. Farmers responded to demands of the world market in the 1970s. Environmental groups criticized the rebreaking of land for the purpose of supplying the export market. Congress responded with the Soil and Water Resources Conservation Act (RCA) of 1977. This set the stage for several conservation provisions that were written into the Agriculture and Food Act of 1981. Other acts followed, which provided an opening for soil conservation provisions in the Food Security Act of 1985. Soil conservation provisions denied program benefits to those who did not comply with the conservation programs.

In 1936 the first effort was made to restrict payments of over $10,000 to farmers because the benefits did not help all farmers equally. In 1973 the Agriculture and Consumer Protection Act tied several bills together and reduced payments to $20,000 per year. Target prices and deficiency payments were introduced. This was the first major political movement toward market orientation. The Food and Agriculture Act of 1977 continued the trend toward market orientation, but it also adjusted to the reality of capital-intense farming by raising the payment levels to $50,000.

In 1982, 1983, and 1984, loan and target prices were reduced because surpluses accumulated due to a reduction in exports. The 1985 Act provided for lower price supports through 1990 based on average market prices. The Farm Act of 1990 continued the same policy, and exports rebounded, which kept surpluses from accumulating. Because of renewed exports since 1987, farmers have experienced record gross and net cash incomes.

The social aspects of farm programs now overshadow the benefits paid to producers. In 1977 the appointment of an Assistant Secretary of Agriculture for Food and Nutrition was a major step in the pro-consumer trend within the USDA. In 1980 food stamps, school lunch programs, food for the elderly, food aid for Indian reservations, and other food assistance programs cost $17 billion compared to $3.6 billion for price supports and subsidies. Food aid made up over half the USDA budget. In 1993 more than 25 million students participated in the school lunch program and 27 million people were on food stamps. From 1990 through 1992 more money was spent on food assistance programs than in the first 50 years of support programs. By 1994 at least 60 percent of the employees of the USDA worked in nonfarm programs. These social programs have been charged to agriculture. They have been a positive benefit to society and will continue regardless of what happens to the support programs.

As the 1967 study pointed out, programs hastened the rise of large farms. This was a positive goal, even though those of the populist tradition did not see it that way. Studies in the 1990s point out that the smallest rural communities were the most adversely affected by acreage reduction programs. Rural communities under 2,500 decreased from 7.0 percent of the population to 4.3 percent. The greatest impact was on communities under 1,000. Since 1950 approximately 50 rural nonfarm people were lost per decade for each 1,000 acres of cropland diverted by the various programs. This amounted to a 7.4 percent loss per decade in the 100 most agriculturally dependent counties. The loss of farmers affected the rural nonfarm population, which caused the decline of the overall rural economy.

Changing Structure

A 1964 USDA study on the contracting and expanding sectors of agriculture pointed out that from 1939 to 1964 the number of farms had decreased by 2.4 million, of which 2.3 million grossed less than $2,500. Most of the remaining operators in that sales category were part-time or semi-retired. Their nonfarm income was nearly $3 for each $1 of gross farm sales. Most of them were 65 years or older. By contrast, only 20 percent of the income of farms grossing more than $10,000 came from off the farm. Because of the change in dollar value, the USDA now uses $40,000 as the dividing line between commercial and residen-

	Household Type			
	Full-Time Commercial Farms		Viable Small Farms	Low Income
	Large	Small		
Number of households	99,651	397,916	1,125,969	448,412
Share of households (percent)	4.8	19.2	54.3	21.6
	------------ Dollars ------------			
Household farm-related income (average)	63,037	17,373	−817	−7,334
Negative farm-related income (percent)	21.7	25.9	63.4	71.2
Off-farm income (average)	27,179	17,198	52,667	11,550
Earned income (average)	18,995	11,620	41,535	27,022
Percent reporting earned income	50.4	53.3	81.0	42.7
Unearned income (average)	8,183	5,579	11,132	5,519
Total household income (average)	90,216	34,572	51,850	4,216
Average farm net worth	1,153,240	484,018	243,163	230,626
Average government payment	17,611	7,221	982	939
Percent share of government payments	28.5	46.7	18.0	6.8

Farm income by classification. (*Outlook Proceedings, 1994*, USDA)

tial farms. About 70 percent of farms with sales under $40,000 have a net loss each year from farming. That group makes up 1.5 million out of the 1.9 million remaining farms.

The National Food and Fiber Commission of 1967 concluded that it was not desirable to raise support prices high enough for the lower-income farms to make a reasonable living. President Johnson declared that farm policies had to look to the future and not to the past. He said in his executive order regarding the study: "New ways must be explored to keep agriculture and agricultural policy up to date, to get the full benefit of new findings and new technology, to make sure that our bountiful land is used to the best of our ability to promote the welfare of consumers, farmers, and the entire economy."

Johnson advocated making the most efficient use of our resources, which implied that farm consolidation had to continue. This was done so that the "highest standard of living for all . . . citizens" could be achieved. He said that the best way to enhance the rural sector was "by increasing the occupational and geographic mobility of the population." Programs had to be developed for the 90 percent of farm youth who

Workers could dig, pick, and haul potatoes at the rate of about 40 bushels per person per day, Maine, 1908. (NAL)

would have to find jobs outside agriculture to improve the conditions of those who remained in farming.

Johnson stressed: "The sum of these individual decisions is a mammoth change in the structure of agriculture." Society could not afford the cost of maintaining those who desired to stay in agriculture. The trend toward larger and fewer farms started in the last decade of the nineteenth century and accelerated after World War II.

The four-row digger harvests four rows of potatoes and lays them in the center of two rows that the harvester will dig and elevate to the truck. In this manner potatoes are harvested at the rate of about 12 bushels per worker per minute. (Larry Monico)

The Commission concluded that change should be encouraged by continuing rural economic development. Because of the constantly rising cost of labor, caused in part by legislation, capital had to be substituted for labor. It was then pointed out that the trend would not be stopped for some time because "agricultural progress serves as the basis for the economic growth of any country." Without progress in agriculture there could be no specialized industry.

With the census of 1920, urban population surpassed rural population. Since that time the per capita income of non-metropolitan counties has nearly always lagged behind that of metro counties, and the unemployment rate has generally been higher in the non-metro counties. During the 1980s, 54 percent of the non-metro counties lost population. A General Accounting Office study on rural development noted that in

1990 about 3.9 million people (1.6 percent of the population) lived on farms. This represented a 31 percent decline in farm population since 1980. Farming represented only 2 percent of the GNP. In 516 (22 percent) of the 2,349 non-metro counties, farming was the dominant occupation, with 10 percent or more of the income. By contrast, manufacturing was the dominant occupation in 945 (40.2 percent) of the non-metro counties, and government activities led in 725 (30.9 percent).

In 1992 the 100,000 designated large full-time commercial farms grossing over $250,000 in annual sales had an average farm-related net income of $63,037. Of that group about 18 percent had a negative net income. Those grossing from $50,000 to $249,999 had an average net of $17,373, and about 24 percent of that group had a negative net. Those grossing $49,999 or less in income had an average net loss of about

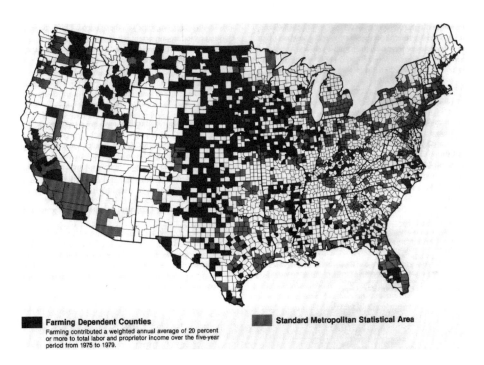

Farming-dependent counties in 1980. (USDA Rural Development Research Report No. 49)

Sources of farm household income, 1992

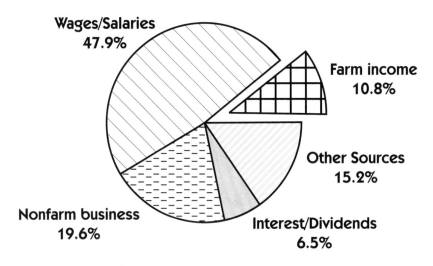

(*Outlook Proceedings, 1994,* USDA)

$3,500. Approximately 70 percent of those 1.5 million lost money. This was offset by average off-the-farm earnings of over $40,000. Nationwide, all farmers received only 10.8 percent of their income from farming. These figures explain in part why the *Yearbook of Agriculture* has been discontinued and why the census no longer counts farmers separately.

Environmentalism

In recent years two emerging views have arisen about agriculture. One view is industrial. This implies large-scale farming that is capital and management intense, uses labor extensively, achieves maximum yields with standardized production techniques, and is run as a business. The other view is sustainable. This implies smaller-scale farming that has lower capital but greater labor needs, uses on-site techniques, is adaptable to individual farms, and has less concern for maximum production. Sustainable agriculture must be run as a business to endure, but there is more emphasis on a way of life. In either case, only the efficient will survive.

Geographer John Fraser Hart contends that certain persons would like to keep the countryside as a museum—i.e., keep it looking the way they want to see it. Hart wrote:

> Some people who are concerned about changes in the countryside want to reject recent technological innovations and go back to an older farming approach. This is variously known as alternative, ecological, environmentally sound, low-input, natural, organic, regenerative, or sustainable agriculture.

Hart stresses that groups with that philosophy have effectively used publicity and have pressed the government to accept their ideas. The majority of the farmers are skeptical, even though a few have been successful with farming systems that use more labor and less chemicals and machinery. Their successes receive wide publicity, but little has been mentioned about the farmers who tried and failed or about the nonfarm income of those who have succeeded.

Individuals favorable to the current system point out that "high-yield farming is more sustainable today than farming has been in 10,000 years." They profess that without our current methods we would have massive famines. High-yield farming reduces soil erosion by 50 to 90 percent because herbicides instead of plowing are used for weed control. GPS, combined with radar and microprocessors, can adjust chemical and seed application rates as often as seven times a second. Top yields with no erosion are a practical reality. High-tech advocates contend that the so-called contaminants in ground water "are less dangerous than the pesticide residues on our foods, which are a trivial risk."

Environmentalism will play an increasingly important role in agriculture, partly because the public is concerned about food safety and has an interest in stewardship of the land. Some environmental constraints

Combining wheat and no-till planting soybeans at once, Maryland, 1982. (NAL)

will inhibit agricultural production. This will likely increase the cost of products. The Clean Water Act passed in 1972 came about because the Environmental Protection Agency (EPA) determined that 70 percent of river impairment was caused by nonpoint sources of pollution and that agriculture was responsible for most of that.

Some farmers have adopted LISA—low-input sustainable agriculture. LISA incorporates ecological, organic, regenerative, and biological ideas of those who prefer not to farm in the traditional manner. Their goal is profitable farming while protecting resources and environmental quality and cutting back on off-farm inputs. Crop rotation, soil-building practices, and more intensive livestock production constitute their plan for success.

The Food, Agriculture, Conservation, and Trade (FACT) Act of 1990 defines sustainable agriculture as an integrated system of plant and animal production practices with a site-specific application that in the long run will satisfy our food and fiber needs, enhance the environment on which agriculture depends, make efficient use of nonrenewable resources, sustain the economic viability of farming, and enhance the quality of life for all involved. An added objective is to increase opportunities for self-employment in agriculture on owner-operated farms.

Sustainable agriculture is part of the new public policy for agriculture. The contention is that there may not be much to gain from continued industrialization of farming, so it is important to work on ecological considerations. FACT and other programs have virtually mandated that price stabilization and supply management be linked with agricultural sustainability. Operations will have to be much more site-specific oriented.

Since 1900, production expenses have increased from 45 percent of gross income to 75 or 80 percent. In 1987 herbicides were used on more than 95 percent of corn and soybean acres and 60 percent of wheat acres. Some individuals feel that there may have been an overapplication of fertilizers and chemicals. The farmers used them because they were so economical. Until the early 1990s there was little incentive to develop the low-input systems, but it appears governmental intervention may change that. The result may lead to reduced efficiency, which could affect farm profits, cut the nation's competitive edge in the export markets, make it more difficult to improve the environment, and add to the cost of food. The intense regulations favor the larger farmers because they have better management staffs to keep up with the changes and can better afford the additional expenses that are incurred. Larger farms already have at least one full-time professionally trained person working on environmental concerns. Small farmers cannot afford that expense.

In 1988 the government funded two studies to learn more about the use and the effect of chemical and nonchemical practices. At that time there were more than 45,000 pesticides on the market derived from 600 chemical compounds that make up the active ingredients. The EPA approved about 350 of those pesticides for use on food products. The other compounds are used in nonfood applications. Another part of the 1988 studies strove to determine the long-range economic viability of some of the systems suggested. As of 1994, comparisons have been so diverse from various experiments that no positive answers have been determined.

The director of the Agricultural Marketing Service's Commodities Scientific Support Division wrote: "Chemical technology has played an important role in this [scientific] revolution. . . . It has brought significant benefits to the American consumer. . . . The public is often not fully informed, and retains a perception that pesticides present a serious health threat." In 1987 the FDA tested 14,500 food samples and found no excessive residues in 95 percent and no detectable residues in 57 percent. Less than 1 percent had residues exceeding established tolerances.

Many farmers have made adaptations in their current methods of operation that retain most of the high-tech methods but are more environmentally satisfying. In 1992 less than 12 percent of the acreage planted to major crops was plowed with the moldboard plow. The Food Security Act of 1985 mandated that for farmers to benefit from the support programs, conservation measures, including residue management, had to be adopted. Tillage that leaves 30 percent or more residue on the surface after planting is defined as conservation tillage. Mulch tillage leaves more than 30 percent of the soil covered, and compared to conventional tillage it saves time, fuel, and labor. In the no-till method the soil is not disturbed from harvest to planting. In 1992, 14 percent

No-till wheat seeding, Washington, 1976. (NAL)

of the major crops were produced by no-till. It can reduce erosion by 75 percent and is the most effective method for doing so.

Because surplus production continues to be a problem and likely will remain so for decades to come, many individuals advocate removing fragile lands permanently from production. Academicians Frank and Deborah Popper have pointed out that about 150 western counties have a population of less than 2 people per square mile. Those counties cover an area of 950,000 square miles, with a total population of slightly over 570,000. If 6 people per square mile is used as a frontier criterion, then 394 counties could be included. Those counties contain about 2.2 million people and cover 1.6 million square miles.

The Poppers suggest that probably the best use for this area should be preservation, recreation, and retirement. Ranching, farming, logging, and mining should be seen as secondary industries. Environmental legislation will accelerate such a movement. It is possible that recreation and retirement may prove more profitable than the current use of this land. The transition to occupations in the above fields would probably not involve a major change for many persons.

Today there about 358.5 million acres of public land in the 48 contiguous states. Indians control another 50 million acres, and Alaska and Hawaii have millions more acres. Most of this land is of lower grade from an agricultural standpoint. Current legislation favors retaining these areas as public land. Most of this land is in 11 western states and is desert, semi-desert, mountainous, or marginal for agriculture. Its greatest value is for forestry, grazing, minerals, wildlife habitat, recreation, and retirement living. Of the nearly 880 million acres used for grazing, about 474 million are privately controlled; the remainder are in federal grazing districts. In some areas the land is so marginal that 100 acres are required to support one cow.

In addition to providing environmental improvement and added long-range benefit to agriculture, the environmental movement will likely turn to agriculture to provide alternative fuels for energy use. The 1990 Clean Air Act amendments will have a major impact on agriculture. They are likely to boost the demand for ethanol and will require a doubling of the size of that industry by 2000. Tax credits could also increase the demand for ethanol, which would create additional demand for corn. Other crops, including some vegetables, are being considered as sources of ethanol. A large project in Minnesota contemplates the production of 175,000 acres of alfalfa, from which the stems will be used for energy production after the leaves have been removed for protein.

Forest preservation, as discussed in previous units, was one of the first major efforts to safeguard natural resources. The movement has continued. In 1964 the Wilderness Act was passed, which opened the way for public and Congressional involvement in the management of the Forest Service. In 1970 the National Environmental Policy Act paved the way for direct public involvement in the daily operation of the Forest Service. In 1973 the Endangered Species Act was passed. This was followed by the Forest and Rangeland Resources Planning Act of 1974, which provided for a long-range program to protect forests and rangeland. In 1976 the National Forest Management Act replaced the Organic Act of 1897 and forced more public involvement in decision making.

A hydraulically powered blade cutting a tree, South Carolina, 1967. (NAL)

The area in forests declined for two centuries, but by 1993 forest acreage totaled 737 million acres, an area equal to that of 1920. About one-third of that is federally owned. With the exception of 1990, regrowth has exceeded removal every year since 1952. By the 1980s annual net growth exceeded harvest by as much as 37 percent. National roundwood supplies increased from 9.5 billion cubic feet in 1952 to 18.2 billion cubic feet by 1991.

There are several reasons for those gains. The first is that the average volume of standing timber per acre has increased by 30 percent since 1952. A second factor is that loss to wildfires has decreased from 20 to 50 million acres in the early 1900s to 2 to 5 million acres in the 1990s. A third cause is that tree planting reached record levels in the 1980s and has continued. In 1993 there were 70,000 certified tree farms on 95 million acres of private land. A final reason is that by 1986 over 28 percent of the paper and cardboard used was recycled.

The Food and Fiber Commission pointed out that the greatest need for improved forestry was on the 300 million acres of privately owned

A 20- to 25-year-old stand of red pine forest that has been trimmed and pruned, Wisconsin, 1975. (NAL)

timberland. Farmer-owned woodlots made up about 35 percent of the nation's commercial timber, and land owned by small holders another 26 percent. Initially, many of those lots provided firewood for home and business use. Farmers often chopped wood for extra income, in addition to chopping it for themselves. The need for this wood disappeared when other sources of energy became more economical.

National forests provide the major source of water for 1,800 western towns and cities and over 600 hydroelectric developments. Forestry-related products make up the fourth largest manufacturing industry, after transportation equipment, machinery, and food processing. Of the over 2 million persons in forest-related work, only about 75,000 work directly in the woods. Approximately 700,000 work in the nearly 40,000 sawmills and in the manufacture of lumber products, 375,000 in the making of furniture and wood fixtures, and 565,000 in pulp and paper mills. The others are in subsidiary industries and the Forest Service. Each timber dollar is worth about $20 by the time it reaches the final consumer. In addition, at least 60 million people use the forests for recreational purposes annually, which provides a major source of employment in western and northern areas. No realistic value can be placed on the worth of water for multiple uses, including power.

The United States remains the world's leading importer of timber products. Softwood lumber, wood pulp, paper, and board lumber are imported from Canada, and veneer and plywood from Asia. We import about one-fifth of our annual consumption. On the other hand, we are the second largest exporter after Canada. Timber products represent about 3.4 percent of our imports and about 4.2 percent of our exports. Our exports are chiefly to Japan, Western Europe, and China.

Fickle Consumers

The greatest challenges facing the food chain, including farmers, are changing consumer demands, environmentalists (including animal welfare advocates), globalization, and public policies. On the positive side, population and income growth will continue. Currently about 90 million people are added to the world's population annually. In the United States and other developed nations, overall income growth will likely slow down as the percentage of two-career households becomes nearer to the maximum. Dual-income homes will still increase food consumption and expenditures because of their demand for more convenience and further-processed foods. More people will eat out more often. High-income consumers lead the way in changing styles for in-home, away-from-home, and take-out food. They want top products and are prepared to pay for them.

The price of food impacts demand, but farmers and processors are very effective in keeping production high, which holds prices down unless there are unforeseen obstacles. Low-income consumers, who represent about 20 percent of the market, stay with traditional foods longer than higher-income, better-educated groups do. About half of them receive aid from the federal government in the form of school lunches, food stamps, and special milk programs. In 1994 the school lunch program was affected by federal dietary changes.

Demographic changes, such as smaller households, older heads of households, and ethnic diversity all affect food demands. Consumers want smaller packages, more fresh foods, foods more appealing to older people, fewer baby foods, and foods from their native lands. At the same time, there is a change in food consumption patterns because less food is needed to fuel the human body than in the past. Lumberjacks who worked in the woods from daybreak to darkness in the cold winter months averaged 5,600 calories daily. Meat, potatoes, and pastries were a large part of their diet, which averaged about 7 pounds of food per day.

Office work does not require as much intake as manual work. So-called health foods are more in demand. At the same time, food has become an increasing part of the social environment. The great variety of foods leads people to expect something new, which makes them inconsistent about the types of foods they eat. Consumers are fickle. It is a challenge to keep pace with their wants. Some want to eat so-called natural foods, which compounds the problem of persons in the food chain. Much of the change depends upon who makes the decision—industry, governmental regulators, farmers, or environmental interests.

Annual food consumption was about 1,533 pounds per capita in the late 1930s. During World War II it peaked at 1,651 pounds and declined steadily until 1972. At that point there was a shift in the type of foods. Fruits, cereals, juices, seafood, vegetables, and poultry increased at the expense of beef, pork, and dairy products. There was a decrease in high-cholesterol foods, sugar, coffee, and eggs, but a contrary trend in the increase in high-fat foods, sweeteners, soft drinks, calories, and fast foods. There is a marked shift from inexpensive to more expensive foods.

Consumers forced many changes through their demands for safe foods, which they wanted to have produced in an environmentally and technologically proper way. They convinced the government to regulate production, which forced a change in procedure as to how food is produced and processed. Once again this ongoing movement is pro large farms, which produce under contract to preserve product identity with the manufacturing firms.

Some of the major regulations introduced in recent years, in addition to those mentioned in the previous topic, are the Federal Insecticide and Rodenticide Act of 1947, in which the FDA was given power after 1953 to inspect a plant following a written notice to the owner; the Food Additives Amendment of 1958; the 1967 Wholesome Meat Act; and the Nutrition Labeling and Education Act of 1990. Also, the National Center for Food Safety was established in 1988.

The changes have been controversial. FDA scientists have determined that pesticide residues "account for less than one one-hundredth of one percent of all fatal cancers. . . . [M]ore than 98 percent of the overall risks from food related cancer was inherent in the foods themselves." Biochemist Bruce Ames claims that deaths by cancer from human-made pesticides in food and water are "close to zero." A Purdue scientist maintains that pesticides consumed by average adults in the past were equal in toxicity to that of an aspirin or of the caffeine in a cup of coffee.

Dennis Avery, an agricultural analyst, asserts that the greatest danger in reducing pesticide use is that organic production could not supply the demand cost effectively. Nonchemical methods could not produce the necessary quantity and quality of fresh produce at competitive prices. Others maintain that because plants contain their own toxins, virtually 99.99 percent of the pesticides people consume are in a natural state in plants. The human body is indifferent as to whether a chemical is synthetic or natural.

A science advisor for the USDA wrote:

> Inorganic or commercial fertilizers contain the same chemical nutrients but in simpler forms. . . . It is not accurate to refer to inorganic fertilizers as "artificial" just because they have not been made by living cells. A plant is not aware of the type of fertilizer, organic or inorganic, that is furnishing the chemicals for its growth. . . .
>
> There is no scientific evidence that plants grown only with organic fertilizers or that meat from animals raised only on organically fertilized feed has greater nutritive value than our regular food produced by the usual agricultural methods.

A New Rural Society

Because of agriculture's constant underemployment, net income in the industry commonly was less than 50 percent of what it was in other sectors. By 1929 the differential had dropped to 43 percent, and during the 1950s and 1960s the income spread between farmers and nonfarmers continued to narrow.

In the boom years of the 1970s, the net income of farmers exceeded that of nonfarm people for the first time in history. During the price decline in the 1980s, net income fell to 72 percent of what it had been in the 1970s. Since the recovery in the late 1980s, combined farm and nonfarm income of farm families has exceeded the income of nonfarm people. The irony of that is that in 1992, farm households received 89.2

percent of their income from nonfarm sources, leaving only 10.8 percent from farming.

Accelerating Change

The rapid change in agriculture will potentially have major effects on our society, but at this writing the consequences are not clear. U.S. agriculture continues to grow more dependent on international markets, which increases its interdependency. Technology has made and will continue to make greater changes in farming. The technological lag is obvious on farms and in rural areas that are unable or unwilling to make the necessary adaptation. Environmentalism will remain an increasingly important part in the restructuring of farming and rural areas.

Isolation and sparse population have been the two biggest brakes on rural growth. Even with modern transportation and communication, it is difficult to make up for being cut off from the mainstream of economic life. The lack of a critical mass hinders the development of the rural economy. The government does not have the resources to help the limited population of the rural areas because of more pressing problems.

During the 1940s, 1950s, and 1960s, the industrial economy created a demand for labor that attracted workers away from farming. At the same time, there was a sharp increase in human skills and knowledge—hence, improved managerial ability. The unprecedented agricultural economy of the 1970s created a short-lived aberration in the decline of farms. The 1970s was the only decade of the twentieth century in which rural America did not have a decline in population.

From 1950 to the 1990s the share of the GNP attributable to the natural resource industries (agriculture, mining, and timber) has declined from 11.7 percent to slightly over 3 percent. This is in keeping with the long-term trend in which natural resources will continue to play a diminishing part in our economy. Rural America needs to find supplemental industry to avoid further loss.

In the 1960s and 1970s, manufacturing went to rural areas and absorbed the surplus rural labor. By the 1970s, $2^{1}/_{3}$ times as many rural people were employed in manufacturing as in agriculture. In 1972 the Rural Development Act was passed to stem the flow of rural population to the urban centers. Then, in the late 1970s, manufacturers found themselves in competition with foreign manufacturers and stopped expanding. This was part of the globalization process, which had a serious impact not only on U.S. rural areas but on rural areas worldwide. The change is the result of well-documented trends in the international marketplace that put less value on what rural America has traditionally produced.

Rural America no longer can focus on an agrarian-centered life, but there is no historical basis to determine what the end result will be. Worldwide economic change challenges any government policy that stands in its way. The best the government can do is to facilitate the transition by seeking new opportunities. In the end, however, the long-range outcome will depend on what rural people decide to do. They must create their own future.

In a market-oriented economy the economic rewards go to the producers who have the competitive edge. We have a commercial, industrialized agriculture with the world's best infrastructure, which enables us to be competitive. Mining, farming, and lumbering have become very labor efficient in recent decades and require far fewer workers to produce more. However, many people in rural areas are so immobilized by what they feel is a tragic loss of their valued past that they lack the adaptability and creativity to cope with change.

The reduction in the number of people necessary to supply our food needs has been a boon to the nation. A decline from 96 percent of the population in agriculture in 1790 to 50 percent in 1900 to 1.5 percent in 1994 has upset those with strong populist leanings. In 1880 nonagricultural labor exceeded agricultural. In 1929 the workers in manufacturing alone outnumbered those in agriculture. Since 1962 the number of unemployed workers (even in times of prosperity) has exceeded the labor force in agriculture.

A 36-row corn planter that can be reduced to 18.5 feet for road travel in five minutes. This rig can plant 11 acres in one round on a half-mile field in 10 minutes. Three trucks are required to supply it with seed, fertilizer, and chemicals. Iowa, 1977. *(Farm Show)*

A Government Accounting Office (GAO) study in 1992 noted that the farm bill of 1990 included the creation of the Rural Development Administration for the purpose of promoting rural America. The study determined that the biggest problem was the lack of an adequate job base. Large manufacturing plants could not locate there because that would only create more problems. Small manufacturing and cottage industries were the apparent solution. The study found that about 20 percent of the work force of rural America was in agriculture. In some midwestern states there was only 1 farmer under 35 years of age for every 20 who were over 55. This meant that agricultural numbers had finally adjusted to long-range needs.

During the 1980s employment in the metro areas grew by 21 percent against only 12 percent in the non-metro areas. Over half the rural counties lost population during the 1980s. Only those bordering metropolitan areas or those with natural amenities for tourists or retirement grew. Since 1982 virtually all the growth in rural counties has been in 500 retirement and recreation counties. The greatest outmigration has been from areas of marginal farms. This has produced rising rural incomes for those who have remained. In recent decades rural areas have had a higher unemployment rate, a higher percentage of poverty, and a greater need to retain population if they were to remain viable communities.

Low density of population and small size prevented rural settlements from having economy of scale in delivering private and public services. Hospitals, schools, churches, and most private businesses had difficulty generating the necessary volume. Most of the food processing and manufacturing industry has moved to larger centers. The GAO found that the most important problem in rural areas was remoteness, which prevented rural people from understanding what was taking place in the national and global economies.

A USDA study pointed out what really happened to change the rural scene. In 1900 it took 983,563 farmers out of 5.74 million to produce 50 percent of the gross sales from agriculture; in 1940, 688,912 out of 6.1 million; and in 1969, 221,690 out of 2.78 million. In 1994, 75,682 farms out of 1.9 million produced half of all sales. Those 75,682 farms had average sales of over $900,000 and average expenses of about $740,000, leaving about a $160,000 average net per farm. The average investment of those farms was nearly $1.7 million.

Former Secretary Orville Freeman summed up the position well: "Lots of people are on farms that are too small and can't grow. Perhaps the people are old and uneducated. . . . No price-support program will do any good for a guy on forty acres with a mule. . . . These people should be dealt with under poverty programs not programs for commercial agriculture."

Part-Time Farming

Part-time farming has been part of agriculture from the beginning. Early farmers had to work off the farm to earn cash to pay taxes and purchase a few staples. Off-farm work was an accepted way of helping the farm survive. When the frontier moved westward the opportunity for part-time farming diminished. However, even on the remotest frontiers some farmers worked for the railroad, in the woods, in the mines, or as trappers and hunters.

One of the problems has been to define what is a farm and what is a part-time farm. Since 1850 the Census Bureau has changed the definition of a farm nine times. It always left the definition very vague so the largest possible number of farms could be counted. In the 1920 census a farm consisted of three acres of land with $250 in sales. This caused an increase of 300,000 farms under 10 acres. Later the definition was any place that sold at least $50 worth of agricultural produce and had 10 or more acres of land. In 1978 the definition was revised to any place that sold $1,000 or more in farm produce. This dropped the number of farms by 302,000. Without the change in definition, only 79,000 farms would have been lost.

Geographer Hart contends that at least one-third of the reported farms today are nonfarms and probably even more than that are not productive agricultural operations. Hart suggests that the loss of those 750,000 farms would not affect the nation's production. It would have only a limited effect on the operators, because they earn so little from farming. In 1992 most of them had net losses from farming.

In 1934 about 11 percent of the farmers worked off the farm 100 days or more. That year 29 percent of all farmers sold less than $500 worth of commodities, and about 40 percent of them had farms of less than 50 acres. Most of those farms were in the East and the South. By 1987, 43.8 percent of the farmers worked off the farm 100 days or more. Of these farmers 36 percent sold less than $5,000 in produce, 28.5 percent farmed less than 50 acres, and 45.5 percent listed another principal occupation other than farming.

Some economists suggested that the census should ignore farms with annual gross sales of less than $5,000 because they are chiefly rural residences. This would have reduced farm numbers by 44 percent. By 1970 less than 15 percent of all farm families earned their entire income from farming. By 1979 that figure fell to 8 percent, and over 70 percent listed a primary occupation other than farming. In 1992 the bottom 448,412 farmers, who grossed less than $15,000 from the farm, lost an average of $7,334. This was offset with an average nonfarm income of $11,550. The 1,123,969 farmers who grossed from $15,000 to $49,999 had an average loss of $817, but they averaged $52,667 in nonfarm income. The lowest group of the farm sector had a poverty level of living, while those in the next category had a higher net income than the commercial farmers in the $50,000 to $249,999 gross farm sales bracket.

What are the reasons 76 percent of our farmers give for their decision to be part-time farmers? A study in the late 1970s indicated that they liked the pastoral setting. They were hobby farmers who had a few acres to continue their hobbies or recreation. Some enjoyed gardening and used the proceeds to reduce family food costs. Others believed in the organic concept and wanted to raise their own food and also to earn some extra income. Some were retired and did part-time farming to supplement their income. Others had a lifelong dream of wanting to be a farmer and were willing to work off the farm or use other income to support the farm. A very large portion were looking for an alternative way of life and were willing to pay the price for that privilege. The USDA found that many eventually were unable to overcome the hardships of part-time farming and returned to the city or at least gave up the idea of farming. If they were lucky, they were able to rent the land to another farmer. If not, they learned that it was less costly to leave the land idle than to farm it. Whatever the reason, part-time farmers are still counted in the census, even though fewer than 400,000 farmers provide the nation with virtually all our produce.

Changing Life Style

Previous discussion documented the role of women in the economics of farming. In spite of their contributions, women made little progress regarding their equality until the Donation Act of 1850 permitted wives of homesteading families to hold 320 acres. Next, the Homestead Act permitted single women to file individual claims. After 1888, women were given political equality in six western states. They exerted those rights and held as much as one-third of the land in those states.

However, it was not until 1978 that the sex of the farm operator was recorded by the Census of Agriculture. Prior to then it was difficult to determine how involved women were in farm management. There was always an undercount, because only one operator was enumerated per farm and that was usually the husband. The reality was otherwise, for as recently the early 1900s, farm women provided about 80 percent of the cash used for family living from the sale of farm commodities they produced and processed.

Beginning in 1930 their contribution took a different form, for from 1930 to 1980, farm women entered the paid labor force at a faster rate than nonfarm women. In 1930 only 12.3 percent of farm women were in the labor force, in contrast to 24.9 percent of nonfarm women. By 1980 the percentages were 45.8 and 50.2, respectively. In recent years at least one-fourth of all farm women were filling "triad roles—homemaker, farmer, and employee." About 57 percent worked because of financial need, 16 percent to keep proficient in their field of training, 18 percent for social reasons, and 9 percent for other reasons.

Improved transportation and communication after the 1930s helped farm families become more urbanized in their outlook. This tended to break down the semi-patriarchal family system of early agricultural society. The family system became more egalitarian because of more education for the woman and more work opportunities. By the 1960s there was little difference in the relative dominance by husbands or wives in decision making in families living on farms, in rural nonfarm areas, in small cities, or in larger cities. By then there were no marked differences among families based on tasks performed.

A study of 1,500 farm families in the 1980s, of whom 89 percent earned some income off the farm, indicated a declining willingness of farm wives and children to work on the farm without being paid. This was especially true for the younger family members, who expected to earn wages and benefits comparable to those from nonfarm work. This is in sharp contrast to the past, when all family members were expected to work without pay to enable the farm to survive.

A 1980 Florida survey indicated that off-farm income, most of which came from women's work, provided over one-half the net income for farms grossing up to $100,000 and more than one-third for farms grossing up to $200,000. That survey pointed out that women also were doing more farm work than they had in the 1920s and the 1930s. In the earlier period, they worked on the farm 11 hours per week, and in the 1980s, 22 hours. Modern appliances enabled a reduction in housework in those decades from 50 to 26 hours. The determination as to who should work off the farm was based on whether the husband or the wife could earn the most income. In a large majority of cases, the women were nurses or teachers or had other professional training that enabled them to earn the higher off-farm income.

Live-bird market where birds are delivered daily to wholesalers and then to retailers, New York, 1986. (NAL)

An Iowa study revealed that in earlier days farm women made income from selling processed dairy or poultry products, catering meals, and selling homemade items. As the farm structure changed with mechanization, the small flock or dairy herd quickly faded out of the picture. Health laws regarding the sale of unpasteurized dairy products or uncandled or ungraded eggs put the small herds and small flocks out of business. Flocks on Iowa farms decreased from 98 percent in 1930 to 4 percent in 1982. Those remaining were chiefly in large commercial enterprises or on farms of older people.

In the post–World War II transition, women had the choice of becoming involved with farm enterprises as bookkeepers, marketers, "gofers," or field workers or taking jobs in town. It depended upon where they could achieve the greatest net income. In Iowa, a ranking rural state with a high portion of small-scale farms, the percentage of farm wives in the paid labor force grew from 8.3 to 39.4 between 1940 and 1980. Some women on economically viable farms gave up their nonfarm jobs when they realized that they could improve their net income by working and helping to manage the farms.

A survey of 160,000 farm families in major farm states indicated that over 77 percent of the women were involved in managing the farm business. Of that group, 71 percent took a part in decisions in buying and selling land; 70 percent in renting; 64 percent in purchasing inputs; 63 percent in selling commodities; and 60 percent in trying new production practices. Of the actual labor activities, 97 percent ran errands, 89 percent did record keeping, 80 percent did tax preparation, 77 percent helped in harvesting by running machinery or trucks, 74 percent cared for livestock, 65 percent supervised other workers, and 65 percent monitored market news reports.

There were reasons other than cash income that led farm women to work off the farm. One of the major factors was securing health insurance. Other benefits were increased social security, paid vacation and/or sick leave, pension, life insurance, employee discounts, and education. Many women were willing to put forth the extra effort because they wanted to raise their children on farms. Some did so because they wanted to help their children begin farming. In many cases women worked off the farm because couples realized that their marriages were less stressful if both spouses were not involved in farming. On the other hand, some couples gained satisfaction from partnering in the farm enterprise.

A 1980s study in Michigan pointed out that on 27 percent of the hobby farms with incomes under $2,500, women did the total workload. On small farms with from $2,500 to $20,000 gross sales, women did 24 percent of the total workload, and on farms with over $20,000 in sales, they did 23 percent of the workload. About 68 percent of the wives on hobby farms worked on the farms, in contrast to 79 percent on small farms, and 85 percent on larger farms.

Women who were classified as farmers or farm managers increased from 2.7 percent in 1950 to 5.4 percent in 1982 to 7.5 percent in 1994. The greatest growth percentagewise came in the South, where the women tended to be white and widowed and where half were over 60 and had a small livestock operation. Farms operated by women averaged 285 acres versus 423 for those farmed or managed by men. Women had far less debt than men, and farm income represented about one-third of the total income. Very few of the farms were carried on as complete units once the woman who was owner/operator no longer farmed. As more women graduate in professional agricultural courses, their role in the total agricultural chain will increase.

Foreign Investments

Corporate farming and ownership of land by foreigners constitute two of the most emotional issues in rural America. They present no real threat to U.S. agriculture, but the public perception, especially among some farm groups, is that these two issues pose a potential danger to our current system. Some farm state politicians have perpetuated the issue to promote their campaigns.

A recent USDA study established that foreign investors own 5.2 million acres of farm, forest, or pasture land, about one-half of 1 percent of the total. Three-fourths of that land is owned by U.S. corporations that have foreign stockholders who hold 5 percent or more of the stock of the corporations. In the three states with the largest percentage of foreign-owned land, 80 percent of that land is owned by such corporations. Of that land, 19 percent is cropland, 23 percent pasture, 14 percent miscellaneous, and 44 percent timberland. Crop and pasture land represent less than 1/10 of 1 percent of all privately held agricultural land. The timberland is held chiefly by U.S. corporations with foreign stockholders.

Over 40 percent of foreign owners are Canadians, who hold 12 percent of the foreign-owned land. That represents about 3.91 million acres. The British hold another one-third of the foreign-owned land by virtue of the fact that they have stock in U.S. paper and timber corporations. The total investment in agricultural land is about $10 billion. In addition, foreigners also have another $32 billion invested in agribusiness firms.

The Japanese became the heaviest new foreign investors in U.S. agriculture and agribusiness during the 1980s. By 1990 they had $3 billion invested in land and agribusiness, making Japan the fourth largest foreign investor. The Japanese have interests in wineries, Alaska fish processing, a U.S. convenience store chain, feed additives, soybean processing, and the grain storage business. This is a positive factor because it stimulates growth and income in the United States. After the 1988 U.S.-Japan Beef and Citrus Understanding, the Japanese invested in those industries to assure themselves supplies for export to Japan. By 1990 they owned 9 ranches, 9 feedlots, and 21 beef processing plants. They also had joint ventures in cattle ranching in at least nine states.

Japanese firms are very interested in the food industry aimed at the Japanese ethnic market in the United States and in Japan. Because of that interest, they invest in grazing land, in forests, and in grape, rice, and citrus production. Nearly 70 percent of Japanese land is owned in joint venture with U.S. citizens or corporations.

Citizens from the United Kingdom remain the largest investors in agribusiness, with a total of $11.044 billion. The next largest block of foreign ownership is by members of the European Community, which have about $27 billion invested in U.S. agribusiness. On the other hand, U.S. investors have $34 billion in holdings in foreign agribusiness. Foreigners see U.S. agriculture as the world's leader and invest here because our nation provides a safe place for investments while enabling them to produce commodities they want to import.

New Farm Groups

After the founding of the AFBF in the early 1920s, no other farm organizations were created until 1955, when the National Farmers Organization (NFO) was established. The NFO initially planned to have farm holidays and to create legislative pressure. In 1958 the NFO decided to try collective bargaining by negotiating contracts with processors to raise and stabilize prices. On September 1, 1962, it had a withholding action that included hogs, cattle, sheep, corn, and soybeans. By September 12 the action weakened because nonmembers were no longer willing to cooperate. On October 2 the action was suspended.

A second major withholding action was attempted in 1964 and a third in 1967 with virtually no success other than calling attention to the plight of some farmers. Most farmers were too independent to organize to the degree needed to make withholding work. The NFO has not released membership figures, but it apparently never achieved the size of the other major organizations. The boom of the 1970s dimmed the activities of the NFO.

In 1977 the American Agricultural Movement (AAM) was formed. Initially, the AAM threatened to strike by not producing a crop or not buying equipment. This was a radical stand, which caused the tactic to fail, as only the most zealous were enthusiastic about not planting. As an alternative the AAM held tractorcades to Washington, D.C., in January 1978 and February 1979. This marked the dramatic emotional height of the organization's protest. "Such actions antagonized most policy makers [including Secretary Bergland], much of the once-friendly media, and scores of one-time AAM activists." The tractorcades caused the organization to become fragmented, because many disagreed with the tactic.

The AAM was caught between the values of technological agriculture and the fact that technology was the cause of overproduction. In desperation to gain more political clout, the AAM associated with smaller statewide groups involved in agricultural protests, such as Nebraska's Center for Rural Affairs, Iowa's Prairie Fire Rural Action, the Rainbow Coalition, and the National Save the Family Farm Coalition. The AAM had to compromise its position. The diverse nonfarm groups were interested in proposals such as protecting family farms, producing crops with minimal chemical use, and saving rural communities. The AAM was not interested in those causes, but it needed the support of the groups and had to stay with them. This opened the door for activist groups to become involved in agricultural regulatory legislation. The rebound of the farm economy hurt the already badly shattered organization.

Declining Rural Communities

Town boomers were commonplace on the frontier. Wherever the railroad established a depot or the river boat firm picked a spot on the bank, someone boomed the area. Towns could be within eight miles of each other without serious competition during the days of horse traffic.

In the East, small rural towns were on the decline prior to the Civil War, but much of the nation was not yet settled by that time, so the institution of the small rural service town continued to survive for decades. By the late 1800s railroads caused the demise of many towns, because it was common for individuals to travel by train to larger nearby towns to shop. The advent of the automobile drastically changed the future of the small town, because people were willing to pass the local village and drive up to 30 miles, even on dirt roads, to shop in larger centers.

By the 1920s books were being written exposing the disillusionment with small towns. The writers challenged the myths about the virtues of the small communities. During the 1920s economist Thorstein Veblen credited the small town as being a molder of public sentiment, but he also pointed out its shortcomings. One of Veblen's chief contentions was that small-town merchants charged the farmers all they could without driving them away. This writer knows of small-town merchants who stated that if they could not operate on a 25 percent margin, they would close their businesses. After the 1930s paved highways provided an escape for the customers.

Vacant buildings and declining population in a large portion of the towns have been characteristic of midwestern history for almost every decade since the beginning of settlement. However, outmigration be-

Typical small-town blacksmith who fixed nearly all farm machines, except tractors, and also shod horses, North Dakota, 1925. (Donald Brandt)

came more pronounced in the 1930s and accelerated in the 1950s. An Iowa farm couple who live on the edge of a small village told this writer that the most depressing part of their daily life was driving down the main street each morning and evening on the way to and from the farm. In that village one-third of the buildings were in ruins, another third were boarded up with plywood, and the other third had owners who wished they could find buyers.

Iowa became a state in 1846, and by 1930 it already had 2,205 abandoned towns, villages, hamlets, and country post offices. There were many reasons for those closures, such as loss of the county seat, abandonment of a military road, exhaustion of coal mines, declining importance of grist and sawmills, and declining importance of river traffic. Early plat books and road maps of any state reveal a large number of villages that no longer exist. The decline has not yet abated.

In 1985 the commissioner of agriculture in a leading agricultural state related to this writer that the decline of the 1980s was so serious that two county-seat towns were in danger of not surviving. In another midwestern state, two county-seat towns have fewer than 50 citizens.

In a northern Minnesota county 41 post offices were opened, most between 1907 and 1915. Within a decade many of those post offices and their communities were in decline. Today this county has 10 post offices remaining, most of which are small fourth-class offices.

In an in-depth study by this writer of 70 counties in three agricultural states, most businesses, such as medical facilities (including nursing homes), elevators, automobile agencies, and machinery dealers, acknowledged that they could not continue if they could not do $2 million in business annually. Elevators could not compete if they could not provide for unitrain service, because of the freight differential. Farmers using semitrailer trucks hauled their grain many miles to unitrain elevators. The other businesses could not achieve the volume to purchase the equipment needed. This study revealed that the smaller the largest town in any given county, the older was the population of that county and the higher were government transfer payments in relation to total income. Government transfer payments were chiefly social security and welfare.

The government has long been aware of the problem of declining rural areas. With improved transportation and communication, rural towns are no longer more immune to change than other sectors of the economy. Farms have declined from a peak of 6.8 million to 1.9 million, of which only about 400,000 are economically viable businesses. Today, 3.5 million people (over half of whom have off-farm jobs) live on farms where once 32 million people farmed. The towns that were once needed to service the farms have lost their reason to survive. The most damaging aspect of this is the decline of the labor pool needed for agricultural production.

The 1986 OTA study on the changing structure of agriculture searched for answers to what could be done. Part of its conclusion was that "Hard-hit communities may need technical assistance to attract new businesses to their areas, to develop labor retraining programs, and to alter infrastructure to attract new inhabitants." Higher wages and better living conditions might draw skilled workers back to farming. A few communities may profit by becoming processing centers for agricultural products, some may find new opportunities, and many others will continue to fade away.

During the 1970s the Family Farm Policy Review conducted 7,000 meetings to determine what the USDA could do to "safeguard and preserve agriculture's strength . . . by protecting the farm families, and our traditional pattern of farming." This study was second only to the Country Life Commission in scope of review of agriculture and rural America. The result was the Rural Development Act of 1972, which was designed to help maintain farm numbers by providing industrial opportunity for off-farm work. The small rural service center necessary as late as pre–World War II days is no longer needed. The business-oriented farmer has become a discriminating buyer and seller. What has caused another major change, particularly since the 1970s, is that hometown loyalty is no longer present.

A modern shop located on the farm site, eliminating the need for the blacksmith shop, Minnesota, 1977. (Drache)

The Fading Myth

In 1980 only 3 percent of the people lived on farms and only 15 percent of the population had been raised on a farm. By 1994 less than 1.5 percent lived on farms and only about 6 percent had been raised on one. But the myth of farming as a way of life, which was created in earlier decades, drove agriculture until the frontier closed. Then farmers had to take a more businesslike approach.

Geographer Hart described all the talk about saving the family farm: "Picturesque poverty for peasants may be fine for city folk who want to enjoy a nice Sunday afternoon drive, but these people would flatly refuse to live and labor on the kinds of hardscrabble farms they romanticize."

Wheat and summer fallow strips provide beautiful scenery for those touring the Great Plains, Montana, 1971. (NAL)

Hart advocated that if the public wanted to keep rural landscapes attractive, it should directly propose landscape conservation as sort of a museum of agriculture and/or nature. By the 1940s, the USDA proposed restricting the growth of farms by limiting program benefits. However, by then the economies of scale brought on by technology were so great that farmers would have continued to expand without the benefit of support-program insurance.

By 1960 the top 3 percent (102,000) of all farms produced as much as the bottom 78 percent. If the top 9 percent of all farmers had used the technologies of the top 3 percent, there would have been no need for the output of the other 91 percent. That year the top 1,200 farms produced as much as the bottom 1.6 million. A writer of that era suggested that there was no need to move people out of agriculture unless there was a food shortage and more production by top-producing farmers was needed. There was no reason to encourage people to leave farming until their labor was needed elsewhere. So the surplus farmers were permitted to continue farming. The President's Food and Fiber Commission of 1967 verified those findings.

By the 1960s it became more difficult to advocate the family farm myth as the pillar of democracy because farm numbers were so small. But the end was not yet in sight. In a 1963 USDA study the writers compared the myth that small farms implied social equality and economic security to the European idea of the 1700s and 1800s that land

Monfort feedlot, the nation's largest, capable of finishing over 250,000 cattle annually, Colorado, 1970. (NAL)

ownership implied the ability to lead in government. In sixteenth- and seventeenth-century Europe, land was a major criterion for qualification for office. By 1963 the farm sector was less than 9 percent of our population, and it no longer had much impact on our society.

The desire to own land, which was once so treasured, declined as people sought other ways to capitalize on economic opportunity. A professor of rural sociology writing for the USDA noted:

> Probably in no other enterprise can a worker continue longer in uneconomic work than farming. An older farmer on a paid-up farm can live off the capital investment, consolidate operations, and feel no great hurt. Assets run down, and he experiences something of the well-being of retiring on the job. Economically this is wasteful.

The chief interest of such operations is subsistence living, which is not conducive to the general welfare. The writer of the above asked how rigid society should be in guarding land from those who are capable of little more than subsistence living. He contended that the nostalgia for rustic rural life was often perpetuated by politicians, clergy, artists, and essayists who wanted to shape an image of farm life for a ready audience. Farm museums, the purchase of old farm tools, bed and breakfasts in the country, and the rental of a farm for a week are all symbolic but not realistic.

Countless studies beginning as early as the 1920s indicated the reluctance of farmers to adapt to the industrial society. One of the most famous studies centered on the slowness of farmers to adapt to hybrid seed corn after it became commercially available. The study noted that the top 2.5 percent were innovators, who kept changing agriculture, followed by the 13.5 percent who were early adopters. Even the most costly government programs could not save those who did not have the management ability or the desire to farm efficiently and at a profit.

What happened to the small manufacturer, the small-town merchant, the blacksmith, the country school, the rural church, and the local creamery? The question often asked is why those institutions were not mourned in their passing. In 1970 the USDA wrote a revised definition of the family farm that answers much of the question. The definition: "The family farm is a primary agricultural business in which the operator is a risk-taking manager, who with his [or her] family does most of the farm work and performs most of the managerial activities." Nothing was said about economic viability or the ability to support a family, because the commercial farms had grown beyond that concern.

The small-farmer concept is contrary to the ideals of the industrialized–urbanized–economically diversified society. The large farmer was never liked by his smaller farmer neighbors, but in the 1990s the large farmer fits into a niche in the industrial food chain. Approximately 99 percent of large farms are still owned by the families that operate them. The greatest threat to the family farm is found within the changing farm family.

Summary

From 1954 to 1994, U.S. agriculture led the world in change. Chemicals, biotechnology, mechanization, computers, and management altered the industry as few would have dared to visualize at the beginning of the era. With the help of the world's best agribusiness infrastructure, the business of farming constantly produced a surplus. Other nations looked to the United States as the supplier of last resort but always expected that a progressive U.S. agriculture would have the produce ready. The rapidly shrinking world will continue to challenge agriculture. U.S. agriculture will continue to export to help balance the nation's trade deficit.

The professional farmer uses management, capital, and technology to produce the lowest-cost food possible. He or she leases land, just as any industry leases physical assets, and no longer assumes that appreciated land value is needed to provide for part of the farm income. The myth that tenant farming is somehow not good for agriculture is no longer valid.

Farming has become a totally integrated part of the dynamic and efficient food chain that supplies consumers with over 700,000 items at less than 9 percent of their disposable income. Included in that percentage is the added cost of eating away from home more than any other civilization has done to date. The greatest challenge faced by agriculturists will be the ability to anticipate and adapt to changes demanded by consumers, who expect an adequate supply of low-cost, high-quality, safe food.

References

Abrahamsen, Martin A. *Agricultural Cooperation—Pioneer to Modern.* Washington, DC: USDA Bulletin No. 1, Reprint No. 4, 1965, 49–71.

Bolling, H. Christine. *The Japanese Presence in U.S. Agribusiness.* Washington, DC: USDA-ERS, Foreign Agricultural Economics Report No. 244, June 1992, 1–42.

Brooks, Nora L., and Donn A. Reimund. *Where Do Farm Households Earn Their Incomes?* Washington, DC: USDA-ERS, Agricultural Information Bulletin 560, February 1989.

Center for Agricultural Business. *Agriculture 2000: A Strategic Perspective.* West Lafayette, IN: Purdue University, 1993.

Daberkow, Stan, and John Parks. *Global Trade in Agricultural Inputs.* Washington, DC: USDA-ERS, Research and Technology Division, Statistical Bulletin No. 812, September 1990, 1–88.

Dunham, Denis. *Food Cost Review, 1992.* Washington, DC: USDA-ERS, Agricultural Economics Report No. 672, September 1993.

Economic Research Service. *Corporations Having Agricultural Operations (A Preliminary Report).* Washington, DC: USDA, Agricultural Economic Report No. 142, August 1968.

Economic Research Service. *Corporations with Farming Operations.* Washington, DC: USDA, Agricultural Economic Report No. 209, June 1971.

Fassinger, Polly A., and Harry W. Schwarzweller. "Work Patterns of Farm Wives in Mid-Michigan," *Home and Family Living.* East Lansing: Michigan State

University, Agriculture Experiment Station Research Report No. 425, January 1982, 1–19.

Gardner, Bruce. *The Impact of Environmental Protection and Food Safety Regulation on U.S. Agriculture.* Arlington, VA: Agricultural Policy Working Group, 1993.

General Accounting Office. "Rural Development: Profile of Rural Areas," *Fact Sheet for Congressional Requesters.* Washington, DC: GAO/RCED-93-40FS, April 1993.

Hart, John Fraser. *The Land That Feeds Us.* New York: W. W. Norton Co., Commonwealth Fund Book Program, 1991.

Hart, John Fraser. "Nonfarm Farms," *Geographical Review* 82, April 1992, 166–179.

Havlicek, Joseph, Jr. "Megatrends Affecting Agriculture: Implications for Agricultural Economics," *Journal of Farm Economics* 68, No. 5, December 1986, 1053–1064.

Heady, Earl O. "Influence of the Stage of Development and Urbanization on the Structure of Agriculture," *Benefits and Burdens of Rural Development: Some Public Policy Viewpoints.* Ames: Iowa State University Center for Agricultural and Economic Development, 1970, 107–134.

Henkes, Rollie. "Tomorrow's Seeds: Patent Pending," *The Furrow* 97, No. 6, September 1992, 10–15.

Hill, Howard L., and Frank H. Maier. "The Family Farm in Transition," *Yearbook of Agriculture, 1963.* Alfred Stefferud, ed. Washington, DC: USDA, 1963, 166–176.

Hill, Walter E. "Making Rural Policy for the 1990s and Beyond: A Federal Government View," *Agricultural Outlook '92 Proceedings.* Washington, DC: USDA, World Agricultural Outlook Board, January 1992, 586–595.

Hoag, W. Gifford. *The Farm Credit System: A History of Financial Self-Help.* Danville, IL: Interstate Publishers, Inc., 1976.

Hostetler, John A. *Amish Society.* Baltimore: The Johns Hopkins University Press, 1980.

Ikerd, John E. "Impacts of Policy on the Economics of Sustainable Agriculture," *Agricultural Outlook '93 Proceedings.* Washington, DC: USDA, World Agricultural Outlook Board, January 1993, 679–690.

Johnson, Doyle C. "Financial Performance Trends and Economic Outlook for the U.S. Greenhouse, Turfgrass, and Nursery Industries," *Agricultural*

Outlook '94 Proceedings. Washington, DC: USDA, World Agricultural Outlook Board, January 1994, 392–394.

Kaiser, H. Fred. "Forest and Timber Supplies in the 21st Century," *Agricultural Outlook '94 Proceedings.* Washington, DC: USDA, World Agricultural Outlook Board, January 1994, 167–178.

Kalbacker, Judith Z. "Women Farm Operators," *Family Economics Review.* Washington, DC: USDA-ERS, Family Economics Research Group. Special Issue: Household Production, 1982.

Kinsey, Jean. *Women in Agriculture: The U.S. Experience.* St. Paul: University of Minnesota, Department of Agriculture and Applied Economics, Staff Paper Series P87-32, October 1987, 1–35.

Kyle, Leonard. *The Economics of Large Scale Crop Farming.* East Lansing: Michigan State University, Agricultural Economics, Staff Paper P7-7, May 1970.

Lipton, Kathryn L., and Alden C. Manchester. *From Farming to Food Service: The Food and Fiber System's Links with the U.S. and World Economies.* Washington, DC: USDA-ERS, Agricultural Information Bulletin No. 640, Commodity Economics Division, January 1992.

Majchrowicz, T. Alexander, and David E. Hopkins. *U.S. Farm and Farm-Related Employment in 1988.* Washington, DC: USDA-ERS, GPO, Agricultural Information Bulletin No. 634, December 1991.

Marking, Syl. "Believe This—And Act," *Soybean Digest* 53, No. 14, December 1993, 10.

Morehart, Mitchell J., James D. Johnson, and David E. Banker. *Financial Characteristics of U.S. Farmers, January 1, 1989.* Washington, DC: USDA-ERS, Agricultural Information Bulletin No. 579, December 1989.

Nikolitch, Radoje. "Our 100,000 Biggest Farms: The Relative Position in American Agriculture." Washington, DC: USDA, Economic Research Service, Report No. 49, February 1964.

Phillips, Michael J., and W. Burt Sundquist. "A New Technological Revolution: How Will Agriculture Adjust?" *Policy Choices for a Changing Agriculture.* North Central Regional Extension Publication 266, June 1987, 6–10.

Popper, Frank J., and Deborah E. Popper. "The Reinvention of the American Frontier," *The Amicus Journal* 13, No. 3, Summer 1991, 4–7.

President's National Advisory Commission on Food and Fiber. *Food and Fiber for the Future.* Washington, DC: GPO, July 1967.

Reimund, Donn A., and Fred Gale. *Structural Change in the U.S. Farm Sector, 1974–87.* Washington, DC: USDA-ERS, 13th Annual Family Farm Report to Congress, Agricultural Bulletin No. 647, May 1992.

Schlebecker, John T. "The Great Holding Action: The NFO in September, 1962," *Agricultural History* 39, No. 4, October 1965, 204–213.

Stauber, Karl N. "Rural Development Issues," *Agricultural Outlook '94 Proceedings.* Washington, DC: USDA, World Agricultural Outlook Board, January 1994, 32–36.

U.S. Congress, Office of Technology Assessment. *Technology, Public Policy, and the Changing Structure of American Agriculture.* Washington, DC: OTA-F-285, March 1986.

Bibliographical Note

There are dozens of sources available for researching agricultural history and related topics. *Agricultural History* and the *Journal of Farm Economics*, later called the *American Journal of Agricultural Economics*, are well indexed and are excellent sources for articles. Most journal articles have good bibliographies. Probably the first place any researcher should look is in the bibliographies published by the Agricultural History Society. They provide a ready reference to a broad field of topics.

No serious researcher can overlook the *Reports* and/or *Yearbooks of Agriculture*, which span 1862 to 1993. They are particularly good in revealing the activities of the USDA and the thinking of the time in which the articles were written. The *Outlook Proceedings*, which report the articles presented to the Agricultural Outlook Conferences held annually since 1923, are an excellent source of data reflecting contemporary challenges to the industry. Most college and large library indexes provide information on extensive governmental publications. For material catalogued since 1970 most college libraries have Agricola, the computer data base, or have access to it.

Index

A

A & P, 118; early leadership, 359; founding of, 158
Abilene, Kansas, 195
Accounting, farm, 143, 144
Act of 1796, 71
Adams Act, 136
Advertising, impact of, 422
Africans, imported, 24
Agrarian, idealism, 273; myth, rise of, 69; revolts, 11
Agrarianism, decline of, 231
Agribusiness, 346; infrastructure, 390; rise of, 352
Agricultural Act of 1948, 320
Agricultural Adjustment Act, impact of, 315
Agricultural, associations, 197; attachés, 138; chemistry, 80, 84; college, 80; committee, 73; ladder, 175, 180, 342; press, 77
Agricultural committee, created, 73
Agricultural courses, elementary, 149; for women, 174; high school, 148
Agricultural Credits Act, 347
Agricultural economics, development of, 144
Agricultural Marketing Acts, 309, 314
Agricultural Outlook Conference, founded, 305
Agricultural policy, change of, 314
Agricultural regions, arid, xxvi; coastal plains, xxv; Great Plains, xxvi; subhumid lands, xxv; summer-dry area, xxvi
Agriculture and Consumer Protection Act, 432
Agriculture and Food Act of 1981, 432
Agriculture, commercial, 58, 94, 115, 152, 221; decline in New England, 60; foundations of U.S., 4; industrialized, 390; literature of, 54; maturity of, 362; promotion by railroads, 197; promotion of, 146; regulation of, 142; scientific, 81, 142; teaching of, 80
Agway, founded, 233
Airplanes, uses of, 337, 338
Alcohol, from crops, 357
Alfalfa, for fuel, 444
Algonquin Tribe, 20
Alliance Israelite Universelle, 210
Alliance of Colored Farmers of Texas, 230
Allis-Chalmers, tractor production, 332
American Agricultural Movement, formed, 461
American Farm Bureau Federation, founding of, 302
American Farm Economics Association, 144
American Farmer, first farm journal, 77
American Farm Management Association, founded, 144

478 / Index

American Federation of Labor, 180; in agriculture, 295
American Forestry Association, formation of, 217
American Jersey Cattle Club, 188
American Revolution, cause of, 14, 59; results of, 60, 71
American Society of Equity, 231, 234
Amherst, 80
Amish, 405
Animal, health, 134; science, 354
Anthrax, inoculant for, 135
Antibiotics, use of, 356
Aquaculture, 387
Arator, first book on southern agriculture, 77
Arbor Day, established, 217
Arkansas Agricultural Wheel, aims of, 230
Artificial insemination, 376; history of, 356
Association of Southern Agricultural Workers, 222
Attitude, of public, 311
Auctions, livestock, 348, 349
Automobile, advent of, 200; impact of, 287
Ayrshires, 86

B

Babcock, S. M., invented butterfat test, 189
Back-to-the-land movement, 182
Bacon's Rebellion, 12
Bailey, Liberty Hyde, 245, 271
Bank for Cooperatives, 347, 414
Bankhead-Jones Act, 281
Bankhead-Jones Farm Tenant Act, purpose of, 325
Bankhead-Jones Farm Tenant Purchase Program, 343
Banks, failure of, 271
Barbed wire, development of, 208
Barley, 26
Battle of Fallen Timbers, 94
Beal, William J., corn breeder, 186
Beans, 7
Bennett, Hugh H., promoted conservation, 320
Bergland, Bob, 398
Berkshire Agricultural Society, 79
Better Farming Associations, 146
Bicycles, need for roads, 199
Binder, corn, 170; development of, 166; self-, 110; twine-tie, 167; use of, 110; wire-tie, 166
Biotechnology, development of, 375
Birdseye, Clarence, 361
Blacks, denied education, 291; leave farms, 223, 267; resettlement of, 327
Black stem rust, 34, 53
Blast, 34
Boll weevil, 189; in Texas, 144
Bonanza farms, advantages of, 344; Dalrymple, 167; Red River Valley, 169; use of factory system, 178
Borden, Gail, 119
Borlaug, Norman, 372
Boston, as livestock market, 94
Botanical Garden, established, 54
Bounties, 17
Boycott, impact of, 412
Boys, work of, 91
Bracero program, 297
Bryan, William Jennings, nomination of, 230
Bulletins, farm, 136, 137

Bureau of Agricultural Economics, 137
Bureau of Animal Industry, 135
Bureau of Forestry, established, 218
Bureau of Indian Affairs, 203
By-product of hog slaughter, 123

C

Cacti, fruit of, 9
California, 43
California Agricultural Experiment Station, 142
California Fruit Growers' Exchange, 234
Campbell, Hardy W., developed "dry farming," 215
Campbell, Thomas M., 146
Canadian investments in agriculture, 460
Canals, for irrigation, 215; transportation on, 100
Cane, 9
Cannery and Agricultural Workers Industrial Union, 295
Canning, of food, 119
Capital, long-term, 273; need for, 116; substituted for labor, 340
Capper-Volstead Act, 310; aided cooperatives, 303
Carey Act, 214
Carolinas, settlement of, 37
Carts, two-wheeled, 47, 59
Carver, George Washington, 143
Catalogs, first mail-order, 228
Catholic Rural Life Movement, created, 278
Cattle drives, 95, 194, 195; end of, 208
Cattle, dutch, 31; importation of, 86; range, 204; shipment of, 195; Spanish, 205; Texas Longhorns, 204
Census Bureau, 454

Census, first agricultural, 74
Centralizers, established, 157
Central Pacific Railroad, employed Chinese, 179
Chain stores, 158; growth of, 358
Change, acceleration of, 450
Chavez, Cesar, 404
Cheese factories, 120, 188
Chemicals, adopted, 339; advantages of, 442; use of, 372, 373
Chemurgic Council, work of, 413
Chemurgy, 412; development of, 357
Cherokees, 202
Chicago, as packing center, 157
Chicago Board of Trade, 154
Chicago Union Stock Yards, opening of, 156
Chickahomineg Tribes, 19
Child Nutrition Act, 323
Chinese, employment of, 179; farmers, 179
Churka, 98
Cider, 26
Cincinnati, 96; as center of packing industry, 122
Civilian Conservation Corps, 321
Civil War, 132; impact on migration from farm, 251; in England, 57; watershed for agriculture, 182, 220
Clay, Henry, 86
Clayton Act, 310
Clean Air Act of 1990, 443
Clean Water Act, 440
Clubs, boys', 145, 248; girls', 145, 248
Cold storage, development of, 198
Collectives, Jewish, 210
Colorado potato beetle, 189
Columbian Agricultural Society, 79
Columbus, Christopher, 19

Combine, acceptance of, 168; development of, 167
Commercial agriculture, 94, 115; development of, 58; growth of, 152; trend toward, 221
Commodity Credit Corporation, 317
Communal farming, 11, 26, 35, 405
Communes, Hutterite, 211
Communist Party of the United States, 296
Composts, 85
Conestoga wagon, 47
Connecticut Board of Agriculture, 142
Connecticut Experiment Station, 352
Conservation, by CCC and WPA, 312; forest, 320; of public lands, 219; promoted, 243; required, 442; soil, 320, 432
Conservationist(s), Pinchot, 140; and farm program, 318
Conservation Reserve Program, 370
Consolidation in the South, 337
Consumers, dual-income, 447; fickle, 446; high-income, 447; low-income, 447; protection of, 317
Contract farming, 399
Convenience stores, development of, 421
Cooperative bull associations, 356
Cooperative Extension Act, 306
Cooperative Extension Service; see Extension service
Cooperative oil associations, 313
Cooperatives, 413; aid to, 310; encouraged by Grange, 228; failure of, 303; heyday of, 236; reasons for decline, 236; scope of, 232; size of, 414; start of, 232
Cooperative shipping associations, 311
Corn, borer, 353; cribs, 5; development of, 185; early production, 51; hybrid, 351, 352; in Louisiana, 44; monoculture, 183; planter, 115, 170; planting of, 26; price of, 352; production cost, 143; purchase of, 19
Cornell University, 80
Cost of living, 272, 310
Costs, operating, 137
Cottage industries, 452
Cotton, demand for, 192; export of, 124; harvesting of, 336; labor, 25; loss of market, 264; machinery, 335; monoculture, 221; plowed down, 316; production of, 97; sea-island, 98
Cotton gin, invented, 98
Cottonseed, demand for, 194
Council of North American Grain Exchanges, 146
Country Life Commission, 301; findings of, 226; plight of farm wife, 242
Country Life Movement, findings of, 244
Country store, decline of, 416
Courses, vocational, 250; vocational agriculture, 306
Cowboys, 19; strike of, 179
Cowherds, 17
Cowpens, 19
Cradle scythe, 50, 103
Creameries, 120
Cream separator, invention of, 188
Cream, test for butterfat, 189
Credit, by dealers, 397; elimination of, 158; need for, 275
Credit system, lack of, 221
Crop lien system, adopted, 221
CRP (Conservation Reserve Program), impact on production, 410

Culpepper Rebellion, 37
Cultivator, straddle-row, 170

D

Dairy centers, 97
Dairy Herd Improvement Associations, 356
Dairying, 188, 354; commercial, 121; cow testing associations, 189; improved production, 355; lease cows, 403
Dairymaids, 121
Dale, Governor Thomas, 11, 19, 35
Dalrymple, bonanza farm, 167
Darwin, Charles, 186
DDT, use of, 373
Debt, reduction of, 268; of planters, 42
Debt-to-asset ratio, in farming, 398
Deere, John, 112, 163; tractor production, 332
Delco light plants, 289
Desert Land Act, 140
Devon cattle, 86
Discount warehouse clubs, 423
Diseases of crops, 34, 53, 189, 353
Distribution system, efficiency of, 244
Diversification in the South, 262
Diverted acres, impact of, 427
Division of Forestry, created, 217
Domesticated animals, dog, 6; turkeys, 6
Donation Act of 1850, 456
Downsizing, need for, 315
Drainage, legislation, 190; tile, 190
Drill, development of, 169; shoe, 169; single-disk, 169; use of, 113
Drought, of 1930s, 323, results of, 225
Drovers, 95
Dry Farming Congress, 215
Dry farming, merits of, 215; problems of, 213
DuPont, E. I., 85
Dutch, as farmers, 30

E

Economics, farm, 136
Economy, impact of global, 298; world, 407
Education, early agricultural, 80; funding of, 249; increase of, 221; opposition to, 249
Eggs, preservation of, 120
Electric cooperative power association, first, 290
Electricity, cost of rural, 290; development of, 289; importance on farm, 285
Eliot, Jared, 54, 59
Embargo of agricultural exports, 412
Emergency Relief Appropriation Act, 290
Endangered Species Act, 444
Energy, cost of, 380
English corn laws, repealed, 124
Environmental groups, 432
Environmentalism, 438, 450
Environmental Protection Agency, 440
Erie Canal, 90, 100
Erie Railroad, 197
Eskimos, starvation of, 203
Esterly, George, 108
Ethanol, 357; demand for, 443; production of, 415
European customs, resistance to, 10
Experiment stations, vii, 141; established, 142; forest, 216; in Alaska, 204
Exports, x, 37; growth of, 150, 224, 300, 369, 407; major agricultural, 123; of

cotton, 98; of hides and tallow, 43; of value-added agricultural products, 409; percent of farm income, 150; revival of, 320; to West Indies, 33, 57; volume of, 299

Extension agents, first Negro, 146; in cooperative movement, 312

Extension programs for farm women, 242

Extension service, vii, 144; created, 146, 306; promoted cooperatives, 237; task of, 248

F

Fahrenheit, work in cooling, 198

Fairs, agricultural, 78, 79; impact of, 247; livestock, 31; trends of, 247

Families, size of, 22

Family farm, advantages of, 277; decline of, 401; defined, 277, 470; myth, 273

Family Farm Policy Review, findings of, 465

Fanning mill, use of, 110

Farmall, in cotton production, 336; row-crop tractor, 331

Farm Bloc, creation of, 307

Farm broadcasters, first, 288

Farm Bureau Management Service, created, 271

Farm Credit System, change in, 413; need for, 346; promoted, 228

Farm earnings, 393

Farm enterprises, changes in, 285

Farmers, age of, 452; aid to, 182, 325; attitude of, 274; classes of, 75; concerns of, 245; health of, 242; income of, 225, 447, 449; living standards of, 3, 242; problems of, 224; ranked by income, 395; reluctant to change, 246, 275; represented in Congress, 319; status of black, 246

Farmers' Alliance, aims of, 229

Farmers' Cooperative Demonstration Group, 144, 181

Farmers' Educational and Cooperative Union, 231

Farmers' Holiday Association, 269, 315

Farmers Home Administration, write-downs, 398

Farmers Union, aims of, 224, 229; promotes cooperatives, 235; *see also* National Farmers Union

Farm income, 393; improvement in, 267

Farming, a way of life, 274; capital-intense, 396; communal, 405; corporate, 398, 400, 401; integrated, 398; large-scale, 183, 345; part-time, 403, 453; start-up cost, 397; tenant, 180, 402

Farm Journal, corn contest, 353

Farm life, description of, 238

Farm Life Studies Section, purpose of, 280

Farm management, 143

Farm numbers, decline of, 369, 431; last increase of, 278

Farm organizations, 226

Farm population, decline of, 251; surplus of, 304

Farm problem, x

Farm production, studies of, 136

Farm programs, 427; payment limits, 432; payments from, 430; permanent, 317, social aspects of, 432

Farms, abandoned, 221, 251; census definition of, 454; challenges of small, 307; consolidation of, 344;

decline of numbers, 223, 277, 433; demise of small, 430; expansion of, 394; experimental, 137; hog, 455; impact of larger, 266; income of, 433, 437; loss of, 271; population on, 466; preservation of small, 429; production of large, 467; size of, 27, 51, 270, 325, 344; small, 274, 276; transfer of, 286; work year, 275

Farm Security Administration, 325

Farmsteads, individual, 90

Farm vehicles, early, 46

Farm Workers Association, 404

Farm youth, programs for, 435

Fast-food industry, 425, 426

Fawkes, Joseph, steam plow, 162

Federal Aid Road Act, 287

Federal Emergency Relief Administration, created, 325

Federal Farm Board, created, 310

Federal Farm Loan Act, 347

Federal Insecticide and Rodenticide Act, 448

Federal Intermediate Credit Banks, 347

Federal Land Bank, 347

Federal Reserve Board, action of, 270

Federal Reserve System, creation of, 347

Federal Trade Commission, 359

Fencing, cost of, 17, 207; rail, 16, 93

Ferguson, tractor production, 332

Fertilizer, commercial, 84, 339, 350, 351; use of, 54, 184

Feudal system, 43; end of, 60

Finances, economy of scale, 396

Financial problems, causes of, 270

Fire wardens, 217

Fish, 8; freezing of, 156; in New England, 56; production of, 388

Flail, 50

Flax, 26; promoted, 123; straw, 358; use of, 53

Floriculture, 384

Florida Fruit Exchange, 233

Florida, settlement of, 43

Flour milling, percent of, 152

FMHA programs, impact of, 430

Food, Agriculture, Conservation, and Trade Act of 1990, 440

Food and Agriculture Act of 1977, 432

Food and Drug Act, 136

Food and fiber system, 392

Food, assistance programs, 433; consumption patterns, 447; convenience, 267; cost of, ix, 155, 159; 426; different products, 421; distribution of, 116; frozen, 361; instant, 417; prepared, 116; processing of, 118, 119; purchased, 28; safe production of, 448; shortage of, 35; toxins in, 448

Food chain, ix, 116, 415; growth of, 358

Food, Drug, and Insecticide Administration, 322

Food industry, changes in, 267

Food processing at home, 116

Food Production Act, 306

Food production, problems of, 27

Food sector, scope of, 426

Food Security Act of 1985, 429, 442

Food stamp program, 323

Forage crops, 26

Ford, Henry, 164; leader in chemurgy, 357; opened supermarket, 360; tractor production, 331

Fordson tractor, 164

Foreign Agricultural Service, 407

Forest and Rangeland Resources Planning Act, 444

Forest fires, Hinckley, 217

Forest, products, 322; reserves, 218

Forestry, exports, 55; restrictions on, 60; schools, 217; service, 217
Forests, 93; cultivation of, 218; protection of, 33; removal of, 56; restrictions on cutting, 56; ownership of, 445
4-H clubs, 248, 276
Franklin, Benjamin, 81
Freighters, 101
Freight rates, 100, 227
French settlements, 42, 43
Fries' Rebellion, 87
Froelich, John, 163
Frontier, cattle, 204; close of, 224; settling on, 88; significance of end of, 301
Frozen food, 361, 416
Fruit, cooperatives, 234; trees, 26
Fruit Workers Union, 180
Fuels, alternative, 443
Furfural, use of, 357
Futures trading industry, development of, 153

G

Galbraith, John, 320
Gardens, experimental, 54
General Agreement on Tariffs and Trade (GATT), 300, 370, 411
General Allotment Act, 203
General Foods, promoted frozen foods, 361
General Land Office, 140
Genetic engineering, potential of, 378
Georgia, establishment of, 40
Girls, work of, 91
Glidden, J. F., 208
Global Positioning System, value of, 378
Golden Age of Agriculture, 241

Gold rush, 91, 99, 205
Government aid to agriculture, 67
Government programs, x, 411; objective of, 362; results of, 369
Grain drill, 113, 114, 169
Grain, exports of, 31
Grand Island Livestock Commission Company, 348
Grange, 201, 227, 233
Granger laws, 228
Grape industry, 214
Grazing, fees for, 219; on commons land, 18; on public lands, 219
Great American Tea Company, 118
Great Atlantic & Pacific Tea Company; see A & P
Great Plains, xxvi, 204
Greenhouse industry, 155; products of, 384
Gross National Product, farm portion, 392, 437; from natural resources, 450
Guano, 84
Gypsum, 83, 84

H

Haines Illinois Harvester, 109
Halladay, Daniel, invented windmill, 161
Hanson, Timothy, 51
Harsford, Eben, 80
Hart, C. W., 164
Hart, John Fraser, Museum of Agriculture, 466
Hart-Parr Company, 164
Harvard, 80
Harvey, Fred, restaurants, 155
Hatch Act, vii, 141
Hawaiian Islands, agriculture of, 187

Hay, 26, 51, 94
Header, 108
Headright, 14, 36
Hebrew Emigrant Aid Society, 210
Hemp, 26, 123
Herefords, 86
Hessian fly, 35, 53
Highways, early, 101; 199
Hoard's Dairyman, 188
Hoe culture, 5, 25
Hog cholera, 135
Hog Island, 16
Hogs, as scavengers, 17; Berkshire, 86; Poland China, 86; Duroc-Jersey, 86; Chester White, 86
Holsteins, 86, 188
Home demonstration agents, first black, 147
Homestead Act, vi, 139, 140, 456
Homestead, by veterans, 223; cost of, 207; problems of, 274
Horner, John, perfected combine, 168
Horse, Chickasaw, 47; Conestoga, 47
Horses, cost of, 166; declining use for, 330; peak numbers, 330; use of, 46, 97
Housing Act of 1949, 281
Housing, condition of, 29
Houston, agricultural secretary, 137
Hudson's Bay Company, 91
Hull, Cordell, 305
Hunting, 9, 10
Hunts Point, wholesale market, 417
Hussey, Obed, 107
Hutterites, 211, 212
Hypermarkets, 421

I

Ice for refrigeration, 199

Illinois Cooperative Act, 313
Illinois Farm Supply Company, 313, 314
Immigrants, as farm workers, 176; avoid agriculture, 223; Chinese, 179; Japanese, 292; Mexican, 297
Immigrant train, 197
Income, farm, 225, 447, 449; off-farm, 457
Indentured servants, 15, 22, 32
Indian(s), v, 16, 19; agriculture, xxv, 3, 4; land policy, 21; policy toward, 68; relations with, 19, 96; schools, 203; trade with, 35; treaties with, 202; villages, 6
Indian Trade Act, 203
Indigo, 44; bounty for, 40; market, 60
Industrial agriculture, 429, 439, 451
Industrialization, aid to trade, 408
Industrial Revolution, 121
Industrial Workers of the World, 180, 295
Infrastructure, agricultural, xii
Innovations, adoption of, 469
Insects, control of, 374
Integrated pest management, goal of, 379
International Harvester Company, 164, 331
International Longshoremen's Association, 412
International Society of Soil Science, formed, 320
Interstate Commerce Act, 228
Investments, foreign, 459, 460
Iowa Agricultural College, 143
Iroquois League, 6
Irrigation, 43, 138; by missionaries, 214; by Mormons, 99, 213; canals, 215; center-pivot, 382; cost of, 380; districts established, 213; early, 9; flood, 381; for rice, 39; in California, 206;

increase of, 383; Indian, 8; problems of, 380; sprinkler, 381; trickle (drip), 384; windmills, 215

Isolation, 29; breakdown of, 286; cause of, 247; hindrance to growth, 450; problems of, 247, 276

J

Jackson, Andrew, Indian policy, 69

Jamestown, 35

Japanese, as farm workers, 179; immigration of, 292; investments in agriculture, 460

Jefferson, Thomas, vi; Indian policy, 68; philosophy of agriculture, 70

Jerseys, 86

Jewish Agricultural and Industrial Aid Society, 182

Jewish Agricultural Society, 211

Jews, as farmers, 209; settlements of, 209, 211

Johnson, Howard, restaurants, 425

K

Kansas Agricultural College, 147

Kansas, settlement of, 223

Kelley, Oliver H., founder of Grange, 227

King Kullen Stores, 360

Kinkaid Act, 140

Knapp, Seaman A., 144

Koch, Robert, 135

Korean War, 320

Kyle, Leonard, findings of, 395

L

Labor, cost of, 21, 160, 166; demand for, 450; excess of underemployed, 292; family, 29; hired, 176; immigrant, 176, 222; importation of, 267; in a changing agriculture, 177, 451; income of farm, 403; indentured, 15, 22, 32; in food chain sectors, 392; in food industry, 419; Japanese, 179; leaving agriculture, 428; management of, 177; move from farm to city, 223; number in agriculture, 423; on larger farms, 294; problems of, 178, 292; productivity of, 371; results of shortage, 267; savings from electricity, 290; shortage of, 21, 120, 175, 192, 221, 245, 336; supply of colonial, 21; surplus of, 403; time needed for production, 333; waste of, 245

Lancaster Turnpike, 101

Land, abandoned in New England, 206; acres of public, 443; allotment of, 32, 57; consolidation of, 27; dispersing of, 14, 72, 206; increase of price, 265; ownership of, 11; seizure of, 20; speculation in, 72, 206

Land-grant colleges, contribution of, 428; for Negroes, 141

Land Ordinance of 1785, 71

Lane, John, 112

Lane, John, Jr., 113

Large-scale farming, 345; resentment of, 183

Legislative acts, Act of 1796, 71; Adams Act, 136; Agricultural Act of 1948, 320; Agricultural Adjustment Act, 315; Agricultural Credits Act, 347; Agricultural Marketing Acts, 309, 314; Agriculture and Consumer Protection Act, 432; Agriculture and Food Act of 1981, 432; Bankhead-Jones Act, 281; Bankhead-Jones Farm Tenant Act,

325; Capper-Volstead Act, 303, 310; Carey Act, 214; Child Nutrition Act, 323; Clayton Act, 310; Clean Water Act of 1990, 440, 443; Cooperative Extension Act, 306; Desert Land Act, 140; Donation Act of 1850, 456; Emergency Relief Appropriation Act, 290; Endangered Species Act, 444; Federal Aid Road Act, 287; Federal Farm Loan Act, 347; Federal Insecticide and Rodenticide Act, 448; Food, Agriculture, Conservation, and Trade Act of 1990, 440; Food and Agriculture Act of 1977, 432; Food and Drug Act, 136; Food Production Act, 306; Food Security Act of 1985, 429, 442; Forest and Rangeland Resources Planning Act, 444; General Allotment Act, 203; Hatch Act, vii, 141; Homestead Act, vi, 139, 140, 456; Housing Act of 1949, 281; Illinois Cooperative Act, 313; Indian Trade Act, 203; Interstate Commerce Act, 228; Kinkaid Act, 140; Land Ordinance of 1785, 71; meat inspection law, 135; Morrill Act, 141; Morrill Land Grant College Act, vi; Mutual Security Act, 268; National Environmental Policy Act, 444; National School Lunch Act, 323; Packers and Stockyards Act, 311; Plant Variety Protection Act, 379; Preemption Act of 1841, 73, 90; Public Law 480, 268, 369, 411; Purnell Act, 281; Reclamation Act, 138, 214; Robinson-Patman Act, 360; Rural Development Act, 451, 465; Sherman Anti-Trust Act, 310; Smith-Hughes Act, 306; Smith-Hughes Vocational Education Act, 322; Smith-Lever Act, 248, 306; Soil and Water Resources Conservation Act, 432; Soil Conservation and Domestic Allotment Act, 319; Steagall Amendment, 320; Stock-Raising Homestead Act, 140; Swamp Land Acts, 190; Timber Culture Act, 140, 217; Trade Expansion Act, 268; Watershed Protection and Flood Prevention Act, 321; Weeks Act, 220; Wilderness Act, 444; for women, 456

Lend-Lease, 268

Libraries, traveling, 251

Liebig, Justus von, 80, 84, 184

Lien laws, 192

Life style, changing, 455

Lime, 82, 84

Lincoln, Abraham, vi, 133; interest in agriculture, 70

Lister, development of, 170

Livestock, 85; commercial packing, 122; destruction of, 28; housing of, 33; imported, 28; improvement in breeds, 96, 339; in Carolinas, 38; killing of, 316; need for, 16; problems of marketing, 348

Livestock industry, commercial, 93, 94; growth of, 16

Livestock shipping associations, 235

Loans, intermediate-term, 413

Lobbyists, 308

Logan, James, 33

Log boat, 46

London Company, 14

Lorain, John, experiments on corn, 185

Lumber products, manufacture of, 152

M

Machinery, adoption of, 165, 221; horse-powered, 160; investment in, 340; restrictions on production, 265

Maize, 7, 8

Management, farm, 136, 182; need for help, 271; professional, 345, 393
Manorial system, 12
Manufacturing, in rural areas, 451; of textiles, 85
Manure, use of, 82
Marketing, control of, 157, 234; direct buying, 400; shipping associations, 350
Market reports, 118; local, 288
Market(s), access to, 252; colonial, 59; decline of terminal, 400; early, 58; forecasting of, 304; foreign, 123, 150; gyrations of, 406; livestock, 122, 348; loss of British, 98; need for, 55; reports by USDA, 305; terminal, 196, 350; urban, 101; wholesale, 417
Marl, 84
Marshall Plan, 268
Marsh brothers harvester, 110
Maryland, settlement of, 36
Mason jars, 120
Massachusetts Bay Colony, 27, 57
Massasoit, 20
Matanuska Colony, 326
Mather, Cotton, experiments on corn, 185
McCormick, Robert, 107
McMillen, Wheeler, 353, 357
Meadows, 7
Meat, inspection of, 135, 156; preservation of, 51; refrigeration of, 157; self-service marketing, 361; shipped in ice cars, 123
Meat packing, 96; change in, 418; leading manufacturing, 152
Mechanical power, rise of, 160
Mechanization, failure to adopt, 334; need for, 176; results of, 333
Meteorological Service, 134

Mexican Mutual Aid Society, 296
Mexicans, immigration of, 297
Michigan, first agricultural college, 80
Middle Colonies, settlement of, 30
Middlemen, 118
Migration, off-the-farm, 251, 271, 278, 279, 281; out of farming, 404; to the farm, 279
Milk, adulteration of, 134, 153; condensed, 120; delivery of, 153; inspector, 134; powdered, 120; production of, 86, 188, 265, 376; unsanitary production of, 153
Minnesota State Forestry Association, 217
Minnesota, University of, 143
Missions, Catholic, xxvii; Spanish, 43
Money, management of, 394
Moore, Hiram, 167
Mormons, in agriculture, 99; settlement of, 212; use irrigation, 213
Morrill Act, 141; second, 141
Morrill Land Grant College Act, vi
Morton, S. S., 164
Mower, use of, 105, 106
Mulch tillage, 442
Mules, 85
Multiple cropping, 4
Mutual Security Act, 268
Myth, family farm, 273

N

Napoleonic Wars, 124
National Board of Farm Organizations, 302, 308
National Catholic Rural Life Conference, 278
National Center for Food Safety, 448

National Electric Light Association, studies by, 289
National Environmental Policy Act, 444
National Farm and Home Hour, 289
National Farmers' Alliance, 230
National Farmers Organization, 461
National Farmers Union, 308
National Food and Fiber Commission of 1967, 428, 434
National Grange; *see* Grange
National road, 101
National School Lunch Act, 323
Navigation acts, 57
Nebraska, settlement of, 223
Nebraska Tractor Test, 164
Negro, extension agent, 146
Net worth of average families, 431
Newbold, Charles, first cast-iron plow, 104
New England Farmer, 77
New England, town, 13; township, 15; settlement of, 25
New Orleans, as market, 95
New South, beginning of, 264
New York Cotton Exchange, 154
Nonpartisan League, 272
North American Free Trade Agreement (NAFTA), 370, 410
North Carolina Farmers' Association, 230
Nursery products, 384; experimental, 54
Nutrition, need for, 322

O

Oats, 26
Occupations, secondary, 3
Offal, use of, 123
Off-farm income, 281, 431, 457
Off-farm work, importance of, 281

Office of Farm Management, 137, 144
Office of Markets, created, 138
Office of Public Road Inquiries, 199, 287
Office of Technology Assessment, findings of, 394, 465
Ohio Canal, 90
Oliver, James, 112
Open range, 16; end of, 205
Oregon Trail, 96
Organic Chemistry in Its Applications to Agriculture and Physiology, 84
Overcapitalization, 271
Overproduction, continuation of, 223
Oxen, use of, 46

P

Pacific Coast, settlement of, 91
Packaging, improvement of, 417
Packers and Stockyards Act, 311
Panic of 1837, 89, 91
Panic of 1873, 227
Paris green, 189
Parity, goal of, 319
Parr, C. H., 164
Patent Office, 133; in charge of agriculture, 74
Patrons of Husbandry; *see* Grange
Patroon system, 13, 31
Peas, 26
Penn, William, 15, 32
Pennsylvania, breadbasket, 31
People's Party, 230
Pesticides, 441
Philadelphia Centennial Exposition, 165
Philadelphia Society for Promoting Agriculture, 76, 216
Piggly Wiggly, 360
Pinchkey, Eliza Lucas, 40

Pinchot, Gifford, 140
Pittsburgh, as export center, 89
Plantations, number of, 42; poverty of owners, 221
Plant breeding, 4, 379
Plant Variety Protection Act, 379
Plough Boy, 77
Plow, development of, 112; double-gang riding, 113; first cast-iron, 103; moldboard, 112; steel, 105; sulky, 113; use of moldboard, 442
Plowing, 46; time required, 332
Plymouth Colony, 15, 26
Policies, early agricultural, 68
Popper, Frank and Deborah, proposals of, 443
Population, growth of, 447; loss of, 225, 436; of rural communities, 433; on farms, 466; surplus of rural, 226
Populist Party, failure of, 230
Porkopolis, 96
Postal system, 29
Potatoes, introduction of, 53
Poultry, industrialization of, 377
Power, 5; early, 46; revolution in, 328
Predators, 17
Preemption Act of 1841, 73, 90
Preemption, problem of, 72
Pressure cooker, 119
Prices, decline of, 176, 224, 268; improvement of, 225; reduction of target, 432; rise of, 267; stabilization of, 317
Price supports, 309, 429
Primogeniture, law of, 12
Proclamation of 1763, 88
Production, boom in, 339; challenges to, 56; cost of, 391, 441; curtailment of, 307, 316; growth of, 333; increased cost of, 441; of small farmers, 222; problems of, 269; reduction of cost, 333; surplus of, 269; time needed, 429
Production Credit Associations, 347
Productivity, decline of, 151; increase of, 371
Profits, obstacles to, 225
Protests, early, 87
Public domain, disposal of, 67, 72
Public Law 480, 268, 369, 411
Pueblo dwellers, 9
Purnell Act, 281

Q

Quitrents, 12, 37

R

Rabies, 19
Radio, adoption of, 288
Rail rates, 101
Railroads, impact of, 224, 462; importance to agriculture, 197; importance to livestock industry, 194; unitrains, 464
Railroad system, 102
Rainfall, unpredictability of, 207
Rake, horse, 105, 106
Ranching in California, 205
Reaper, 103, 107
Reclamation Act, 138, 214
Reconstruction Finance Corporation, 347
Reforestation, 322
Refrigeration, importance of, 157; mechanical, 198; of fruit, 198
Rehabilitation, rural, 279, 326
Reid, Robert, corn breeder, 186
Reindeer, 203

Reno, Milo, farm leader, 269
Rensselaer Institute, 81
Research for agriculture, 371
Resettlement Program, 324; for blacks, 327
Restaurants, first, 154; Fred Harvey, 155; Howard Johnson, 425
Revival of South, 192
Revolutionary War, impact of, 59
Revolution in agriculture, xi
Rice, aerial seeding of, 338; production of, 39
Riley, Glenda, 91
Rivers, transportation on, 100
Roads, building of, 101; centralization of planning, 200; development of, 199; need for, 287; plank, 101; promotion of, 228
Roanoke Island, 19, 34
Robinson-Patman Act, 360
Rogin, Leo, 105
Rolfe, John, 35
Roosevelt, Theodore, encouraged conservation, 243; promoted cooperatives, 232
Rotation of crops, 82, 84
Ruffin, Edmund, 83
Rural, decline of communities, 462; housing, 281; problems, 453; schools, 249; sociology, 280
Rural Development Act, 451
Rural Development Act of 1972, 465
Rural Development Administration, 452
Rural Electrification Administration, created, 291
Rural electrification, need for, 228
Rural free delivery, 201, 228
Rural Organization Service, created, 280
Rye, shift to, 53

S

Salmon, 8
Saltzburgers, 41
Salvation Army, 182
San Jose scale, 189
Sapiro, Aaron, 303
Schidler, James, 275
School(s), consolidation of, 200, 243, 250, 281; forestry, 217; Indian, 203; lunch program, 323; rural, 149; tractor, 164
Schultz, Theodore, 266
Scientific agriculture, lack of, 227
Seafood, 10
Sedalia, Missouri, 194
Seeder, broadcast endgate, 114; use of, 113
Seed, importation of, 138; wild, 9
Seeding, hand, 113
Self-sufficient farming, 55
Sharecropping, 177, 191, 267
Shays' Rebellion, 87
Sheep, cost of, 94; Merino, 85; production of, 94
Sheep Shearers Union of North America, 179
Shenandoah Valley, 18
Sherman Anti-Trust Act, 310
Shipbuilding, 56
Short courses, developed, 148
Shorthorns, 86, 96
Sickle, capacity of, 49; use of, 105
Silk, production of, 41
Silo filler, 170
Silver, Grey, 308
Skinner, John, 77
Slaughterhouses, 31, 58
Slavery, 24; abolished, 98

Slaves, 12; cost of, 24; emancipation of, 177; importation of, 41
Small-farm myth, 273; in Europe, 468
Small farms, 15, 274, 276; challenges of, 307; demise of, 430; preservation of, 429; production of, 222
Smallpox, 20
Smith, Captain John, xxv, 20
Smith-Hughes Act, 306
Smith-Hughes Vocational Education Act, 322
Smith-Lever Act, 248, 306
Smithsonian Institution, 134
Social status of farmers, 241
Soil, analyzed, 141; compaction of, 190; conservation of, 432; exhaustion of, 83, 160; fertility, 5, 83; productivity of, 84; rebuilding of, 54; types, 83
Soil and Water Resources Conservation Act, 432
Soil bank, 369, 410, 431
Soil Conservation and Domestic Allotment Act, 319
Soil Conservation Districts, established, 321
Soil Conservation Service, created, 320
Soldiers' grant, 72
South, revival of, 192; settlement of, 34
Soybeans, 353, 354
Spanish settlements, 42
Spillman, W. J., 183
Spinning wheel, 26
Squanto, vi, 20, 26
Squash, 7
Squatters, 206
Stagecoaches, 103
Standard of living, 3, 242
Standard Oil, experiments with ethanol, 357
Starvation, v; of livestock, 19

Steagall Amendment, 320
Steam engine, self-propelled, 163
Steam power, 162
Stock-Raising Homestead Act, 140
Stone boat, 47
Stores, country, 117; self-service, 360
Strikes, by cotton pickers, 295; by farmers, 269; early agricultural, 294
Subsidies, cost of farm, 319
Subsistence farming, 55
Subsoiling, value of, 191
Sugar, beets, 186; cane, 43, 185; development of industry, 186; Hawaiian production, 187
Summer-dry area, xxvi
Sunflowers, 9
Sunkist Growers, Inc., 234
Supermarkets, development of, 360; impact of, 420; size of, 423
Surplus, problems of, 270
Surveying of soil, 138
Sustainable agriculture, 439, 440
Sutter, John, 91
Swamp Land Acts, 190
Sweeps, horse-powered, 111
Swift, G. F., 157

T

Tariffs, change in, 409; demand for protective, 224; harm of, 305; impact on agriculture, 299
Taxes, real estate, 272
Tax system, harmed farmers, 224
Taylor, Henry C., 144, 305, 342
Technology, rate of adoption, 396
Telephone, aid to farmers, 247; impact of, 286
Teller Reindeer Station, 203

Index / 493

Tenant farming, 13, 15, 180, 181, 192, 341, 342, 402
Tennessee Association, 222
Texas fever, 135
Thanksgiving, 20
Thresher, development of, 110; flail, 110
Tiling, 84
Timber Culture Act, 140, 217
Timber, production of, 216–218
Timothy hay, 51
Tobacco, 8, 35; destroyed, 36; export of, 36; in Louisiana, 44; labor, 25; loss of market, 60; price of, 36
Tongs, associations, 179
Tools, inventory of, 48
Towns, abandoned, 225, 464
Township, 13
Trace minerals, value of, 375
Tractor(s), adoption of, 275; demonstrations, 164; development of, 163; four-wheel-drive, 164, 370; impact of, 285; improvement in, 332; increased use of, 330; number of, 165; replaced families, 332; schools, 164; small, 164
Trade balance, 151
Trade Expansion Act, 268
Trade Union Unity League, activities of, 296
Trains, agricultural instruction, 197; immigrant, 197
Transportation, improvement of, 199; of perishables, 198
Treadmills, 111
Trees, lack of, 207
Trucks, 163; impact of, 337; number of, 165
Tuberculosis, test for, 135
Tugwell, Rexford, 317
Tumbrels, 47

Turkeys, 6, 9
Turnpike road system, 101
Tuskegee Experiment Station, 143
TV dinners, 417

Underemployment in farming, 427, 449
Union Cold Storage and Warehouse Company, 198
Unions, 180, 295, 296, 404
Union Stock Yards Company, of Chicago, 122; opening of, 156
United Farm Workers Organizing Committee, 404
United States Agricultural Society, 74, 133
United States Department of Agriculture, (USDA citations generally under specific departments), vi, 143; attempts to create, 73; created, 133; employees of, 393, 433; goals after 1900, 184; social programs of, 322
United States Department of the Interior, 133
Unitrains, impact of, 464
University of Minnesota, 143
Urban, Thomas, views of, 390

Van Rensselaer, Stephen, 81
Veblen, Thorstein, findings of, 463
Vertical integration, 398
Village green, 18
Vitamins, value of, 375

W

Wages of farm women, 172, 173
Wagon, Conestoga, 47

Wallace, Henry A., 430; causes of farm problems, 270; hybrid corn work, 352; philosophy of, 308
Wallis Cub tractor, 164
Ward, Aaron Montgomery, first mail-order catalog, 228
War of 1812, 89, 124
Warren, G. F., 166, 345
Washington, Booker T., 143
Washington, George, vi, 133; farewell address, 73; interest in agriculture, 69; land survey, 190
Waterloo Gasoline Traction Engine Company, 163
Watershed Protection and Flood Prevention Act, 321
Watson, Elkanah, 79
Way-of-life farming, 427
Weather Bureau, created, 134; first weather reports, 288
Webster, Daniel, 112
Weeks Act, 220
Wesleyan University, 142
West Indies, 33
Westward movement, 89
Wheat, 44; farming in California, 205; monoculture, 224; production areas, 90
Wheatland Riot, 180
Whiskey Rebellion, 87, 95
White House Conference of Governors, on conservation, 220
Wholesale markets, 417
Widows, inheritance, 30
Wilderness Act, 444
Wild rice, 8
Wilson, James, 137

Windmills, demand for, 161; used for irrigation, 215
Wind power, 161
Winnowing, 110
Winter wheat, 53
Women, as farm managers, 459; as homesteaders, 91, 208; as mediators, 286; black in fields, 335; changing roles, 267, 282, 458; courses for, 147; did field work, 220, 335; efforts to help, 243; equality for, 208; fewer in agriculture, 223; importation of, 22; in the Grange, 227; legislation for, 456; life on farm, 242; on farms, 171, 239; on the frontier, 91; rights of, 174, 208; work of, 5, 6, 22, 34, 121
Women's Christian Temperance Union, 229
Women's suffrage, 208
Wood, for fencing, 93; for fuel, 93
Wood, Jethro, 104, 112
Wood preservatives, first used, 219
Woolen mills, 94
Work, changing patterns, 120; off-the-farm, 266
Works Projects Administration, 321
World Food Conference, 408
World War I, impact of, 262
World War II, impact of, 265

#

Yearbook of Agriculture, discontinued, 438
Yellowstone Timberland Reserve, 217
Yeoman myth, belief in, 246